TOURO COLLEGE LIBRARY
Kings Hwy

WITHDRAWN

Computer-Supported Collaborative Learning in Higher Education

Tim S. Roberts
Central Queensland University, Australia

TOURO COLLEGE LIBRARY
Kings Hwy

IDEA GROUP PUBLISHING
Hershey • London • Melbourne • Singapore

KH

Acquisitions Editor:	Mehdi Khosrow-Pour
Senior Managing Editor:	Jan Travers
Managing Editor:	Amanda Appicello
Development Editor:	Michele Rossi
Copy Editor:	Jane Conley
Typesetter:	Sara Reed
Cover Design:	Lisa Tosheff
Printed at:	Yurchak Printing Inc.

Published in the United States of America by
    Idea Group Publishing (an imprint of Idea Group Inc.)
    701 E. Chocolate Avenue
    Hershey PA 17033
    Tel: 717-533-8845
    Fax: 717-533-8661
    E-mail: cust@idea-group.com
    Web site: http://www.idea-group.com

and in the United Kingdom by
    Idea Group Publishing (an imprint of Idea Group Inc.)
    3 Henrietta Street
    Covent Garden
    London WC2E 8LU
    Tel: 44 20 7240 0856
    Fax: 44 20 7379 3313
    Web site: http://www.eurospan.co.uk

Copyright © 2005 by Idea Group Inc. All rights reserved. No part of this book may be reproduced in any form or by any means, electronic or mechanical, including photocopying, without written permission from the publisher.

Library of Congress Cataloging-in-Publication Data

Computer-supported collaborative learning in higher education / Tim S. Roberts,
editor.
     p. cm.
  Includes bibliographical references and index.
  ISBN 1-59140-408-8 (hardcover) -- ISBN 1-59140-409-6 (pbk.) -- ISBN 1-59140-410-X
(ebook)
  1. Group work in education. 2. Education, Higher--Computer-assisted instruction.
3. Education, Higher--Effect of technological innovations on. I. Roberts, Tim S.,
1955-
  LB1032.C575 2005
  378.1'758--dc22

                          2004003755

British Cataloguing in Publication Data
A Cataloguing in Publication record for this book is available from the British Library.

All work contributed to this book is new, previously-unpublished material. The views expressed in this book are those of the authors, but not necessarily of the publisher.

3/6/08

Computer-Supported Collaborative Learning in Higher Education

Table of Contents

Preface

This book brings together 14 contributions from researchers and practitioners actively involved in the field of computer-supported collaborative learning (commonly referred to as CSCL). The authors describe a variety of different learning situations, some undoubtedly very successful and some perhaps not quite so. Taken as a whole, the work presented here is richly illustrative of the wide diversity of research and practice currently being undertaken in this rapidly expanding field.

In an earlier book (Roberts, 2003), I expressed the view that online collaborative learning was an idea whose time had come. My confidence in the truth of this statement has grown stronger over the intervening period. Not only are students across the globe now coming to expect that their courses will be supported by online web-based materials and resources (and becoming indignant if they are not), but there is also a growing recognition among educators and students alike that the provision and enhancement of generic skills that can be used in "the real world" outside of academia is of vital importance. Among the most highly regarded of these skills can be counted the ability to work productively in teams, in both social and work settings, especially in situations where the various team members may have diverse backgrounds, experiences, and opinions. Indeed, it is in just such an environment that collaborative work can bring the greatest benefits.

Each of the authors represented in this volume has much to contribute to the central questions of how students can learn collaboratively using the new technologies, the problems that can be expected, and the benefits that may ensue. In their various ways, they examine how computer-supported group work differs from face-to-face group work, and the implications for both educators and students.

The aim of the first chapter, *Computer-Supported Collaborative Learning in Higher Education: An Introduction*, is to act as a jumping-off point for both researchers and practitioners interested in exploring this area for the first time. The chapter has three main objectives: first, to describe some of the benefits and problems that can be expected in a CSCL environment; second, to give an outline of some of the practical steps that need to be considered for CSCL to be successful; and third, to provide pointers to some of the more recent research reported in the literature. No attempt has been made to provide an exhaustive list of all of the research in this area – there is far too much! The selection of what gets a mention is, therefore, highly subjective. A list of references is provided at the end of the chapter for those seeking to pursue particular issues in more depth.

In Chapter 2, *Online Group Projects: Preparing the Instructors to Prepare the Students,* Valerie Taylor makes the excellent point that while group projects are often included in on-campus classes, instructors teaching computer-supported courses are frequently reluctant to attempt similar group projects online. She stresses the importance of staff development — if the process of integrating group work into an online environment is to be successful, staff must be trained appropriately. The chapter outlines lesson modules for teaching online instructors to use group projects in their courses.

In Chapter 3, *Time, Place and Identity in Project Work on the Net*, Sisse Siggaard Jensen and Simon Heilesen identify some of the fundamental conditions and factors that affect collaborative project work on the net. Understanding them is fundamental to developing key qualities in net-based collaborative learning such as confidence, reliability, and trust. They argue that collaboration and social interaction develop in continuous oscillations between abstract and meaningful frames of reference as to time and place, and that such oscillations condition the creation of a double identity of writer and author modes in social interaction. Further, they argue that collaborative work creates an ever-increasing complexity of interwoven texts and that strategies must be developed for organizing these.

In Chapter 4, *The Collective Building of Knowledge in Collaborative Learning Environments,* Alexandra Okada investigates how collaborative learning environments (CLEs) can be used to elicit the collective building of knowledge. This work discusses CLEs as lively cognitive systems and looks at some strategies that might contribute to the improvement of significant pedagogical practices. The study is supported by rhizome principles, whose characteristics allow us to understand the process of selecting and connecting what is relevant and meaningful for the collective building of knowledge. A brief theoretical and conceptual approach is presented, major contributions and difficulties about collaborative learning environments are discussed, and new questions and future trends about collective building of knowledge are suggested.

In Chapter 5, *Collaboration or Cooperation? Analyzing Small Group Interactions in Educational Environments*, Trena Paulus illustrates how computer-mediated discourse analysis (CMDA) can be used systematically to investigate online communication. She argues that intended outcomes of learner interactions, such as meaningful dialogue and joint knowledge construction, must be identified and analyzed to better understand the effectiveness of online learning activities. The CMDA approach is illustrated through analysis of a synchronous chat held by a three-person graduate student group as it completed a course assignment at a distance.

In Chapter 6, *Mapping Perceived Socio-Emotive Quality of Small-Group Functioning,* Herman Buelens, Jan Van Mierlo, Jan Van den Bulck, Jan Elen, and Eddy Van Avermaet demonstrate the influence of the socio-emotional quality of small-group functioning in a collaborative learning setting. They report a case study from a sophomores' class at a Belgian university, where the subjects were 142 undergraduates subdivided into 12 project groups of 12 students each. The aims of the study were to map group members' perception of the socio-emotive quality of their own group functioning and to examine if and how problems in groups of learners can be detected as soon as possible. Having demonstrated that dysfunctionalities within groups can be detected rather early, the authors suggest that corrective interventions should be implemented when they can still have an effect.

In Chapter 7, *A Constructivist Framework for Online Collaborative Learning: Adult Learning and Collaborative Learning Theory*, Elizabeth Stacey reviews and discusses theoretical perspectives that help to frame collaborative learning online. The chapter investigates literature about the type of learning and behavior that are anticipated and researched among participants learning collaboratively and discusses how these attributes explain computer-supported collaborative learning. The literature about learning is influenced by perspectives from a number of fields, particularly philosophy, psychology, and sociology. This chapter describes some of these perspectives from the fields of cognitive psychology, adult learning, and collaborative group learning.

In Chapter 8, *The Real Challenge of Computer-Supported Collaborative Learning: How Do We Motivate ALL Stakeholders?*, Celia Romm Livermore starts from the premise that to be effective, computer-supported collaborative learning has to be intrinsically motivating. In contrast to much of the literature in the field, which focuses almost exclusively on the needs of students, the chapter discusses three groups of stakeholders whose concerns and motivation have to be considered: students, instructors, and institutions. She introduces a paradigm that integrates the needs of the above three stakeholders. This is followed by a description of the Radical Model, an innovative approach to computer-supported collaborative learning that is an example of applying the proposed paradigm in practice. The chapter concludes with a discussion of the research implications arising from the model.

In Chapter 9, *Use and Mis-Use of Technology for Online, Asynchronous, Collaborative Learning*, William Klemm suggests that online learners are typically considered to be isolated learners, except for occasional opportunities to post views on an electronic bulletin board, and that this is not the team orientation that is so central to collaborative learning theory. So why does formal collaborative learning receive so little attention in online instruction? First, the teachers who do value collaborative learning generally are traditional educators and not involved in online instruction. Second, online teachers often have little understanding or appreciation for the formalisms of collaborative learning. In this chapter, the inadequacies of electronic bulletin boards, which, although universally used, do not readily support collaborative learning, are explained. As a better alternative, shared-document conferencing environments that allow learning teams to create academic deliverables are discussed. Finally, examples are given of well-known collaborative learning techniques and how these are implemented with shared-document conferencing.

In Chapter 10, *The Personal and Professional Learning Portfolio: An Online Environment for Mentoring, Collaboration, and Publication*, Lorraine Sherry, Bruce Havelock, and David Gibson describe the Personal and Professional Learning Portfolio (PLP), a software application designed to provide a flexible learning environment suitable for group collaborative work. After giving a description of the PLP's origins, structure, and pilot implementations across a range of educational settings, they detail two higher education sites to illustrate the key issues involved. The primary intent of the chapter is to bring awareness of the PLP to new audiences and expand consideration of its potential applications, while at the same time shedding light on the factors that influence adoption of collaborative technologies in institutional settings.

In Chapter 11, *Problems and Opportunities of Learning Together in a Virtual Learning Environment,* Thanasis Daradoumis and Fatos Xhafa explore new ways of collaborative learning in a virtual learning environment based on acquisition of knowledge from previous experience. They identify both the problems faced in real collaborative learning practices and the ways these problems can be overcome and turned into opportunities for more efficient learning. These issues concern pedagogical, organizational, and technical elements and constraints that influence the successful application of collaborative learning in distance education, such as efficient group formation, the nature of collaborative learning situations that promote peer interaction and learning, the student roles and tutor means in supervising and guiding the learning process, and an effective assessment of group work. They argue that the proposed methodology not only achieves better learning outcomes but also contributes to the tutor's professional development in a networked learning environment that facilitates social interaction among all participants, while building on existing skills.

In Chapter 12, *Web-Based Learning by Tele-Collaborative Production in Engineering Education,* Amiram Moshaiov deals with the need and the potential of reforming design projects into web-based learning by tele-collaborative production in engineering education. The chapter provides an overview of related topics including the impact of computer-mediated communication (CMC) on engineering and engineering education, the role of social creativity and dominance of multi-disciplinary thinking in modern engineering, assessing designers and the design process, and more. In addition to discussing the need and the potential of reforming engineering design projects, two major strategies for web-based learning by collaborative production in engineering education are discussed. It is concluded that short projects focusing on early design stages should be encouraged for the current assimilation of tele-collaboration, whereas long and complex design tasks may currently be better handled in a local framework.

In Chapter 13, *Relational Online Collaborative Learning Model*, Antonio Santos Moreno describes an instructional online collaborative-learning model that addresses the phenomenon from a systemic human relations and interaction perspective. Its main purpose is to aid students in their social building of knowledge when learning in a CSCL environment. The model argues that knowledge building in a networked environment is affected by the communication conflicts that naturally arise in human relationships. Thus, the model is basically proposing a way to attend to these communication conflicts. In this line, it proposes a set of instructional strategies to develop the student's meta-communication abilities. The concepts and instructional suggestions presented are intended to have a heuristic value and are hoped to serve as a frame of reference to: 1) understand the complex human patterns of relationships that naturally develop when learning in a CSCL environment, and 2) suggest some basic pedagogical strategies to the instructional designer to develop sound online networked environments.

In Chapter 14, *Online, Offline, and In-Between: Analyzing Mediated-Action Among American and Russian Students in a Global Online Class*, Aditya Johri argues that computer-supported collaborative learning is a situated activity that occurs in complex settings. This study proposes a sociocultural frame for theorizing, analyzing and designing online collaborative learning environments. The specific focus of this study is: learning as situated activity, activity theory as a theoretical lens, activity system as an analytical framework, and activity-guided design as a design framework for online learning environments. Using data gathered from a naturalistic investigation of an online collaborative learning site, this study reveals how these lenses and frameworks can be applied practically. The study also identifies the importance of design iterations for learning environments.

It is the editor's hope that the 14 chapters that comprise this book prove to be both stimulating and thought-provoking for readers interested in the field of computer-supported collaborative learning. If some of the information presented here inspires teachers to experiment with new ways of teaching, while perhaps other material provokes controversy and discussion, this book will have fulfilled a useful purpose.

Computer-supported collaborative learning is still very new. Researchers and practitioners alike still have much to learn. For all of us in this field, it is an exciting time.

Reference

Roberts, T.S. (2003). *Online collaborative learning: Theory and practice.* Hershey, PA: Information Science Publishing.

Acknowledgments

I would like to acknowledge the many people who have made this book possible. First and foremost, I would like to thank all of the authors of the individual chapters. Without exception, their contributions to this book have provided unique and thought-provoking insights into the increasingly important world of computer-supported collaborative learning.

Many thanks also to all of those who served as reviewers, without whose critical but always constructive comments this book would never have seen the light of day. Special mention in this regard must go to Sue Bennett, Dianne Conrad, John Dirkx, Richard Ferdig, Charles Graham, Albert Ingram, Agnes Kukulska-Hulme, Sebastian Loh, Hanni Muukkonen, John Nash, Rod Nason, Sabita d'Souza, Lesley Treleaven, Raven Wallace, Leigh Wood, and Ke Zhang.

Extra special thanks must be extended to Joanne McInnerney, who not only assisted in the review process, but also played a very significant role in ensuring that the finished product was of a uniformly high quality throughout.

I would also like to express my gratitude to Central Queensland University for providing me with the very necessary time and resources to see the book through to completion.

Last but very definitely not least, I would like to thank the very helpful and professional staff at Idea Group, Inc., whose patience and assistance throughout the whole process has been of the highest order.

This book is dedicated to my wife Jane, and my children Rickie-Lee and Mitchell, who teach me new and wondrous things every day.

Tim S. Roberts
Central Queensland University
Bundaberg, Queensland, Australia
May 2004

Chapter I

Computer-Supported Collaborative Learning in Higher Education:
An Introduction

Tim S. Roberts
Central Queensland University, Australia

Abstract

The rapidly increasing use of computers in education, and in particular the migration of many university courses to web-based delivery, has caused a resurgence of interest among educators in non-traditional methods of course design and delivery. This chapter provides an introduction to the field of computer-supported collaborative learning (CSCL). First, some of the major benefits are listed. Then, some of the common problems are described, and solutions are either given or pointed to in the literature. Finally, pointers are given to some of the more recent research in this area.

Introduction

It is interesting that collaborative learning methods were experimented with, and found to be successful, at least as early as the late 18th century, when George Jardine employed them for his philosophy classes at the University of Glasgow. He came to believe that *"...the teacher should move to the perimeter of the action...and allow the students freedom to...learn from one another"* (Gaillet, 1994). However, it is only recently, with

Copyright © 2005, Idea Group Inc. Copying or distributing in print or electronic forms without written permission of Idea Group Inc. is prohibited.

the advent of the new technologies, that many academics and instructors have become interested in exploring possible alternative methods of course design and delivery.

This renewed interest is evidenced by, among other things, the increasing number of conferences devoted to this topic; the number of papers submitted to both conferences and journals; the formation of numerous research groups around the world; and the number of web-sites devoted to providing resources in this area, such as the Online Collaborative Learning in Higher Education website (Roberts, 2002).

Benefits

The importance and relevance of social interaction to an effective learning process has been stressed by many theorists, from Vygotsky (1978) through advocates of situated learning such as Lave and Wenger (1991), and many other recent researchers and practitioners. Computer-supported collaborative learning (CSCL), if implemented appropriately, can provide an ideal environment in which interaction among students plays a central role in the learning process.

Ted Panitz, a Professor of Mathematics and Engineering at Cape Cod Community College, has written extensively about collaborative and cooperative education, mainly as it relates to the K-12 (Kindergarten to Year 12) sphere. However, much of his writing is equally applicable in higher education. He lists a substantial number of benefits to collaborative learning (Panitz, 2001); the list here is slightly abbreviated and amended:

Academic Benefits

Collaborative learning:

- promotes critical thinking skills

    Under this dot point Panitz suggests that collaborative learning develops higher level thinking skills; stimulates critical thinking; helps students clarify ideas through discussion and debate; enhances skill building and practice; develops oral communication skills; fosters metacognition in students; and improves students' recall of text content through cooperative discussions;

- involves students actively in the learning process

    And here, that it creates an environment of active, involved, exploratory learning; encourages student responsibility for learning; involves students in developing curriculum and class procedures; provides training in effective teaching strategies to the next generation of teachers; helps students wean themselves away from considering teachers the sole sources of knowledge and understanding; fits in well with the TQM and CQI models of effective management; promotes a learning goal rather than a performance goal; fits in well with the constructivist approach; and allows students to exercise a sense of control on task;

Copyright © 2005, Idea Group Inc. Copying or distributing in print or electronic forms without written permission of Idea Group Inc. is prohibited.

- improves classroom results

  Panitz suggests that collaborative learning promotes higher achievement and class attendance; promotes a positive attitude toward the subject matter; increases student retention; enhances self management skills; increases students' persistence in the completion of assignments and the likelihood of successful completion of assignments; helps students stay on task more and be less disruptive; and promotes innovation in teaching and classroom techniques;

- models appropriate student problem-solving techniques

  Collaborative learning fosters modeling of problem solving techniques by students' peers; allows assignment of more challenging tasks without making the workload unreasonable; can help weaker students improve their performance when grouped with higher achieving students; provides stronger students with the deeper understanding that comes only from teaching material (cognitive rehearsal); leads to the generation of more and better questions in class; provides a safe environment for alternate problem solutions; and addresses learning style differences among students.

Social Benefits

Collaborative learning:

- develops a social support system for students

  For example, it promotes student-faculty interaction and familiarity; develops social interaction skills; promotes positive societal responses to problems and fosters a supportive environment within which to manage conflict resolution; creates a stronger social support system; fosters and develops interpersonal relationships; and helps students to develop responsibility for each other;

- builds diversity understanding among students and staff

  Collaborative learning builds more positive heterogeneous relationships; encourages diversity understanding; fosters a greater ability in students to view situations from others' perspectives (development of empathy); and helps majority and minority populations in a class learn to work with each other (different ethnic groups, men and women, traditional and non-traditional students);

- establishes a positive atmosphere for modeling and practicing cooperation

  Collaborative learning establishes an atmosphere of cooperation and helping; helps students learn how to criticize ideas rather than people; helps to model desirable social behaviors necessary for employment situations that utilize teams and groups; helps students practice modeling societal and work related roles; fosters team building and a team approach to problem solving while maintaining individual accountability; creates environments where students can practice building leadership skills; increases leadership skills of female students; develops learning communities; provides the foundation for developing learning communi-

Copyright © 2005, Idea Group Inc. Copying or distributing in print or electronic forms without written permission of Idea Group Inc. is prohibited.

ties within institutions and in courses; helps to promote social and academic relationships well beyond the classroom and individual course; and helps teachers change their roles from their being the focus of the teaching process to becoming facilitators of the learning process (they move from teacher-centered to student-centered learning).

Psychological Benefits

Collaborative learning:

- can increase students' self esteem

  Group learning can help to reduce anxiety; enhance student satisfaction with the learning experience; promote a mastery attribution pattern rather than helpless attribution pattern; and encourage students to seek help and accept tutoring from their peers;

- develops positive attitudes towards teachers

  Such an environment can create a more positive attitude on the part of students towards their instructors; and create a more positive attitude by instructors toward their students; and set high expectations for students and teachers.

Even if one quibbles with some of the items in Panitz' list and notices a little duplication in places, the benefits – academic, social, and psychological – are substantial. Other benefits have also been noted by a large number of other researchers (e.g., Graham & Misanchuk, 2003; Johnson & Johnson 1996).

The benefits of collaborative learning within a computer-supported environment can be at least as great as those within a classroom or lecture hall.   In an asynchronous environment, students do not need to meet at a regular place at regular times, so "missing a session" assumes less importance. Fruitful and constructive discussion and dialogue can take place at any time of the day or night, whenever inspiration or enthusiasm strikes. Good ideas are less likely to be lost, and thoughts can be followed through without regard to the normal time constraints. Opinions can be considered on their merits, without some of the stereotypical assumptions that may be superimposed in a face-to-face environment based on the speaker's gender, physical appearance, or mannerisms.

Employers - whether private or public corporations, government organizations, or small business - are today more than ever ranking generic skills at least as highly as subject-content knowledge when they select graduates. Among the most frequently listed and highly prized of those generic skills are computer literacy and the ability to work effectively in a team environment. If introduced appropriately into the curriculum, CSCL has the potential to provide students with both.

Copyright © 2005, Idea Group Inc. Copying or distributing in print or electronic forms without written permission of Idea Group Inc. is prohibited.

Problems and Solutions

If the benefits are so numerous (and they are), why is the adoption of collaborative learning techniques not widespread? A number of problems are apparent, among the most prominent of which is the simple problem of inertia. It is often the case in higher education institutions that it is easier for educators to follow accepted practices than to carve out new paths. Those brave enough to attempt to replace the traditional lecture-tutorial model with something as radical as group work may risk finding themselves subject to abuse from superiors, colleagues, and students, who may regard non-traditional methods of instruction with suspicion and distrust. The problems can therefore be broken down into three categories: those occurring because of the influence of other stakeholders; those affecting the instructors; and those directly related to the students and the learning process.

With regard to other stakeholders, those seeking sources of information so as to persuade colleagues, administrators, and managers that the benefits of collaborative learning outweigh the problems would be well-advised to explore the writings of Ted Panitz (1997, 1999, 2001). His articles deal with many fundamental topics, such as why teachers often resist collaborative learning techniques, why students resist collaborative learning, and the reactions of other stakeholders (such as parents and administrators). The list of benefits of collaborative learning given above is his. He has also described 18 policy issues that need to be considered if the introduction of collaborative learning techniques is to be successful (Panitz, 1997, 1999; Panitz & Panitz, 1998).

With regard to the instructors, it is typically the case that teaching staff are most comfortable using the traditional methods by which they themselves were taught. Further, the majority of teachers and lecturers will not have had any training in delivering collaborative classes via a computer. Mason (1970), as cited in Bruffee (1999), says that:

"Redesigning an education system is a relatively easy exercise. Changing one's own method of teaching, especially when it has been acclaimed as successful by all the old standards, is very much harder."

Salmon (2000) has suggested that in any computer-supported session, the instructor might need to be a chair, host, lecturer, tutor, facilitator, mediator of team debates, mentor, provocateur, observer, participant, co-learner, assistant, community organizer, or some combination of these! This clearly points to the fact that the skills required on the part of the instructor are more complex and diverse than those required for a face-to-face lecture. Thus, instructors need to be prepared for the different roles they will have to assume. In some cases, this may mean formalized training programs.

With regard to the students, those coming to CSCL courses for the first time can be apathetic or sometimes openly hostile. This may be because of the "CS," or the "CL." The solution to both of these problems is to ensure that students are computer-literate and used to the idea of working in teams prior to the commencement of the course. This can be best achieved by having computer skills and teamwork introduced as core compo-

Copyright © 2005, Idea Group Inc. Copying or distributing in print or electronic forms without written permission of Idea Group Inc. is prohibited.

nents of an introductory course, and making students aware that CSCL may be a feature of future courses. However, since this requires cooperation from program administrators and academic managers, it may not be possible in all cases.

Much of the literature in this area would indicate that the two most important steps for the instructor to take in cases where students are new to CSCL are, first, to inform students in advance of the multiple benefits to be obtained from group work, and second, to acquaint all students with their responsibilities as team members.

Graham and Misanchuk (2003) have suggested that there are three general stages that are important if using groups in a CSCL environment is to be successful:

- creating the groups,
- structuring the learning activities, and
- facilitating group interactions.

They recommend a series of practical steps that can be undertaken in each of these three areas to maximize the chances for successful learning to occur. Other excellent sources of good advice are to be found in Kemery (2000), Paloff and Pratt (1999), and McConnell (2000).

Initial resistance to the idea of working in groups is quite common. Salomon (1992), among others, has pointed out that despite the mass of literature praising collaborative learning, teams very frequently do not work well, and lists some common problems:

- the "free rider" effect, where one or more students do not do their fair share (Kerr & Bruun, 1983),
- the "sucker" effect, where one or more members is left to do all of the work (Kerr, 1983),
- the "status sensitivity" effect, where cliques form within the group (Dembo & McAuliffe, 1987), and
- the "ganging up on the task" phenomenon, where subtasks are divided among individual members of the group without much (if any) collaboration taking place (Salomon & Globerson, 1987).

All of these problems have effective solutions, however. The most common complaint about any form of group work is that one or more members of the group have not contributed. This can never be completely eliminated, but can be mitigated by a variety of techniques, including collecting regular reports from team members as to each member's responsibilities and how fully they are being accomplished. Non-contributing members can be reassigned to other groups or be awarded reduced marks, according to the circumstances.

Copyright © 2005, Idea Group Inc. Copying or distributing in print or electronic forms without written permission of Idea Group Inc. is prohibited.

In a computer-supported environment, hostile or bullying students are likely to be less disruptive and less intimidating to others in the group since they are only virtually, rather than physically, present. However, such an environment can be prone to other problems, such as flaming (the sending of deliberately inflammatory email), spamming (the sending of bulk, unwanted email), etc.

Davis (1993) has supplied excellent solutions to a number of problems of collaborative learning, broken down under the following headings:

- general strategies,
- designing group work,
- organizing learning groups,
- evaluating group work,
- dealing with student and faculty concerns about group work, and
- setting up study teams.

Another highly-recommended guide to effective strategies for cooperative learning has been provided by Felder and Brent (2001); they deal with

- forming teams,
- dealing with dysfunctional teams,
- grading,
- distance learning, and
- avoiding discouragement.

The second of these, dealing with dysfunctional teams, is discussed at some length.

Collaborative learning is sometimes seen as a means of assisting the less able students to achieve better grades than might otherwise be expected. This view carries with it the implication that this usually occurs at the expense of exceptional grades for those more able.

However, it is quite possible for collaborative learning to benefit all students. It is often said by academics and instructors generally that the best way to learn a subject is to teach it, and for good reason – when teaching, one needs to gain a thorough knowledge of the subject, not only to prepare material, but also to be able to answer questions confidently. No matter how good the preparation on the part of the instructor, further questions will almost inevitably arise during the course of instruction, thus leading to an even better understanding.

It seems naïve in the extreme to assume a similar process will not occur when the students themselves take on the unofficial role of instructors to other students within their group.

Copyright © 2005, Idea Group Inc. Copying or distributing in print or electronic forms without written permission of Idea Group Inc. is prohibited.

Thus, it can be expected that in many cases the learning will increase for all students within the group, and not just those who are least able to learn for themselves. This conclusion seems to be supported by research in this area.

For example, Webb and Sugrue (1997) report that *"among groups with above-average students. . .the higher level of discussion translated into an advantage in the achievement tests for the below-average students (in those groups),"* both when they were tested on a group basis and also individually; on the other hand, *"high ability students performed equally well in heterogeneous groups, homogeneous groups, and when they worked alone."* Both of these results have also been shown in different contexts by other researchers (Azmitia, 1988; Hooper & Hannafin, 1988; Hooper, Ward, Hannafin, & Clark, 1999).

Many examples of successful implementation of CSCL have been reported in the literature. One particularly noteworthy case in an asynchronous learning environment is the so-called "radical model" (Romm & Taylor, 2000). The radical model dispenses with traditional face-to-face teaching almost entirely and places the emphasis on the students themselves to learn within a group setting, using the Web for resource material and email discussion groups for communication and presentation of assessment items, with the instructor providing guidance and feedback as required.

At the beginning of the course, the instructor randomly assigns students into groups. Each group is assigned one of the weekly topics and has to make a single online presentation. Students are assessed not just for their group's presentation but also for their comments about other presentations. Each group presentation is also assessed on the quality of the discussion that follows. Typically, by the end of semester, students will have received over 100 inputs on their work from other students in the group, other groups, and the instructor.

In the last week of term, students are invited to submit a recommendation in writing on each other's group performance. The instructor considers any such recommendations when allocating individual marks for group performance to members of the group. A student who a group decides did not contribute sufficiently may suffer a reduction in mark as a result.

Different assessment criteria may be used – for example, for the electronic presentation, clarity and structure of presentation, originality of ideas, and ability to substantiate arguments by relevant data; for other contributions, understanding the arguments that are made by other presenters, linking them to the relevant literature, and making pertinent critical comments about these arguments.

The students' final marks are based on a combination of their group work throughout the semester and their performance in an end-of-semester examination.

In common with some other forms of collaborative learning, the radical model points the way towards other possible forms of assessment in an asynchronous learning environment. However, as presently constituted, the model still represents an example of a fairly traditional model of assessment, since the grade awarded is based on the standard paradigm of attempting to assess the individual's own efforts, even within the context of an online collaborative learning environment.

Copyright © 2005, Idea Group Inc. Copying or distributing in print or electronic forms without written permission of Idea Group Inc. is prohibited.

Research

Literally thousands of researchers around the world are engaged in some aspect of computer-supported collaborative learning at any given moment; what follows, then, can be at best a brief sketch of some of the more prominent conferences, books, research groups, journals, and articles concerned with CSCL.

Conferences

The number of conferences devoted to computer-supported collaborative learning is still small. Of pre-eminence is the CSCL conference itself, held in the U.S. in 1991 (Illinois), 1995 (Indiana), 1999 (California), and 2002 (Colorado). It has been held outside of the U.S. twice, first in Toronto, Canada, in 1997, and most recently in Bergen, Norway, in 2003. A European version – Euro-CSCL – was held in the Netherlands in 2001 (Maastricht). At the time of this book publication, the next CSCL conference is scheduled for Taiwan, in June 2005.

Europe has hosted eight conferences, at roughly two-year intervals, in the ECSCW series on the closely related field of computer-supported cooperative work, the most recent in 2003, in Helsinki, Finland.

Other conferences frequently attract papers relating to computer-supported collaborative learning. Among the most prominent are the Networked Learning Conferences in the UK, run by the University's of Lancaster and Sheffield; the IEEE International Conferences on Advanced Learning Technologies (ICALT); the International Conferences of the Learning Sciences (ICLS); the International Association for the Study of Cooperation in Education (IASCE) Conferences; the International Association of Science and Technology for Development (IASTED) Conferences on Web-Based Education (WBE); and the International Conferences on Computers in Education (ICCE).

Research Groups

Many groups around the world are actively engaged in research into CSCL, and many valuable resources can be found at their sites.

Among the most prominent are Tim Koschman's group at the Southern Illinois University School of Medicine; Daniel Suthers' Laboratory for Interactive Learning Technologies (LILT) at the University of Hawaii; the TECFA group, until recently led by Pierre Dillenbourg at the University of Geneva; Jeremy Roschelle and Roy Pea at SRI International; the computer-based collaborative group work project led by David McConnell at the University of Sheffield; Gerry Stahl's group at the University of Colorado at Boulder; and Yrjo Engestrom's groups at the University of California at San Diego and at the University of Helsinki in Finland. Mark Guzdial and his group at the Georgia Institute of Technology are especially concerned with collaborative learning in an asynchronous environment.

Copyright © 2005, Idea Group Inc. Copying or distributing in print or electronic forms without written permission of Idea Group Inc. is prohibited.

Formal groups with excellent resource sites include the Collaborative Learning Environment (CLE) group in South Carolina, NJIT's WebCenter for Learning Networks Effectiveness Research group in New Jersey, Pennsylvania State University's Center for Excellence in Learning (CELT) group, and the University of Illinois' Sloan Center for Asynchronous Learning Environments (SCALE).

The University of Minnesota's Cooperative Learning Center (CLC), maintained by David T. Johnson and Roger W. Johnson, is primarily aimed at learning at the K-12 level.

The International Society of the Learning Sciences (ISLS) is a relatively new professional society founded to help unite and support scientific and educational work in the study of learning and education. ISLS builds on the traditions developed and solidified by the International Conferences for the Learning Sciences (ICLS), the Computer-Supported Collaborative Learning (CSCL) conferences and the Journal of the Learning Sciences (JLS).

The International Association for the Study of Cooperation in Education (IASCE) supports the development and dissemination of research and inquiry that foster the understanding of cooperative learning.

Books

Although many books on educational theory make reference to various forms of group learning, relatively few books have been published that specifically focus on computer-supported collaborative learning.

Koschmann (1996) has twelve chapters outlining current research in instructional technology; on the back cover, the hope is expressed that *"...it will help to define a direction for future work in the field."* Dillenbourg (1999) brings together a variety of differing perspectives on collaborative learning from twenty scholars from the disciplines of psychology, education and computer science. McConnell (2000) looks at how communication and distributed advanced learning technologies can be used to support group learning, and considers the importance of the sociocultural dimension of learning. Roberts (2003) brings together thirteen varied perspectives from leading researchers and practitioners in the area of CSCL.

Bonk and King (1998) detail the tools for computer conferencing and collaboration and the learning theories grounding their use. Eisenstadt and Vincent (2000) describe examples of leading-edge research projects from the Knowledge Media Institute at the UK Open University. Littleton and Light (1999) outline experimental studies of process and product, naturalistic studies of computer-based collaborative activities, and contexts for collaboration.

Paloff and Pratt (1999) provide proven strategies for taking learning beyond the classroom and into the online environment, focusing on the critical task of creating a sense of community among learners. Smith and McCann (2001) examine the experiences and lessons from over 20 different institutions pioneering new approaches for more effective teaching and learning.

Copyright © 2005, Idea Group Inc. Copying or distributing in print or electronic forms without written permission of Idea Group Inc. is prohibited.

Journals

There are over a hundred journals that occasionally carry articles directly related to computer-supported collaborative learning. A list can be found at Roberts (2002).

Recent Research

It is not possible to list – let alone detail – all of the research currently being undertaken in the field of computer-supported collaborative learning. The following has the modest aim of selecting a small sample, in order to give a flavor of the types of research being undertaken.

Many different theories of learning have contributed to our current understanding of CSCL, and all of these continue to attract researchers interested in building on the foundations set down by others. Among some of the more prominent of these theories are

- sociocultural theory, e.g., Vygotsky (1978), which emphasizes that interaction within a social environment is vitally important to learning, and that much cognitive development takes place within a certain *zone of proximal development*;

- constructivist theory, e.g., Bruner (1966) and Sherman (2000), which essentially views knowledge of the world as being constructed by the learner;

- distributed cognition theory, e.g., Oshima, Bereiter, and Scardamalia (1995), which emphasizes the interactions between learners, their environment, and cultural artifacts;

- situated cognition theory, e.g., Lave and Wenger (1991), and Brown, Collins, and Duguid (1989), which emphasizes sharing and doing within the context of a social unit.

Of course, this is very far from an exhaustive list. In addition to those mentioned above, Kearsley (2001) has listed an additional forty theories of learning worthy of inclusion in a learning and instruction database.

Koschmann (1999) has proposed a new theoretical framework for understanding learning as a socially-grounded phenomenon based on the writings of the Russian philologist M. M. Bakhtin. Stahl (2002) has proposed a theoretical framework for CSCL incorporating models of knowledge building, perspectives, and artifacts. His writing refers to *building collaborative knowing*, a term derived from the work of Scardamalia and Beireiter (1996). Harapnuik (1998) has discussed how a new learning approach he has termed *Inquisitivism* can be implemented in the development of learning environments catering to adult learners.

The role of computer software in enabling learners to construct and manipulate visual representations of their emerging knowledge has been extensively studied. The term

Copyright © 2005, Idea Group Inc. Copying or distributing in print or electronic forms without written permission of Idea Group Inc. is prohibited.

"representational guidance" has been coined by Suthers and Hundhausen (2002) to refer to how software environments can be used to facilitate the expression and inspection of different kinds of information.

Collaborative learning is of necessity a social activity. Treleaven (2003) has proposed a new taxonomy for evaluation studies of CSCL, with emphasis on three models with sociocultural perspectives. Wegerif (1998) has discussed the social dimensions of asynchronous learning networks, and argued that the social dimension needs to be taken into account in the design of courses. Arias, Eden, Fischer, Gorman, and Scharff (1999) looked at the issues of empowerment and informed participation and concluded that if their importance goes unrecognized, we will not address the challenges faced by authentic real-world learning situations. The importance of feedback mechanisms has been examined by Zumbach, Hillers, and Reimann (2003).

There is a substantial body of work, from both inside and outside of the CSCL arena, pointing to the value of interaction to the learning process, e.g., Anderson (2003). Much has also been written about the subject of vicarious learning. McKendree, Stenning, Mayes, Lee, and Cox (1998) focus on the distinction between exposition and derivation in discourse, discuss how this might be used to describe what happens in learning dialogues, and find benefit to the vicarious learner. That there might be substantial value in vicarious learning has also been pointed out by others, e.g. Sutton (2001) and Fulford and Zhang (1993).

The problem of collaborative learning perhaps not catering sufficiently to individual differences has been noted by a number of researchers, including Huang (2002) and Westera (1999). An eloquent defense of the solitary learner, including a description of *the darker side of collaborative learning*, has been provided by Hopper (2003).

Dillenbourg, Baker, Blaye, and O'Malley (1996) have described the evolution of research into collaborative learning, and argue that empirical studies have recently started to focus less on establishing parameters for effective communication and more on trying to understand the roles variables such as group size, group composition, nature of the task, etc., play in mediating interaction. They argue that the shift to a more process-oriented account requires new tools for analyzing and modeling interactions.

As could be expected, there is a significant amount of research in the literature that reports the results of particular case studies. Contexts range from a small class at an Eastern Pentecostal Bible College (Lavellée, 1999) to a large undergraduate class in organic chemistry (Glaser & Poole, 1999). An examination of collaborative learning for a postgraduate MBA class has been provided by Stacey (1999). Agostinho, Lefoe, and Hedberg (1997) have also described how a postgraduate course was implemented on the Web. The interactions that took place among the students and between the students and instructor are discussed to illustrate how collaborative learning and problem solving can be facilitated and supported. Renzi and Klobas (2000) describe first steps taken at an Italian business university to use CSCL to enhance the quality of teaching and learning for students in large classes, and conclude that plans for wider implementation should recognize differences in the potential contribution of computer-supported teaching and learning across disciplines, and differences in teachers' needs for training. These differences may result in different times for diffusion of computer-supported initiatives throughout a course, unit, or university.

Copyright © 2005, Idea Group Inc. Copying or distributing in print or electronic forms without written permission of Idea Group Inc. is prohibited.

Muffoletto (1997) has suggested that collaboration works well with a professional or graduate course where the level of homogeneity among students is much higher. Many researchers, for example, Ragoonaden and Bordeleau (2000) have emphasized that autonomous, highly independent students generally prefer working alone. Hopper (2003) has supplied a comprehensive defense of the solitary learner.

Benbunan-Fich and Hiltz (1999) tested the effectiveness of using an asynchronous learning network versus traditional manual methods, with individuals and groups discussing and solving a case study. Findings indicated that groups working in an asynchronous networked environment produced better and longer solutions to the case study, but were less satisfied with the interaction process. More recently the same authors have presented similar results in Benbunan-Fich, Hiltz, and Turoff (2003).

A relatively new area receiving increased attention in the last few years is the use of artificial intelligence techniques to aid the learning situation. Dillenbourg et al (1997) propose the use of new types of artificial agents that compute statistics regarding interactions and display them to human or perhaps artificial tutors, or to the learners themselves. Ogato and Yano (2000) have described a knowledge awareness filtering technique to assist efficient collaborative learning using individual user agents.

A group at the Open University of the Netherlands has proposed an intelligent CSCL environment based upon embedding certain properties in the environment that act as social contextual facilitators, with the aim of initiating and sustaining student interaction (Kreijns, Kirschner, & Jochems, 2002). They point out that while there is some positive research on asynchronous CSCL environments, *"...(t)here is also research that shows that contemporary CSCL environments do not completely fulfill expectations on supporting group learning, shared understanding, social construction of knowledge, and acquisition of competencies."* This is probably an understatement.

The use of collaborative learning specifically within the higher education sphere has been researched by David McConnell's Computer-Based Collaborative Group Work (CBCGW) group at the University of Sheffield. An overview of the CBCGW project is given in Lally and Barrett (1999). Bowskill, Foster, Lally, and McConnell (2000) have described a rich professional development environment (RPDE) for university staff to explore and develop networked collaborative learning. Allan, Barker, Fairbairn, Freeman, and Sutherland (2002) have described the use of tutor-less groups, their advantages and disadvantages, from a first-hand standpoint.

Still other research focuses on the advances in technology that may transform CSCL. Roschelle and Pea (2002), for example, have described how wireless handheld computers may have a dramatic impact on the learning environment. Iles et al (2002) have examined the interactive dialogues arising from classroom trials with wireless handhelds.

An excellent summary of some of the more important literature on learning in virtual teams published throughout the decade of the 1990s entitled Learning in Virtual Teams: A summary of current literature by Regina Smith can be found at http://www.msu.edu/~smithre9/Project12.htm.

A list of articles published in the last five years, together with links to other resources, can be found at Roberts (2002).

Copyright © 2005, Idea Group Inc. Copying or distributing in print or electronic forms without written permission of Idea Group Inc. is prohibited.

Summary

This chapter has attempted to provide a brief summary of the benefits and problems of computer-supported collaborative learning, to describe some of the steps that need to be taken if CSCL is to be successfully employed, and to detail some of the current research in this area. In doing so, it has been necessary to omit a huge amount of research that may ultimately prove to be of great worth. Nevertheless, it is the author's hope that some may find the effort useful.

References

Agostinho, S., Lefoe, G., & Hedberg, J. (1997). Online collaboration for effective learning: A case study of a post-graduate university course. Available at: *http://cedir.uow.edu.au/CEDIR/flexible/resources/lefoe.html.*

Allan, B., Barker, M., Fairbairn, K., Freeman, M, & Sutherland, P. (2002). High level student autonomy in a virtual learning environment. *Proceedings of Networked Learning* 2002 conference, Sheffield, UK. Available at: *http://www.shef.ac.uk/nlc2002/proceedings/papers/01.htm.*

Anderson, T. (2003). Modes of interaction in distance education: Recent developments and research questions. In M. Moore (ed.), *Handbook of distance education,* pp. 129-144. Mahwah, NJ: Erlbaum.

Arias, E.G., Eden, H., Fischer, G., Gorman, A., & Scharff, E. (1999). Beyond access: Informed participation and empowerment. In C. Hoadley (ed.), *Proceedings of the Computer Supported Collaborative Learning (CSCL '99) Conference*, Stanford, pp. 20-32. Available at: *http://www.cs.colorado.edu/~l3d/systems/EDC/pdf/cscl99.pdf.*

Azmitia, M. (1988). Peer interaction and problem solving: When are two heads better than one? *Child Development 59*, 87-96.

Benbunan-Fich, R. & Hiltz, S.R. (1999). Effects of asynchronous learning networks: A field experiment. *Group Decision and Negotiation*, 8:409-426.

Benbunan-Fich, R., Hiltz, S. R., & Turoff, M. (2003). A comparative content analysis of face-to-face vs. asynchronous group decision making. *Decision Support Systems.* 34(4):457-469.

Bonk, C. J. & King, K. S. (eds.) (1998). *Electronic collaborators: Learner-centered technologies for literacy, apprenticeship, and discourse.* Hillsdale, NJ: Lawrence Erlbaum Associates.

Bowskill, N., Foster, J., Lally, V., & McConnell, D. (2000). Networked professional development: Issues and strategies in current practice. *International Journal of Academic Development*, 5(2): 93-106.

Brown, J., Collins, A., & Duguid, P. (1989). Situated cognition and the culture of learning. *Educational Researcher*, 18(1): 32-42.

Copyright © 2005, Idea Group Inc. Copying or distributing in print or electronic forms without written permission of Idea Group Inc. is prohibited.

Bruffee, K.A. (1999). *Collaborative learning: Higher education, interdependence, and the authority of knowledge.* Baltimore, MD: The John Hopkins University Press.

Bruner, J. (1966). *Toward a theory of instruction.* Cambridge, MA: Harvard University Press.

Davis, B.G. (1993). Collaborative Learning: Group Work and Study Teams. Available at: *http://teaching.berkeley.edu/bgd/collaborative.html.*

Dembo, M.H., & McAuliffe, T.J. (1987). Effects of perceived ability and grade status on social interaction and influence in cooperative groups. *Journal of Educational Psychology, 79:* 415-423.

Dillenbourg, P. (ed.) (1999), *Collaborative learning: Cognitive and computational approaches.* New York: Pergamon Press.

Dillenbourg, P., Baker, M., Blaye, A., & O'Malley, C. (1996). The evolution of research on collaborative learning. In E. Spadia & P. Reman (eds.), *Learning in human and machine: Towards an interdisciplinary learning science,* pp.189-211., Oxford, UK: Elsevier.

Eisenstadt, M. & Vincent, T. (eds.) (2000). *The knowledge web: Learning and collaborating on the net.* Open and Distance Learning Series. London; Kogan Page Ltd.

Felder, R. M. & Brent, R. (2001). Effective strategies for cooperative learning. Available at: *http://www2.ncsu.edu/unity/lockers/users/f/felder/public/Papers/CL Strategies(JCCCT).pdf.*

Fulford, C.P. & Zhang, S. (1993). Perceptions of interaction: The critical predictor in distance education. *American Journal of Distance Education, 7*(3): 8-21.

Gaillet, L. L. (1994). An Historical Perspective on Collaborative Learning. Available at: *http://www.cas.usf.edu/JAC/141/gaillet.html.*

Glaser, R.E. & Poole, M.J. (1999). Organic chemistry online: Building collaborative learning communities through electronic communication tools *Journal of Chemical Education,* 76: 699-703.

Graham, C.R. & Misanchuk, M.. (2003). Computer-mediated learning groups: Benefits and challenges to using groupwork in online learning environments. In T.S. Roberts (ed.), *Online collaborative learning: Theory and practice.* Hershey, PA: Information Science Publishing.

Harapnuik, D. (1998). *The HHHMMM? What does this button do? Approach to learning.* Available at: *http://www.quasar.ualberta.ca/nethowto/publish/inquisitivism. htm.*

Hooper, S. & Hannafin, M.J. (1988). Cooperative CBI: The effects of heterogeneous versus homogeneous grouping on the learning of progressively complex concepts. *Journal of Educational Computing Research,* 4: 413-424.

Hooper, S., Ward, T.J., Hannafin, M.J., & Clark, H.T. (1989). The effects of aptitude composition on achievement during small group learning. *Journal of Computer-Based Instruction,* 16:102-109.

Copyright © 2005, Idea Group Inc. Copying or distributing in print or electronic forms without written permission of Idea Group Inc. is prohibited.

Hopper, K. B. (2003). In defense of the solitary learner: A response to collaborative, constructivist education, *Educational Technology,* 43(2): 24-29.

Huang, H.-M. (2002). Toward constructivism for adult learners in online-learning environments. *British Journal of Educational Technology.* 33(1): 27-37.

Iles, A., Glaser, D., Kam, M., & Canny, J. (2002). Learning via Distributed Dialogue: Livenotes and Handheld Wireless Technology. In G. Stahl (ed.), *Proceedings of CSCL2002.* Hillsadle, New Jersey: Lawrence Erlbaum Associates.

Johnson, D.W. & Johnson, R.T. (1996).Cooperation and the use of technology. In D.H. Jonassen (ed.), *Handbook of research for educational communications and technology,* pp 1017-1044. New York: Macmillan Library Reference.

Kearsley, G. (2001). Explorations in learning & instruction: The theory into practice database. Available at: *http://tip.psychology.org/.*

Kemery, E.R. (2000).Developing online collaboration. In A. Aggarwal (ed.), *Web-based learning and teaching technologies: Opportunities and challenges.* Hershey, PA: Idea Group Press.

Kerr, N.L. (1983). Motivation losses in small groups: A social dilemma analysis. *Journal of Personality and Social Psychology,* 45: 819-828.

Kerr, N.L. & Bruun, S.E. (1983). Dispensability of member effort and group motivation losses: Free rider effects. *Journal of Personality and Social Psychology, 44:* 78-94.

Koschmann, T. (ed.) (1996). *CSCL: Theory and practice of an emerging paradigm.* Hillsdale, NJ: Lawrence Erlbaum Associates.

Kreijns, K., Kirschner, P.A., & Jochems, W. (2002). The sociability of computer-supported collaborative learning environments. *Educational Technology and Society,* 5 (1). Available at: *http://ifets.massey.ac.nz/periodical/vol_1_2002/kreijns.pdf.*

Lally, V. & Barrett, E. (1999). Building a learning community on-line: Towards socio-academic interaction. *Research Papers in Education,* 14(2): 147-163.

Lave, J. & Wenger, E. (1991). *Situated learning: Legitimate peripheral participation.* Cambridge, UK: Cambridge University Press.

Lavellée, N. (1999). An evaluation of computer-supported collaborative learning in distance education courses offered by Eastern Pentecostal Bible College. Available at: *http://www.epbc.edu/nnlavallee/Collaborative.htm.*

Littleton, K. & Light, P. (eds.) (1999). *Learning with computers: Analyzing productive interactions.* New York: Routledge.

Mason, E. (1970). *Collaborative learning.* London: Ward Lock Educational.

McConnell, D. (2000). *Implementing computer-supported cooperative learning,* 2nd Edition. London: Kogan Page.

McKendree, J., Stenning, K., Mayes, T., Lee J., & Cox, R. (1998). Why observing a dialogue may benefit learning: The vicarious learner. Journal of Computer-Assisted Learning, 14(2): 110-119.

Copyright © 2005, Idea Group Inc. Copying or distributing in print or electronic forms without written permission of Idea Group Inc. is prohibited.

Muffoletto, R. (1997). Reflections on designing and producing an Internet-based course. *TechTrends*, 42(2): 50-53.

Ogata, H. & Yano, Y. (2000). Combining Knowledge Awareness and Information Filtering in an Open-ended Collaborative Learning Environment. *International Journal of Artificial Intelligence in Education*, 11: 33-46.

Oshima, J., Bereiter, C., & Scardamalia, M. (1995). Information-access characteristics for high conceptual progress in a computer-networked learning environment. In J. L. Schnase & E.L. Cunnius (eds.), *Proceedings of CSCL '95: The first international conference on computer support for collaborative learning*. Mahwah, NJ: Lawrence Erlbaum Associates.

Palloff, R. M. & Pratt, K. (1999). Building learning communities in cyberspace: Effective strategies for the online classroom. San Francisco, CA: Jossey-Bass.

Panitz, T. (1997). Collaborative versus cooperative learning: Comparing the two definitions helps understand the nature of interactive learning. *Cooperative Learning and College Teaching*, 8(2).

Panitz, T. (1999). Is cooperative learning possible 100% of the time in college classes? *Cooperative Learning and College Teaching*, 9(2): 13-14.

Panitz, T. (2001). The case for student-centered instruction via collaborative learning paradigms. Available at: *http://home.capecod.net/~tpanitz/tedsarticles/coop benefits.htm.*

Panitz, T. & Panitz, P. (1998). Ways to encourage collaborative teaching in higher education. In *University teaching: International perspectives*, James J.F. Forest edition, pp.161-202. New York: Garland Publishers.

Ragoonaden, K. & Bordeleau, P. (2000). Collaborative learning via the Internet, *Educational Technology and Society*, 3(3). Available at: *http://ifets.massey.ac.nz/periodical/vol_3_2000/v_3_2000.html.*

Renzi, S. & Klobas, J. (2000). Steps toward computer-supported collaborative learning for large classes. *Educational Technology & Society 3*(3).

Roberts, T.S. (2002). The Online Collaborative Learning in Higher Education web site. Available at: *http://clp.cqu.edu.au.*

Roberts, T. S. (ed.) (2003). *Online collaborative learning: Theory and practice.* Hershey, PA: Information Science Publishing.

Romm, C. T. & Taylor, W. (2000). Thinking creatively about on-line education. In M. Khosrowpour (ed.), *Challenges of information technology management in the 21st century.* Hershey, PA: Idea Group Publishing.

Roscelle, J. & Pea, R. (2002). A walk on the WILD side: How wireless handhelds may change CSCL. In G. Stahl (ed.) *Proceedings of CSCL2002.* Hillsadle, New Jersey: Lawrence Erlbaum Associates.

Salmon, G. (2000). *E-moderating: The key to teaching and learning online.* Sterling, VA: Stylus Publishing.

Copyright © 2005, Idea Group Inc. Copying or distributing in print or electronic forms without written permission of Idea Group Inc. is prohibited.

Salomon, G. (1992). What does the design of effective CSCL require and how do we study its effects. *ACM Conference on Computer Supported Collaborative Learning*, 21(3). New York: ACM Press.

Salomon, G. & Globerson, T. (1987). When teams do not function the way they ought to. *International Journal of Educational Research,* 13: 89-100.

Scardamalia, M. & Beireiter, C. (1996). Computer support for knowledge-building communities. In T. Koschmann (ed.), *CSCL: Theory and practice of an emerging paradigm,* pp. 249-268. Hillsdale, NJ: Laurence Erlbaum.

Sherman, L.W. (2000). Postmodern constructivist pedagogy for teaching and learning cooperatively on the web. *CyberPsychology and Behavior: Special Issue.* 3(1).

Smith, B.L. & McCann, J. (eds.) (2001). *Reinventing ourselves: Interdisciplinary education, collaborative learning, and experimentation in higher education.* Bolton, MA: Anker Publishing Co.

Stacey, E. (1999). Collaborative learning in an online environment. *Canadian Journal of Distance Education, 14*(2), 14-33.

Stahl, G. (2002). Contributions to a Theoretical Framework for CSCL. In G. Stahl (ed.) *Proceedings of CSCL2002.* Hillsadle, New Jersey: Lawrence Erlbaum Associates.

Suthers, D. & Hundhausen, C. (2002). The Effects of Representation on Students' Elaborations in Collaborative Inquiry. *Proceedings of CSCL 2002,* Boulder, Colorado, January 7-11, pp. 472-480.

Sutton, L. (2001). The principles of vicarious interaction in computer-mediated communications. *Journal of Interactive Educational Communications, 7*(3): 223-242.

Treleaven, L. (2003). A new taxonomy for evaluation studies of online collaborative learning. In T.S. Roberts (ed.), *Online collaborative learning: Theory and practice.* Hershey, PA: Information Science Publishing.

Vygotsky, L.S. (1978). *Mind in society: The development of higher psychological processes.* Cambridge, MA: Harvard University Press.

Webb, N. & Sugrue, B. (1997). Equity issues in collaborative group assessment: Group composition and performance. *CSE Technical Report 457,* University of California, Los Angeles.

Wegerif, R. (1998). The social dimension of asynchronous learning networks. *Journal of Asynchronous Learning Networks,* 2 (1). Available at: *http://www.aln.org/publications/jaln/v2n1/v2n1_wegerif.asp.*

Westera, W. (1999). Paradoxes in open, networked learning environments: Toward a paradigm shift. *Educational Technology,* 39(1): 17-23.

Zumbach J., Hillers, A., & Reimann, P. (2003). Supporting distributed problem-based learning: The use of feedback mechanisms in online learning. In T.S. Roberts (ed.), *Online collaborative learning: Theory and practice.* Hershey, PA: Information Science Publishing.

Copyright © 2005, Idea Group Inc. Copying or distributing in print or electronic forms without written permission of Idea Group Inc. is prohibited.

Chapter II

Online Group Projects:
Preparing the Instructors to Prepare the Students

Valerie Taylor
De Anza College, USA

Abstract

This chapter provides an overview of the literature on group projects in online learning and outlines lesson modules for teaching online instructors to use group projects in their courses. The lessons themselves are structured to be an example of online staff development for distance learning faculty. While group projects are often included in on-campus classes, faculty teaching online courses are reluctant to use group projects for these classes. The technology and the students' acceptance of the online learning environment should be used to extend the pedagogical benefits of group work. With adequate staff development, online instructors can successfully integrate group learning into online classes.

Introduction

Preparing instructors to prepare students to participate in online group projects is an important precursor to successful collaborative projects in computer-supported courses. Lesson modules developed in *"The Group Project Project"* provided instructors with specific guidance in applying techniques and teaching strategies for collaborative online projects. Interviews with instructors and students participating in online collaboration and group projects, as well as reviews of published research, were influential in determining the content of the instructor preparation modules.

Copyright © 2005, Idea Group Inc. Copying or distributing in print or electronic forms without written permission of Idea Group Inc. is prohibited.

Student collaborative learning and the resulting learning communities are important elements in online teaching, both in principle and in practice. It is the vibrant sense of community of learners that makes successful online courses so rewarding for participants. Group projects need to be considered in the overall instructional plan for usefulness, timeliness, and instructional quality. However, many instructors teaching online classes are themselves new to online teaching and learning. They need guidance in setting up and delivering instruction for their students to fully engage in an online collaborative learning experience. Providing this guidance was the goal of *The Group Project Project*.

The Group Project Project incorporates theories and methods learned, and applies them to online lessons targeting instructors. The project focused on the development of lesson modules (or learning objects) that are intended to be used as part of a larger course. These modules include guidelines and specific "how to's" for instructors, based on reported research in collaborative group projects in online learning. These lessons are designed to prepare instructors to prepare students to participate in online group projects and to apply techniques and teaching strategies for collaborative learning to online group projects. With this set of flexible instructional modules, the basic elements of online group projects can be passed on to instructors either individually or as a group. These modules instruct faculty members on the theory and process for including group projects in their own online course work. Preparation of instructors and students for online collaboration and group projects are critical to their success.

Background

Online collaboration and group projects can provide important learning experiences and are appropriate for inclusion in most online courses (Graham, Cagiltay, Lim, Craner, & Duffy, 2001). Through group project work, students are presented with opportunities to use multiple learning styles, practice communication skills, and engage in critical thinking. Students come together, work through issues and plans, agree to division of labor, and share ideas. When students are adequately prepared for collaborative work and the task or project assignment is appropriate, students can accomplish the project activities successfully and deliver a product that fulfills a broad range of learning objectives. Online learning offers significant benefits of an asynchronous, on-demand, just-in-time learning environment. However, these benefits add a level of complexity to collaboration and group project work. Project work methods must be learned and applied to online group project work in any discipline.

Considerable research is available in the field of collaboration in teaching and learning. Dillenbourg and Schneider (1995) describe collaborative learning as a situation where two or more students interactively build a joint solution to some problem. Tinzmann et al. (1990) provide guidelines for "a thinking curriculum" that includes "in-depth learning; involving students in real-world, relevant tasks; engaging students in holistic tasks; and utilizing students' prior knowledge." An important component of collaboration is the discussion that occurs during project work, since verbal exchanges among the group participants provide the cognitive benefits of collaborative learning (Pressley & McCormick, 1995).

Copyright © 2005, Idea Group Inc. Copying or distributing in print or electronic forms without written permission of Idea Group Inc. is prohibited.

Verdejo (1996) emphasizes a "conversation or dialogue paradigm" in collaborative learning. The shared approach to tasks, student interdependence, and greater student autonomy are key elements of collaborative learning (Henri & Rigault, 1996). Conversation is essential to experiential learning (Barker, Jensen & Kolb, 2002). These conversations contribute to the process of interactively building a joint solution to a problem even if the conversations are asynchronous and electronic.

Many students do not like group projects. Online group projects are perceived as even more challenging than on-campus group projects. More structure, planning, and individual commitment are required online. Concerns include distribution of work, project planning, and work product dependencies (such as "I can't do my part until A finishes his part, and we are running out of time."). However, online collaboration and group project participation are important elements in education and day-to-day life for many people. Students must become proficient online collaborators and group project participants. Instructors need knowledge and guidelines to facilitate this learning.

The lesson modules of *the Group Project Project* were developed to provide instructors with the background, pedagogy, and activities for instructing online classes and preparing students for online group project collaboration.

Student-Centered Learning

"We learn from experiencing phenomena (objects, events, activities, and processes), interpreting these experiences based on what we already know, reasoning about them, and reflecting on the experiences and the reasoning. Jerome Bruner called this process meaning making." -David H. Jonassen (2002), quote in Dunn & Marinetti (2002).

"When you make the finding yourself -even if you're the last person on Earth to see the light -you'll never forget it." -Carl Sagan (1997, p. 413).

Student-centered learning requires active input from students and requires intellectual effort and aids retention. Students must build their own knowledge through activities that engage them in active learning. Effective learning happens when students take stock of what they already know and then move beyond it. The role of the teacher in student-centered learning is to facilitate the students' learning by providing a framework (i.e., activities for students to complete) that facilitates their learning (Hiltz, 1993).

Following the Constructivist Learning Approach, online group project activities are collaborative, conversational, intentional, and reflective (Lum, Mebius, & Wijekumar, 1999). Collaborative work, joint assignments, and learning resources shared among class members and the instructor are integrated (Mason, 1998). To succeed, students are self-disciplined, intrinsically motivated, willing to learn, comfortable with basic technology, have access to a computer with an Internet connection and have adequate computing skills (McCormick & Jones, 1998).

The group will not have all the skills or knowledge necessary to complete the activities and will need to work through a series of trial and error attempts. Experimenting is an

Copyright © 2005, Idea Group Inc. Copying or distributing in print or electronic forms without written permission of Idea Group Inc. is prohibited.

important activity within the project. Depending on the skills within the group, the instructor may have to provide additional instruction or guidance or direction to ensure that the groups will be successful in bridging the knowledge gap before or during the project work.

Practicing skills through project activities ensures that learners have the opportunity to acquire knowledge and move toward the expected learning outcomes. The group work necessitates using and refining skills in many areas of group working, relationship building, and the specific content-related tasks.

Group project work usually involves some individual work and the synthesis of the group deliverable. In an online environment, these activities usually require reading and summarization of the source information. Using online communications -discussion, email, chat-requires students to engage in reading and summarization.

Depending on the project task, the depth of research and analysis can be extensive or relatively minor. Conducting research and analysis online is a natural extension of the project. Articulating (writing, drawing) appropriate to the project should be included. Each student is required to contribute through articulation, informing, and, in some cases, persuading team members. Online, more forms of expression such as images, animation, video, audio, may be possible and encouraged.

Instructional Approach

In online teaching and learning, technology can provide new and challenging avenues for addressing a variety of learning styles. It is important to strive for a balance of instructional methods. Students can be taught in a manner they prefer, which leads to an increased comfort level and willingness to learn. Some learning in a less preferred manner provides practice and feedback in ways of thinking and solving problems. Students may not initially be comfortable with this, but with practice, they will become more effective learners. Teaching designed to address all dimensions on any of the models is likely to be effective (Felder, 1996). While each learning style model has its advocates, all models lead to more or less the same instructional approach.

Traditional instruction focuses almost exclusively on formal presentation of material (lecturing), a style comfortable only for learners who prefer information presented in an organized, logical fashion and who benefit from time for reflection. To reach all types of learners, instruction should explain the relevance of each new topic, present the basic information and methods associated with the topic, provide opportunities for practice in the methods, and encourage exploration of applications (Kolb, 1984).

Starting from Stirling's (1987) three categories of visual, aural, and kinesthetic, Fleming and Mills (1992) found that the categories did not account for the more detailed differences noted among students. Even though students are used to taking in all visual information, the information itself differs. Visual preference was divided into two perceptual modes-visual and read/write. Visual (V) learners have a preference for graphical and symbolic ways of representing information, whereas Read/Write (R) learners exhibit preferences for information printed as words. By presenting information

Copyright © 2005, Idea Group Inc. Copying or distributing in print or electronic forms without written permission of Idea Group Inc. is prohibited.

visually (V), aurally (A), in a read/write fashion (R), and in kinesthetic (K) form, the integrative and real nature of the information are conveyed to the learners.

Distance learning styles, or learning preferences, change over time and by situation (Diaz, 2002). Student characteristics change constantly. A model that continuously monitors student characteristics and determines which characteristics facilitate favorable outcomes is more appropriate than traditional static learning style models. This student- and learning-centered approach in educational practice can be accommodated in an online learning environment by providing information -student tracking, captured discussions, work products-for increasing faculty sensitivity to the individual learner.

Cultural adaptation is essential for online learning to include a culturally diverse student population. This may be just as necessary for a community college course serving local residents as for a course anticipating global enrollments. Dunn and Marinetti (2002) describe cultural differences in learning style. Learning comes from experiencing, interpreting these experiences based on what the learner already knows, reasoning about them, and reflecting on the experiences and the reasoning. This research raises some interesting questions for higher education institutions that attract international students, both online and on-campus. Community colleges in metropolitan areas often have a culturally diverse student population even though they serve only a small geographical area. Just as community colleges are expected to be accessible to students with various disabilities, they need to be accessible to students with diverse cultural backgrounds and learning styles, as well.

There are a broad range of potential problems that may arise in collaborations, including conflict or disagreement, internalization, appropriation, shared cognitive load, mutual regulation, and social grounding (Dillenbourg & Schneider, 1995). Effective collaborative learning requires group composition of optimal heterogeneity. Some difference of viewpoints is required to trigger interactions, while maintaining mutual interest and understanding without triggering conflicts.

Education for understanding develops a family of interrelated abilities (Wiggins & McTighe, 1998). Students who possess a mature understanding of a subject are capable of explaining, interpreting, and applying the subject. They have perspective, empathy, and self-knowledge. Students with an understanding of a subject can explain the subject providing thorough, supported, and justifiable accounts of phenomena, facts, and data. They can tell meaningful stories, offer apt translations, provide a revealing historical or personal dimension to ideas and events, and make subjects personal or accessible through images, anecdotes, analogies, and models. They can effectively use and adapt what they know in diverse contexts. These students have perspective. They can see and hear points of view through critical eyes and ears and see the big picture. They can empathize, finding value in what others might find odd, alien, or implausible. They perceive sensitively on the basis of prior indirect experience. They have self-knowledge and perceive the personal style, prejudices, projections, and habits of mind that both shape and impede their understanding. They are aware of what they do not understand and why understanding is so hard.

Copyright © 2005, Idea Group Inc. Copying or distributing in print or electronic forms without written permission of Idea Group Inc. is prohibited.

Benefits of Group Projects in Online Collaborative Learning

There are significant benefits that can be derived from collaborative learning and project work (Tinzmann et al., 1990). The principle benefits of group projects in online collaborative learning include but are not limited to: building self-esteem, reducing anxiety, encouraging understanding of diversity, fostering relationships, stimulating critical thinking, and developing skills needed in the workforce.

Building self-esteem is an important benefit of online collaboration and group project activities. Students are simultaneously working alone and in an intense community of learners. Students must develop and rely on their own efforts. There is little opportunity to be swept along with the rest of the group. Either they actively participate or they do not. There is no escaping the personal accountability. By contributing to the group effort, students take personal credit for their role in the activity. This visible effort is concrete evidence of participation and learning and contributes to building students' self-esteem.

There is no question that online collaboration and performing group work are challenging for students at all levels. This work is important for stimulating critical thinking. Working out the logistics for forming the group, defining and allocating tasks, actually doing the work, and coming together to present the group's product represent a significant body of work for all members of the group. The group work requires each team member to contribute on many levels. Many of the tasks and interactions necessary to perform those tasks may be new to the students. Students are encouraged to learn and improve a broad range of skills including critical thinking.

Online collaboration and group work require students to develop specific study skills and life skills (Bates, 2000). These include: good communication skills, ability to learn independently, social skills, teamwork skills, ability to adapt to changing circumstances, thinking skills, and knowledge navigation. All these skills have practical application within online learning through collaboration and group work.

Although these benefits may be derived from other forms of learning, the group project success depends on students' mastery of these skills in a short space of time and reinforces that learning with practice that is rarely matched in other learning environments. Online collaboration and group project work require students to become proficient in skills that will serve them well throughout their educational and work lives.

Group Project Learning Model

There are a number of collaborative learning models described in the literature. For developing these lesson modules, important elements from several prominent models were combined to provide a sound pedagogy and a manageable breakdown of the overall process that could be implemented in a short group project within the context of a community college, 12-week quarter-long, distance learning class.

The Group Project Learning Model was derived from the work of Riel (1993), Reid, Forrestal, and Cook (1989), and Tuckman (1965). Other learning models and applicable

Copyright © 2005, Idea Group Inc. Copying or distributing in print or electronic forms without written permission of Idea Group Inc. is prohibited.

research in online teaching and learning, collaboration, group dynamics, and learning styles were investigated. The content of the instructor lesson modules was based on this information.

Riel's (1993) 5 Steps in a Learning Circle was chosen as the base model for developing the lesson modules for instructors. The work in each step is distinct and the sequencing presents a natural progression that is easy to follow. The five steps include forming the Learning Circle, planning the Learning Circle projects, exchanging work on the projects, creating the publication (or deliverable), and evaluating the process. In the Collaborative Learning Model described by Reid et al. (1989), there are five phases for designing instruction for collaborative learning: engagement, exploration, transformation, presentation, and reflection. The phases in the Collaborative Learning Model described by Reid, et al. correspond to Riel's 5 Steps in a Learning Circle. The steps for layout and accomplishing the group project work from these two models provide a strong framework for project activities and describe the process and learning.

Another important element needs to be factored into the online collaboration and group project model-group dynamics. It is well understood and documented (Tuckman, 1965) that the process for learners coming together to work as a group is a critical element in the success of online group project activities. The groups must be directed through the process and given time, opportunity, and specific skill-building tasks to successfully complete the group project. The Tuckman model describes stages that teams go through, from Forming to Storming, through Norming and Performing. Although Adjourning (Clark, 1997) was not in the original model and does not rhyme, it is a reasonable addition. Having a stage for reflection and closure is also important. My apologies to Tuckman.

Online and asynchronously, groups need to be more aware of the individual steps and the transitions between them to provide a solid foundation for the rest of the project activities (Waugh, Levin & Smith, 1994). In Harris' (1995) 8 Steps in Organizing Telecollaborative Projects, the up-front planning and closure are emphasized as separate elements. As students work in groups building a community of learning, it is important to finish the process with closure, especially if the group community is disbanded at the completion of the project.

The Group Project Project modules are viewed from the instructors' perspective, leading a group of students through an online collaborative activity. Considerable effort and knowledge are required to prepare for the group project activity. The six modules describe the instructor preparation phase and the five Riel Learning Circle stages (Riel, 1993). The corresponding elements from the Reid (1989) Collaborative Learning Model and the Tuckman (1965) Small Group Model are shown in square brackets []. These stages are: preparation, forming the Learning Circle [engagement, forming], planning the Learning Circle projects [exploration, storming], exchanging work on the projects [transformation, norming], creating the publication [presentation, performing], and evaluating the process [reflection, adjourning]. These six steps are the subjects of the lesson modules or learning objects in the instructor training for *The Group Project Project*. Each of these steps are described in considerable detail in the following sections.

Copyright © 2005, Idea Group Inc. Copying or distributing in print or electronic forms without written permission of Idea Group Inc. is prohibited.

Instructional Design

There is considerable interest in efficiency and productivity in developing training within corporations and higher education. Producing content for online delivery is extremely labor-intensive. Many instructors interested in online delivery can not spend the time and the resources to produce the necessary course materials. If learning can be broken down into units that have some common applicability, instructional units can be developed once and reused in multiple courses. IBM and Cisco (Barron, 2000), along with several publishing applications vendors, have demonstrated the power and viability of creating, reusing, and sharing lesson modules or learning objects. This trend is likely to expand as the benefits are quantified.

Throughout curriculum development and delivery communities, there is a movement to design and develop libraries or repositories of learning objects. Each three-part learning object consists of a learning objective, the lesson content, and an assessment. Reusable lessons that are applicable to many different learning activities are very appealing. The "develop once, use many" model that characterizes learning object theory would greatly extend the quality and quantity of online content. There is broad applicability of collaboration and online group project learning to a wide range of online classes. Creating reusable, sharable lesson modules leverages the research and development of curriculum. *The Group Project Project* lesson modules were developed for instructors and are intended to help instructors prepare students with knowledge, skills and guidelines that address the what, how, and why of online group project collaboration.

Instructor Training

"In theory, there is no difference between theory and practice. But, in practice, there is." -Jan L.A. van de Snepscheut.

As more instructors enter the field of teaching in an online environment, there will be an increasing need to provide continuing education and professional development. What better way to fulfill this need than by using the tools at hand. *The Group Project Project* instructor modules were created to provide an instructor with a group learning experience in an online environment. The modules can be used as a stand-alone introduction to group projects or combined into a more comprehensive online instructor training. Ideally, several instructors will form a cohort and work through the lessons together. Alternatively, an instructor can use these lesson modules as a self-study course.

The instructor lesson modules focus primarily on practice—practical guidelines and suggestions for development and execution. Lessons reflecting the elements of the Collaborative Learning Model (Reid, Forrestal & Cook, 1989) include discussions and intermediate tasks for group members and suggestions for collaborative refinement and delivery.

The learners participating in *The Group Project Project* online learning experience are themselves instructors. The instructor/learner develops group project activities and

Copyright © 2005, Idea Group Inc. Copying or distributing in print or electronic forms without written permission of Idea Group Inc. is prohibited.

Figure 1

Model	Constructivist	Reid	Kolb
Warm up	Recalling knowledge	Engagement	-
Learn	Summarizing and reading, articulating	Presentation	What
Apply	Constructing, practicing	Transformation	How
Explore	Experimenting, conducting research and analysis	Exploration	What if
Evaluate	Reflecting	Reflection	Why

supporting lesson material while completing the lessons in the modules. These lessons are intended to provide instructors with an online learning experience similar to that of their online students. Many faculty teaching online have never had the opportunity to be online learners or to participate in collaborative group projects online.

The learning models suggest the format for the individual lesson modules or learning objects as illustrated in Figure 1. A standard lesson outline was used to structure lesson modules in the curriculum. The lessons all contain these elements. A constructivist instructor begins a lesson by asking students to recall what they already know about the subject. Then the students are involved in an activity that takes them beyond what they currently know. The student must actively engage in the learning process by doing something. Constructivist activities include the primary activities: constructing, experimenting, practicing, summarizing and reading, conducting research and analysis, and articulating through writing or drawing. These are included in group projects. This format is followed in the lesson module design.

Warm Up

Warm Up [engagement] serves as the attention grabber. Starting with a quote from a student about needs relating to the topic is an effective beginning. Through a series of questions, the Warm Up gets learners thinking about what they already know about the topic. In general, learners will come to these lessons with considerable related knowledge. They need to be reminded of what they already know. The list of goals, objectives, and outcomes for the topic are also provided. Listing objectives sets the learners' expectations for the information, skills, and deliverables that will be accomplished in this module.

Learn

Learn [presentation] provides direction and focus. It describes the underlying instructional design principle for the topic. Tools and methods used to achieve objectives are also described. The Learn section is equivalent to the lecture or instructor-lead portion of the course. Markel (1999) says the notion that instructors in traditional classes spend

Copyright © 2005, Idea Group Inc. Copying or distributing in print or electronic forms without written permission of Idea Group Inc. is prohibited.

most of their time lecturing is a myth; what they really do is help students organize information, help them with their projects, give students a chance to meet with their teams, and motivate the students. And that is exactly what needs to be done in a distance learning environment as well. The Learn section provides text and activities analogous to the lecture in face-to-face learning.

This is the Presentation stage in the Collaborative Learning model described by Reid et al. The Learn section guides the learner through the primary topics of the module and provides a frame of reference for the other learning activities that follow. In the learning and subsequent activities within the module, new online collaboration-specific information is presented and joined to previous knowledge. The Learn section includes a discussion of the design principles and describes the underlying instructional design for the topic. Tools and methods to be used to achieve the objectives are also described in this section.

Apply

Apply [transformation] section describes activities to do to reinforce and extend learning, practical application of theory, and the next steps for developing a group project module for an actual course. Transformation is the primary focus of the application activities. In the Application section, the learner is presented with tasks or activities that require actual performance. In some cases, the activities or tasks are simulations or special test cases set up for the learner to practice. In other cases, these are authentic activities or tasks that will produce a "real" product that the learner can keep and use after the course is complete.

Explore

Explore [exploration] gives the learners the opportunity to expand their own learning. Suggested readings and a list of online references that support and expand on material presented in the Learn section are provided. Demonstrations, case studies, and examples may also be included as appropriate. The Explore area encourages the learner to investigate the topic further. Stories and examples illustrate how others have interpreted and used the information.

The material in the lessons reflects the research and observation on understanding and methods that are important to successful student outcomes in online collaboration and group projects, regardless of content area. As specific questions arise about student project work, the research and readings listed provided direction.

Great care was taken to suggest a few really good references. The target audience is community college faculty including many with full course loads. Readings are intended to augment the summarized and pragmatic materials provided in the module. However, instructors may not have time to read more than one or two of these, so the references must be relevant and directly applicable.

Storytelling is an important technique used to engage students in online learning. The format for the lessons for *The Group Project Project* include an exploration and

Copyright © 2005, Idea Group Inc. Copying or distributing in print or electronic forms without written permission of Idea Group Inc. is prohibited.

demonstration area. This provides an opportunity to include examples or stories about each lesson. We construct knowledge (implicitly) based on our own experiences with the information being presented. If students can tell a story of what they have come to know, they are not only teaching others, but are demonstrating that they truly understand what they "know."

Humans interpret complex data in terms of some underlying story (Mislevey, Steinberg, Almond, Breyer & Johnson, 2001):

We weave some sensible and defensible story around specifics. Such a story addresses what we really care about at a higher level of generality and a more basic level of concern than any of the particulars. A story builds around what we believe to be the fundamental principle and patterns in the domain. (p. 5)

Evaluate

The Evaluate [reflection] section includes problem-based discussion and learning community interaction. The tasks and activities described in the Evaluation section are intended to help the learner with reflection on the learning activities. The learner should come away from the Evaluation activities with a good understanding of the process, the skills, and knowledge acquired. The learning outcomes, set out by the instructor in the planning and preparation phases, are reviewed to ensure that the learning has been as anticipated. This is also an opportunity for assessing the success of the learning opportunity.

The Evaluation may also include actual assessment that may be something as simple and easy to administer as a multiple-choice test. We are moving toward more sophisticated forms of online assessment that might include simulations and other complex performances. These assessments indicate achievement level and offer proficiency inferences with clear instructional implications (Bennett, 2002).

Anderson and Garrison (1998) talk about the teacher-teacher interaction and how it leads to improvements in teaching as it stimulates reflection and communication. Access and adoption of content created by other teachers is instrumental in fostering teacher learning. There are a number of references that provide practical advice and guidelines to address the needs of instructors. These range from self-help advice books (Hanna, Glowacki-Dudka, & Conceicao-Runlee, 2000) to online instruction for developing online activities (Hildreth, Masterson, & Wallace, 2000). Instructors, new to online collaboration and group project activities can draw on the experience and advice of their peers, as the information is readily available.

Group Project Process Lesson Modules

The six lesson modules that define the group project process are described here in detail. The justification for the material included in each step is also provided.

Copyright © 2005, Idea Group Inc. Copying or distributing in print or electronic forms without written permission of Idea Group Inc. is prohibited.

Preparation

Before assigning a group project to students, instructor preparation is necessary if the project is to achieve a successful outcome. Instructor preparation can be segmented into primary components — learning outcomes, interactions, instructional media and tools, social relationships, assessment and measurement, and support systems and services (Miller et al., 1998). Student orientation is also necessary to ensure that all students are adequately prepared to work in groups and produce the expected project deliverables. If technology and tools to be used are new to the students, the orientation must include adequate instruction for team members to successfully use them. These elements are covered in the Preparation module.

Determining goals and objectives, learning outcomes, instructions, and evaluation criteria are essential. The learning outcomes serve as a "contract" between instructor and student. The instructor must effectively communicate these expectations. The learners must understand them to achieve the most effective learning experience. Instructional design strategies appropriate to the distance education experience are needed to support the intended outcomes.

Creating a variety of highly interactive learning experiences is an important step in preparing for online group projects. Learners interact with one another, with an instructor, and with ideas. New information is acquired, interpreted, and made meaningful through interactions. Learner participation is critical to the learning process and must be considered in establishing and maintaining interactions necessary for an effective educational experience.

Designing an instructional experience for any learning environment requires careful consideration of the available tools and media that could be used by learners within that environment. Technologies are tools, and their selection must be guided by the goals and objectives of particular learning programs, the specific characteristics of the learners served by those programs, and the realities of the costs, utility, and benefits to learners that are associated with the technologies that could be employed.

The projects for the online collaborative groups should be developed to engage learners in authentic learning tasks (Brown, Collins, & Duguid, 1989). Anderson and Garrison (1998) raise the concern that a potential problem with online projects is the learner's ability to make sense of overwhelming amounts of information. Application assignments, such as well- constructed group project activities, facilitate the move from theory to practice.

A task that is suited to collaboration sets the framework for the project as participants come together. The task description helps the project team understand how to proceed to define and execute the tasks associated with reaching the project objectives. Structured activities should be designed so that interdependence is essential to successfully completing the assigned task.

The instructions, directions, introductory information, and references should all be readily accessible to the students before and during the project activity. Depending on the online delivery system — web pages, email, listserv, or course management system, students need to be able to locate this information and refer back to it throughout the

Copyright © 2005, Idea Group Inc. Copying or distributing in print or electronic forms without written permission of Idea Group Inc. is prohibited.

group project activity. A comprehensive syllabus or specific project information pages are likely the best way to make this information available to students.

Instructions should include the purpose of each task or activity within the project work and give an approximate time to complete the work. For each task or activity, the goal, objectives, assignments, and due dates should be provided.

In some cases, it may be appropriate to assign roles to students within groups. For more experienced online collaborations, these roles can be suggested without explicit assignments. Team Leader, Encourager, Re-teller, Recorder, and Spokesperson are suggested job titles (Tinzmann et al., 1990). To facilitate high quality group interaction, it may be necessary to teach and practice roles, rules, and relationships for group interaction.

Instructor involvement through the project activity may be limited to defining and presenting clear instructions, providing prompt feedback and clarification, and guiding online discussions — key components in online collaboration facilitation (Espinoza, Whatley, & Cartwright, 1996). The effective facilitator moves out of the middle and has strategies for stimulating real student collaboration and guiding the conversation toward important content (Collison, Elbaum, Haavind, & Tinker, 2000). The instructor must establish and shape intellectual and emotional norms, model appropriate behavior, and admonish harmful input.

For any online learning experience, the student support services should address technical support, instructional resources, faculty development, instructional design and development, and policy and administration to create an environment conducive to distance education. The support systems and services for a distance learner must be as complete, as responsive, and as effective as those provided for the on-campus learner. These services must be in place during the preparation for student group project activity.

Forming the Group Project Teams

Social relationships form the foundation for a community of learners. Group project work depends on informal conversation, trust-building experiences, the interjection of humor, the opportunity to share personal and instructional goals, and interactions among participants. Students who feel they are part of a community of learners are more motivated to successfully work out solutions to problems. The instructor must design strategies and techniques for establishing and maintaining learning communities among learners working asynchronously at a distance.

Once the students are prepared for group work, they are divided into their group project teams. There are a number of ways to select teams with or without student input. Factors such as group size, heterogeneity, experience in group projects, and skills required for project completion are discussed in this module.

In the Engagement phase of the group project activity, participants come together and begin to learn about other participants. Everyone is working to understand the project requirements and to apply current knowledge to the project. The group members identify areas where additional information, skills and clarification are required.

During the initial or forming stage of the group, structure is developed. Roles are assigned or claimed (both implicitly and explicitly). Shared values are discovered.

Copyright © 2005, Idea Group Inc. Copying or distributing in print or electronic forms without written permission of Idea Group Inc. is prohibited.

Relations between members of the group are established, and norms begin to emerge. The general procedures for decision making and problem solving are agreed upon. This may take place very quickly, or it may be long and drawn out. However, it is essential that the group work through all these issues early in the project activity. Shared knowledge and authority, mediated learning, and heterogeneous groups of students are essential characteristics of collaborative groups (Tinzmann et al, 1990).

Asynchronous discussions and group problem solving among students in threaded discussion groups are less expensive, more thoughtful, and easier to schedule, particularly across time zones, than are synchronous alternatives such as chat (Curtis & Lawson, 2001). The number of participants in an online discussion group needs to be limited to 20 to 25 participants per group for general discussions. Groups of two or three participants work well for intense collaboration requiring extensive production. Other group sizes are appropriate depending on the nature of the assigned group project activity. Larger courses can be divided into subgroups. There are a number of techniques that can be used to help students get to know one another and work through the forming process. These techniques can be used for large or small groups (Goddard, 2002).

Group assignments for collaborative activities and the fundamental understanding of group dynamics can be applied to college-level online collaborations (Collis, Andernach, & van Diepen,1996; Foote, 1997). Ideally, each cooperative group should include students with a complete range of ability, learning style, personality, and gender. (Ossont, 1993). Everyone learns from everyone else. All students have the opportunity to make contributions and appreciate the contributions of others. It is critical that students are not segregated according to ability, achievement, interests, or other characteristics (Tinzmann et al, 1990).

In order to appropriately support learners in the online environment, Granger and Benke (1998) suggest that instructors must know their learners. Where are they? Who are they? What resources are available to the learner? What are the learners' needs and limitations? Some of this information can be gathered directly from the learners through questionnaires or open-ended discussion questions. Learner readiness for online learning is an important factor in the successful learning outcomes for collaborative group projects. Maturity, independence, motivation, and inability to attend regular on-campus classes are characteristics of successful online learners.

What are the knowledge goals of the program and what knowledge do learners already have? It is important to build on what learners already know. Learner orientation gets students ready to learn. It ensures that they have the skills, knowledge, tools, and instructions necessary to embark on their group project activities. Maus (2002) states:

If we expect our students to know how to use these tools, we must teach them how to do so... We cannot assume that young people today have had equal access to technology tools and know how to use them in the educational environment. Nintendo is not a stepping-stone to practical computer skills any more than toy cars are a preparation for real driving.

Independent learning should not be seen as ideal. Anderson and Garrison (1998) strongly advise against sacrificing collaborative learning experiences in the name of individual

Copyright © 2005, Idea Group Inc. Copying or distributing in print or electronic forms without written permission of Idea Group Inc. is prohibited.

choice and freedom. In this competitive educational era, many learners demand to be released from group projects. The reasons vary, often so that their grades will not be "pulled down" by the ineptness of team members. As the stronger, smarter member of the group, some students believe that they will be required to instruct the others or carry a disproportionate share of the workload. In some cases, individuals may have commitments that would prevent them from making timely contributions to the group activities and communication. Wherever possible, including all learners in some collaborative activities is essential to achieving the best and most lasting learning outcomes.

No matter how much mutual support, coaching, and encouragement they receive, students must be individually responsible for their own academic achievements. Students must understand that they will play a greater role in their own learning. Students are expected to participate in goal setting for the group and planning their learning activities. Students learn to take responsibility for monitoring, adjusting, and questioning. Students learn to evaluate their own learning and to assess group work, including the effectiveness of learning strategies, the quality of products, the usefulness of materials used in a task, and how future learning might be realized. Because decisions are shared in a collaborative learning environment, students are freer to evaluate their own performance as well as that of the group.

Ideally, online teaching will be inclusive and accommodating to the point where there is no special accommodation needed for students with disabilities. Technology has the power to both create and remove barriers. Awareness is a start. Research shows that a learning environment that includes asynchronous discussions is inclusive and supportive of students with disabilities (Hsi & Hoadley, 1997).

The critical importance of community in education and intellectual development is demonstrated by the phenomenal growth of online communities of interest and purpose facilitated by the Internet and ubiquitous access to email (Lipman, 1991; Turkel, 1995). These virtual communities develop and thrive without physical proximity. They meet the diverse social and intellectual needs of widely distributed individuals. Learner-learner interaction and collaborative learning enhance the quality of learning and need to be included in an online learning experience. Group project activities provide an ideal context for including this experience that is directly related to expected learning outcomes (Anderson & Garrison, 1998). Community, with its sense of both cooperation and critical judgment, contributes to meaningful, deep learning.

Planning the Projects

The planning phase of the project work constitutes the exploration activities. Students work to define the scope of the project and plan the deliverables. They assign roles and responsibilities to team members based on the needs of the project and the skills and knowledge of the team members as determined during these early interactions. The project teams work out a plan for completing the project work. The instructor may provide specific guidelines and instructions or may leave this to the group to work out among the team members. Instructors will be able to assess the degree of formality. Student-directed work can be based on the complexity of the project assignment and students' level of understanding of the work need to complete the activities.

Copyright © 2005, Idea Group Inc. Copying or distributing in print or electronic forms without written permission of Idea Group Inc. is prohibited.

The Storming activities are the interactions to resolve conflicts in values, perspectives, goals, power, and information that are discovered and identified (Tuckman, 1965). This is often a creative stage and should not be avoided or shortened. Many students find this stage difficult and even unpleasant, as they prefer to avoid conflict and confrontation. However, if the students are adequately prepared with strategies, procedures, and skills for dealing with this normal and necessary stage in project development and group formation, the team can transition through this stage and continue with the more directed, task-oriented project work.

Group project participation requires process skills: planning, performing, communicating, and interpreting (Toh & Woolnough, 1993). Giving instruction in planning and communicating is necessary for helping students work on open-ended investigations. In some cases, it is beneficial to assign roles with job titles to members of groups. For other groups with more project experience, the team members can identify roles and assign responsibilities within the group. The task of developing a defined focus helps group members come together to form a community of purpose, and each member understand individuals' roles and responsibilities within the group.

Collaborative learning (Harasim, Hiltz, Teles, & Turoff, 1995) in an online environment has the capacity for active learning, interaction (both quality and intensity), access to group knowledge and support, democratic learning environment, convenience, and motivation to complete tasks. Education depends on acts of communication (Salomon, 1981). At its best, educational communication should be reciprocal (two-way), consensual (voluntary), and collaborative (shared control) (Anderson & Garrison, 1998). These attributes must be in place for the process of constructing meaningful and worthwhile knowledge to take place. Damon (1984) noted that "intellectual accomplishments flourish best under conditions of highly motivated discovery, the free exchange of ideas and reciprocal feedback between mutually respected individuals."

By the end of the Planning phase, students should have worked out roles and responsibilities. Division of work to accomplish the project work should be well underway. Depending on the skills and maturity of the learners, the group may have completed a formal project planning activity and produced task lists, timelines, objectives, and work product outline. However, this level of formal project management may be beyond the scope of the group project activities.

Exchanging Work on the Projects

In the Transforming phase of the project lifecycle, team members complete their independent activities and report back to the group. Research, development, organization, and categorization are all activities that are consistent with transformation of information gathered. Within the group, the nature of communication and interaction is likely to be transformed as well. In this Norming phase, the group's approach to communication and problem solving is more firmly established (Tuckman, 1965).

In this phase, the instructor helps project teams communicate and exchange their work and introduces them to tools and procedures available. Moore (1989) proposed three fundamental transactional relationships in education between the Teacher, Learner, and Content. Garrison (1989) expanded on this by identifying the relationships at the

Copyright © 2005, Idea Group Inc. Copying or distributing in print or electronic forms without written permission of Idea Group Inc. is prohibited.

intersections of these transactions, specifically Support, Independence, and Proficiency, with Meaningful Learning at the intersection of all relationships. It is important to establish and maintain a collaborative culture throughout the project. However, it is especially critical at this stage as it determines the willingness of participants to work collaboratively toward a shared vision and goal. A collaborative environment provides the tools and the resources necessary to conduct a collaborative effort (Riel, 1993; Waugh, Levin & Smith, 1994).

Johnson, Johnson, and Smith (1991) describe positive interdependence as cooperation that results in participants' striving for mutual benefit so that all members of the group benefit from each other's efforts. All group members sink or swim together. Individual effort and team work are essential.

Group interaction usually increases significantly in this phase of the project. There are some important advantages to online collaboration at this stage. Asynchronous interaction provides an opportunity for reflection that is unique to online learning. Document sharing communicates project focus and demonstrates progress toward the final deliverable. Autonomously intra-group interactions —discussions, problem solving, sharing, revising, reviewing, and commenting are all important learning opportunities.

When instructors intervene to assist working groups, even when requested to do so by students, the intervention usually ends with the instructor giving directions. The intervention produces more instructor talk than student talk (Oakley & Crocker, 1977). Too much guidance does not help the students. Even in a hands-on, problem-solving environment, an instructor's desire for students to get the right answer will produce instructor behaviors that eliminate opportunities for problem solving (Martens, 1992). The roles of instructors and students change when online collaboration and group project activities become major features of the teaching-learning experience (Berge, 1995). Instructors become expert questioners rather than providers of answers. Students refine their own questions and search for their own answers. Instructors provide only the initial structure to student work, encouraging increasing self-direction. There is more emphasis on students and groups of students as autonomous, independent, self-motivated managers of their own time and learning process.

As an educational facilitator, the instructor can use questions and probes that focus discussions on critical concepts, principles, and skills. Creating a friendly, social environment that promotes learning is essential for successful online teaching. Promoting relationships, affirming and recognizing students' input, providing opportunities for students to develop a sense of group cohesiveness, maintaining the group as a unit, and in other ways helping members to work together in a mutual cause are all critical to success of any group activities. The group members may create this social atmosphere themselves. However, the instructor may have to become actively involved to promote and model appropriate socialization.

The instructor is responsible for setting the timetable, procedural rules, and decision-making norms. Unobtrusively managing the flow and direction of the discussion without stifling the participants is a key facilitation goal. The instructor must be comfortable and proficient with the technology so the learners may concentrate on the project tasks.

In order for the collaboration to function in an online environment, learners must also be proficient users of the technology. Training and support are essential to sustain the

Copyright © 2005, Idea Group Inc. Copying or distributing in print or electronic forms without written permission of Idea Group Inc. is prohibited.

collaboration. Ideally, learners have an opportunity to gain proficiency in a stress-free, non-graded learning context. Having adequate technical support available, using technology that is available to all team members, and having adequate training in the use of technology prior to starting the project are also important considerations.

Creating the Publication or Product

Group projects normally conclude with the production of a product or deliverable. The Presenting phase provides an opportunity for the group to bring together all the individual efforts to revise and discuss the work and to make decisions about the importance and representation of the work they are completing and how it fits with the objectives and assessment criteria. The processes for creating and delivering the product are covered in this module.

Having established roles, processes, and procedures, the group's time, attention, and energy is directed at the group task. This is the Performing stage of the project lifecycle. Problems in performing may often be traced back to insufficient storming and norming where the group distributed responsibilities, discovered common values, and established procedures. Laying the groundwork and getting a firm foundation are essential to the group's overall success.

To help the group focus its efforts, working to deliver a product is key. As part of the project objectives, the students are given instructions for the preparation and presentation format of the project product. The deliverables for online projects often include text reports in PDF (Portable Document Format) or Rich Text Format (RTF) file formats, web pages, or PowerPoint slides. These products can be published to the web and are easily accessible by students and instructors. It is important to communicate explicit expectations for both individuals and groups. As part of the orientation, considerable time needs to be spent ensuring that all students understand these expectations. There needs to be prompt feedback to students asking for clarification, as the successful formation of the group and its eventual production of the deliverable depends on a complete and shared understanding of these expectations. Potential problems can be avoided if the group members are working toward a common goal.

Evaluating the Process

Evaluation includes the grading or assessment of the project product. In keeping with constructivist theory, students individually and in their groups reflect on the learning of content and process through the group project work. The evaluation criteria should be part of the project orientation, so that students know how the project will be assessed. Assessment and measurement serve several valuable purposes for both instructors and students. Assessment provides information on student achievement for grades and can be used by the students to monitor their own progress and adjust their learning strategies. Feedback from students can also help instructors adjust instruction to better meet students' needs. The final, reflective phase of the project gives team members a chance to evaluate their learning experience, review the process, and think about their roles in the project.

Copyright © 2005, Idea Group Inc. Copying or distributing in print or electronic forms without written permission of Idea Group Inc. is prohibited.

Another important aspect of this project phase is closure or wrap-up that closes out the project for the participants. Although Tuckman (1965) does not include this final stage, it might be labeled adjourning. Others who have studied group dynamics and project lifecycles include a last step and believe it is important and necessary to complete the project team's learning experience. For online group project teams, closure is especially important.

The use of technology in education is inexorable and inevitable (Bennett, 2002). As technology becomes an integral part of what and how students learn, the means we use to document achievement must keep pace. Online assessment is still mostly multiple-choice tests. More sophisticated forms of online assessment are being developed and implemented —simulations and other complex performances that indicate achievement level, linking proficiency measures to actual instruction. As technology is integrated into teaching and learning, the method of assessment should reflect the use of these tools.

By including group projects and assessing the performance, process, and products, assessment is moving toward these assessment goals. Groups of students working together to submit a single product can eliminate or reduce some of the differences in individual student outcomes that would have arisen from differences in technical experience, computer equipment, Internet connection, and other differentiators inherent in distance learning.

Clear objective assessment and evaluation in online teaching and learning are critical (Gellman-Danley & Fetzner, 1998). To be effective, assignments must be related to assessment, with appropriate discussion topics and the effective use of illustrations and visuals. Continual assessment is crucial in online courses because student identification is hard to implement. Through frequent assessments, the instructor can learn to recognize each student's response and participation style. Group project interactions and final reflection are important components for assessment.

In the behaviorist approach to education, the end product or the outcome of learning takes precedence over the process. Often the process is neither evaluated nor considered important (Stahl, 2002). However, in a collaborative group project, the instructor can see how students got from group discussion to the final deliverable. The instructor in the online environment can assess the process, by reading and evaluating the students' online chats and discussions. The instructor may offer assistance to those veering too far from the path or design remedial lessons for the virtual classroom. The advantage of the online approach is having a record of the student discussions; the disadvantage may be the amount of work it presents to the instructor.

Even simple, basic metrics such as enrollment, completion, and success have merit (Bersin, 2002). What is the impact of including a group project in a course? Are retention and completion affected by the inclusion of group project? Are student test scores improved with student collaboration and engagement? While many advocates of online collaborative learning and group project activities believe there are strong academic benefits, collecting actual course data is needed to validate these assumptions. There are many factors that must be considered when developing assessments for online collaboration and group project experience, including the need to assess responsibly, preserve validity, fairness, utility, and credibility of the measurement (Mislevy, Steinberg, Almond, Breyer, & Johnson, 2001, p. 5). In determining the success of the group project

Copyright © 2005, Idea Group Inc. Copying or distributing in print or electronic forms without written permission of Idea Group Inc. is prohibited.

activity and the roles of individual students in the project work, assessment, alignment, accountability, access, and analysis must be fully developed.

Observations and Interviews

These comments were selected from the responses from students in my Web Page Development class taught Spring quarter, 2002, and the JavaScript class I taught in Spring 2003. The group project assignment was to develop an online tutorial for one of the topics covered in class, such as Tables, Images, or Frames. The students worked in groups in the last three weeks of the quarter. Each group made a final presentation to the class. Students decided on the topics and formed groups (limited to five students per group).

Because the subject is technical, all students already had experience using some of the technologies required for online collaboration. The questions in the student survey were intended to look at the attitudes of community college students towards group projects. Their feedback provides valuable insight into their collaborative learning experience.

- *What was your overall reaction to the group project?*

  - First I was not happy about it, because group work need too much time and it also requires dealing with other people. But, in the business world, as I understand, no one works everything by himself/herself; a team of different professional people work on one project together. So, it was a good opportunity to experience this system here while I am in school. Generally, I liked it.

  - I'm not a fan of group projects, so I did not expect much. It turned out to be a pretty good experience.

- *Was the experience beneficial to you?*

  - Yes indeed! I learned new things from my teammates, and I have also learned applying the tags I knew before.

  - Yes, we all had our "assignments" to do ... played a vital role as site designer, while I seemed to be the guy that nudged the team along.

- *Did your group assign roles to team members?*

  - Yes, first we decided what our project should look like. Then we broke it up into five pieces, and divided it to five of us.

- *Did all team members contribute to the final product?*

  - Yes, each of us did our own part as our individual ability.

  - Yes, some more than others.

Copyright © 2005, Idea Group Inc. Copying or distributing in print or electronic forms without written permission of Idea Group Inc. is prohibited.

- *What was the worst part of your group project experience?*

    - Some team members couldn't finish their part on time.

    - The worst part was relying on others to get their part done. Some waited till the last minute. Very stressful!!

- *What was the best part of your group project experience?*

    - Some team members were very concerned and helped their teammates to make the project complete.

    - The day of the presentation. It just all seemed to come together.

- *What could the instructor do to improve the group project experience?*

    - Ensure that each member would be graded for his or her individual effort as well as the group's. Maybe even have inter-group evaluations.

- *Have you worked on group projects in other courses?*

    - Yes, for math courses, but this one is the biggest one I have ever done.

    - None that I can recall.

- *Does the subject of the course affect the success of the group project? How?*

    - No, not at all.

    - Yes. I think a subject that appeals to a wide variety of people would spark the interest of the group members. A group project in a required class, for instance, may lack some of that spark.

- *Did your group use online collaboration - email, chat, web pages? How effective was this?*

    - Using email and web pages, we could save a lot of time.

    - We used email. It seemed to work fairly well.

- *What preparation would help the project team?*

    - Preformatting the project would help the team to focus and finish the project on time.

    - Just getting started earlier. We still had some minor "kinks" in our final draft.

- *What advice would you give to project teams next quarter?*

    - First, they need to agree on what topic they want to work on, and then every one of the group members contributes ideas of his/her best of the topic. After collecting everybody's idea, divide the tasks as every individual's ability.

    - Finally, they need to talk about their final draft and exchange constructive ideas to others.

    - Start as soon as possible and communicate with their team.

Copyright © 2005, Idea Group Inc. Copying or distributing in print or electronic forms without written permission of Idea Group Inc. is prohibited.

As the instructor, I was able to "see" the work of some of the groups as they chose to use a WebCT private group discussion topic for all their communication. One group exchanged 80+ messages in less than two weeks, some time-stamped between 2a.m. and 5a.m. The project was much more complex than those from groups using only email to communicate. In an evaluation survey, students were asked to rate their individual effort compared with that of others in their group. This was a good way to address the inequality of effort, rather than having students rate the effort of others in the group.

Streaming Media

I was a participant in the Streaming Media course pilot offered through California Virtual Campus in July 2002. Each participating campus was expected to provide a team of staff to work through the online course materials and activities. Several individuals also tackled the project activities on their own. The experience was mixed. The project was dubbed "screaming media," reflecting the frustration and excessive effort required when individuals without adequate previous experience tried to complete the work independently even though others on the team had the knowledge. Unfortunately, some of this also reflects a breakdown in the group participating in the group projects. The experience illustrated the importance of team work, communication, collaboration, motivation, and commitment of individuals in the project teams.

Instructional Design Intensive

Foothill-De Anza District Professional Development included a hybrid course called Instructional Design Intensive. Full-time and part-time faculty were invited to participate. There were two on-campus meetings, at the beginning and end of the three-week course period. Assignments included participation in online discussions and online submission of lesson plans for peer review. Participants were divided into facilitated large groups of ten. Within these groups, participants self-selected smaller groups of two or three to complete several activities.

The non-completion was fairly high (approximately 40%), but all participants rated the experience as valuable, regardless of completion. The group work was a good experience for those who participated. Breaking the "class" of 50 into teams of ten provided an interesting but manageable amount of online discussion. The very small teams worked well for the short duration of the course and the individual nature of the lesson plan deliverable.

Mass Communication

These are quotes from student evaluations gathered from a Mass Communication class. This class was an online class. Students were assigned to groups of four or five by the instructor. The students worked in the same groups throughout the six-week summer

Copyright © 2005, Idea Group Inc. Copying or distributing in print or electronic forms without written permission of Idea Group Inc. is prohibited.

quarter. Each of the eight assignments required a group response to several questions pertaining to specific video lessons and/or textbook chapter material.

Several changes are being considered for next quarter. The first assignment will be an introduction to group discussion and formulating a group response, so students become familiar with the concepts and process before working on content-based assignments. Students will be encouraged to use online discussions, rather than depend on email. Students will receive more direction on how to do the work. More specifics about individual and group assessment and grade allocation will be provided in response to student concerns voiced in Summer quarter feedback. Deadlines and time issues are less problematic in regular 12-week quarters.

- Sometimes small groups don't work because I found that there is always someone who does most of the work while the others do hardly anything. I didn't want to be in a group because I like to work independently.

- I did not participate in the small groups, because I found it extremely difficult to coordinate with my group members. If one person sent an email message out to the rest of the group, it would take days for most of the group members to respond. Most only cared to respond immediately before the assignment was due, which makes it impossible to get anything done in a timely manner. I also feel that with group work, everyone should be contributing to every part of the assignment, rather than each person taking responsibility for only a portion and speaking for the others without review. This seemed to be the method that my group members desired. I'm not sure how group work could fair better in a distance learning class - interacting online is difficult when you don't know someone at all. It becomes especially difficult when everyone is on a different schedule and coordination efforts seem to fail because of that. Perhaps a suggestion would to hold periodic, optional class meetings to allow group members to meet and work together in person.

- Yes, I participated in the small group work and it was for the most part helpful. I liked the fact that this made the discussions easier.

- I tried to get in contact with people in my small group, but only one person responded, so we both decided to do all the discussion questions on our own. If there is a better way to get in contact with group members, then the small group projects should be no problem.

- I think the small groups were helpful because it gave everyone a starting point from which to communicate ideas and opinions. We were also able to make comments or ask questions. The feedback from peers wasn't always there but help was available if you tried hard enough. This was my first distance-learning experience and it was a pretty positive one. I think that the groups could have been more efficient but with our hectic individual schedules and short time restraints (Note: Summer quarter is a six-week session), it was difficult to become more intimate as a discussion group. Maybe there could have been less discussions but more thorough dialog between the members for each questions so that there is more intellectual discourse instead of worrying about deadlines.

Copyright © 2005, Idea Group Inc. Copying or distributing in print or electronic forms without written permission of Idea Group Inc. is prohibited.

Instructor Interviews

I spoke with many instructors on the subject of online group projects. They were happy to share lots of interesting and useful information informally. I have captured and transcribed the information. Interviewees are exceedingly generous and thoughtful.

Reading, Writing, and Thinking – Poetry

In a Reading, Writing, and Critical Thinking class based on Poetry, the instructor includes group projects in online and on-campus courses. There are usually four or five students per group, and students are assigned to groups by instructor or at random using WebCT (which worked out very well). The project groups could produce paper, website or PowerPoint presentations about a poem. In the poetry assignment, the presentation included poet info, thesis, and links. Students choose the deliverable format. Interestingly, nobody did a paper.

Job descriptions are provided by the instructor. The instructor defines several roles -e.g., web master, editor-and students select roles from the choices. Directions for the project are clearly spelled out in assignment information -expectations, deliverables, and format. Help is available from the instructor. Online groups tend to be more self-sufficient. WebCT discussions are set up for each group. There have been as many as 80 posts for one group including introductions, project planning, exchanging work products, suggestions, and edits. The instructor also uses Manila's weblog and calendar to provide regular updates and reminders to students for due dates and suggestions.

The duration of project is about three weeks and is assigned in Week 8 of the 12-week quarter. Several groups created PowerPoint presentations. Students are required to include some prose as well as bullet points. This is not good PowerPoint, but it is important in a Reading, Writing, and Thinking course.

Students were reluctant to participate in group projects before they started. Once they got started, they worked very hard, produced great work, and had a good learning experience. The results are well worth the time and effort. Assessing content is important. It is easy to be impressed by the technology. Students are very creative and some put a lot of effort into the "bells and whistles." In *Growing Up Digital*, Don Tapscott (1999) describes how students need collaboration, seek relevancy, and want information now. Educators are focused on assessing fluency with emphasis on critical thinking and process. The instructor is moving to more asynchronous peer review of group deliverables rather than presentation to whole class.

Mass Communications / Journalism

The instructor participated as a student in a very good group project in an online graduate school course at UCLA. The students were motivated graduate students with good time management skills, who were responsible, and engaged in good communication with group members. Groups were four or five people, and students could request the same

Copyright © 2005, Idea Group Inc. Copying or distributing in print or electronic forms without written permission of Idea Group Inc. is prohibited.

group as acquaintance. The instructor assigned students to groups based on bio information. Groups lasted for the whole semester, and several projects were completed by the group, including preparing and presenting papers based on research and personal experience of group members. Group members took turns being the primary writer. Other group members contributed input, edited, and critiqued group paper for submission.

The instructor uses some group projects in her Journalism classes with limited success. It is hard to keep groups together and committed to completing assignments. Younger students are not sufficiently responsible for group projects to be very satisfactory. Group projects in online classes require students to be clearer in communication in text format, providing a record of information exchanges. The asynchronous communication works well for some students, as they can think about their input and responses. Discussions replace casual student interaction. They are more thoughtful and intellectually more interesting than chat. Online group projects provide an opportunity to interact within small group frequently, to get to know one another better, to work together, and to be part of a team.

Business Law I

In his distance learning Business Law class, the instructor uses large group discussions. Students respond with personal opinions to broad class topics — What did you think about the test yesterday? What do you think about testing generally? He divides these large classes of 60+ into several groups. The instructor is just getting started with online groups and discussions but wants to move in that direction.

Business

In his Business course, the instructor found that students were proud of their online group work, and it was a good learning experience preparing a presentation. Combining images and text with oral presentation was very powerful. Asynchronous collaboration is a critical job skill, for it may be the only way for students on-campus as well as online to work on group projects. It is often too hard to find convenient time outside class. Students focus on the task when working online and figure out how to divide work into chunks, then combine individual pieces into group project deliverable. Students working together demonstrate greater creativity than individual projects.

Using PowerPoint has been very successful. Most students have it. The outline structure is good for helping students think about sequence, understand the importance of specific topics, and build projects. Stepwise refinement is easy — add, remove, modify, rearrange, delete content. The flexibility is very helpful. There are minor problems with students focusing on technology and not spending enough time on content. However, pride in product more than outweighs any drawbacks. Students pick their own groups of four or five, which seems to be the best number of students per group.

Copyright © 2005, Idea Group Inc. Copying or distributing in print or electronic forms without written permission of Idea Group Inc. is prohibited.

Financial Accounting

The instructor has been teaching online accounting courses for many years. He included online group projects several years ago when classes had 40 students. He had some success with group projects, but technical problems caused considerable frustration and detracted from the benefits of using collaboration online. Currently, he is teaching classes of 200+ students which are too big to use group projects.

Nutrition

The instructor really likes distance learning. She thinks time spent in class is wasted for most students and instructors, as there is too much emphasis on entertainment, not enough on learning. Most students are not well served by classroom delivery rate or modalities. For her distance learning Nutrition course, the instructor does not use group projects. Students do not like them, and she rarely finds the results satisfactory. Distance learning is hard for students. Students need to be well organized, have good time-management skills, be highly motivated, and have good study skills. Successful distance learners are more likely to be older students.

Reading, Writing, and Thinking 1A

The instructor observed that students' attitudes and expectations are changing. Education needs to be interesting, fast-paced, and visually engaging. Students expect A-grades for showing up to class. In this environment, students do not want to participate in group projects. For this distance learning Reading, Writing, and Thinking class, the instructor primarily uses individual projects, and some peer reviews.

Reading, Writing, and Thinking 1B

The instructor refers to her distance learning Reading, Writing, and Thinking class as a low-tech online course. She uses a simple listserv, and students exchange email with others in their group to work on projects. Projects are due about two weeks after they are assigned. Student interaction is required to produce the deliverable. Students are assigned a partner and students work in groups of two for the online class (compared with groups of four or five in on-campus classes). Requirements are well defined, and students understand what deliverable is required. Students work on several projects per twelve week quarter.

Copyright © 2005, Idea Group Inc. Copying or distributing in print or electronic forms without written permission of Idea Group Inc. is prohibited.

Summary

In the initial proposal for *The Group Project Project*, several areas of investigation were described. Questions to be addressed included: How does group project work fit into an online course? Does online group project work change the students' understanding and application of material used in the online group project? What are the "housekeeping" tasks that must be considered for preparing students for online group projects? What needs to be in the student preparation instruction to help students have a good learning experience?

Research supports the inclusion of collaborative activities and group projects in online learning to provide important learning opportunities to students. Within the constructivist, student-centered models, student-student interaction is a key component to learning. Students are engaged in many different ways — exploration, transformation, presentation, and reflection.

Among instructors teaching in the online environment, there is considerable interest in incorporating group project activities, and the process for planning, developing, and facilitating these projects. The literature cited provides a good overview of group project process. The studies and observations yield helpful information and provide online instruction practitioners with helpful directions to use in the virtual classroom setting.

Initially, low-stakes, low-risk group projects are appropriate. Many instructors and students are just learning to teach and learn online. By starting slowly, there is an opportunity to develop the skills and experience necessary to take on larger, more complex projects. Many faculty teaching online have never had the opportunity to be online learners or participate in collaborative group projects online. Instructors need to be learners, too.

The online projects that are discussed in this chapter are geared to short but significant secondary and community college class projects. Further investigation will yield interesting findings and will impact the definition and development of collaborative activities within distance learning. The extent of instructor involvement is one such topic. The online instructor is the "guide on the side," but there is still considerable debate about how that is manifest with regard to group projects. Is it better to let groups work things out for themselves, or should the instructor provide direction through the project process?

Student motivation is another issue. How can group work be structured to bring out the best in all students? How should the instructor and the other team members handle under-contributors? How is the group leadership determined? What are the direct benefits of group project work for students as individuals?

This leads to the concern about assignment grading. Are all team members awarded the same grade for the project? Is contribution recognized and rewarded in individual grades on a group project activity? How best to determine individual grades? Is peer input reliable? How are team members held accountable? All these questions are important. Best practices around these issues are only beginning to emerge.

More online collaboration is likely in higher education as means of working on more sophisticated projects, extending research, sharing expensive and specialized equip-

Copyright © 2005, Idea Group Inc. Copying or distributing in print or electronic forms without written permission of Idea Group Inc. is prohibited.

ment, and including more geographical dispersion of project teammates. Job skills and vocational training will require more emphasis on real-world activities that include online collaboration in many sectors of the economy. Globalization and geographic diversity are common in businesses today and necessitate online collaboration and group project work.

Specialization creates situations where an organization may have many specialists in different fields. For example, a community college may have only one distance-learning faculty coordinator or instructional designer. Having these specialists meet and work with peers is important to share information and to develop a network of contacts for professional development. These virtual groups are essential to maintaining and improving the specialization. Working collaboratively at a distance may be the only viable solution. The importance of online collaboration and group project activities is expanding into all aspects of online teaching and learning.

College students are technologically literate, information savvy, and very aware of options and opportunities that are available to post-secondary education, and they are more than willing to share this information with faculty and administrators willing to listen. They want help learning how to learn. Actual observation helped identify needs that were not being met and that were not articulated by students. Many of the learners do not know that they do not know how to work collaboratively online. Since instructors are no longer the source of information, of truth, they can take a more useful role as facilitator of learning, not the source. The challenges of developing and facilitating online collaborations and group projects in distance education are more than offset by the benefits of the students' learning experience.

Online group projects need to be considered in the overall instructional plan for usefulness, timeliness, and instructional quality. Preparation of instructors and students for online collaboration and group projects are critical to their success.

References

Anderson, T. & Garrison, D.R. (1998). *Learning in a networked world: New roles and responsibilities. Distance Learners in Higher Education.* Madison, WI: Atwood Publishing.

Baker, A., Jensen, P., & Kolb, D. (2002). Conversation as experiential learning. Retrieved online June 1, 2002 from: *http://makeashorterlink.com/?M372224E.*

Barron, T. (2002). Learning object pioneers learning circuits. Retrieved online June 1, 2002 from: *http://www.learningcircuits.org/mar2000/barron.html.*

Bates, A.W. (2000). *Managing technological change: Strategies for college and university leaders.* San Francisco, CA: Jossey-Bass Publishers.

Bennett, R. E. (2002). Inexorable and inevitable: The continuing story of technology and assessment . *Journal of Technology, Learning, and Assessment.* Retrieved online August 9, 2002 from: *http://www.bc.edu/research/intasc/jtla/journal/pdf/v1n1_jtla.pdf.*

Copyright © 2005, Idea Group Inc. Copying or distributing in print or electronic forms without written permission of Idea Group Inc. is prohibited.

Berge, Z.L. (1995). Facilitating computer conferencing: recommendations from the field. Educational Technology, 35(1), 22-30.

Bersin, J. (2002). *LiNE Zine*. Retrieved online January 30, 2003 from: *http://www.linezine.com/7.2/articles/jbsmyelpn.htm.*

Brown, J., Collins, A., & Duguid, P. (1989). Situated cognition and culture of learning. *Educational Researcher*, 18(1), 32-42.

Clark, A. & Pitt, T. (1997). Creating powerful online courses using multiple instructional strategies. Retrieved October 25, 2002, from the World Wide Web from: *http://leahi.kcc.hawaii.edu/org/tcc_conf97/pres/pitt.html.*

Clark, D.R. (2000). Big dog's leadership page. Retrieved online February 5, 2003 from: *http://www.nwlink.com/~donclark/leader/leadtem2.html.*

Collis, B., Andernach, T., & van Diepen, N. (1996). The web as process tool and product environment for group-based project work in higher education. Paper presented at WebNet'96, San Francisco, October 15-19, 1996. Retrieved online July 12, 2002 from: *http://texas-extension.tamu.edu/agcom/ho...griall/webnet96.gr/webnet96/html/378.htm.*

Collison, G., Elbaum, B., Haavind, S. & Tinker, R. (2000). *Facilitating online learning: Effective strategies for moderators.* Madison, WI: Atwood Publishing.

Connery, B. A. and Vohs, J.L., Group Work and Collaborative Writing. Teaching at Davis, 1988, 14(1), 2-4. (Publication of the Teaching Resources Center, University of California at Davis) Retrieved July 23, 2002 from: *http://www-honors.ucdavis.edu/vohs/sec04-2.html.*

Curtis, D., & Lawson, M. (2001). Exploring collaborative online learning. *JALN*, 5(1), June. Retrieved online July 12, 2002 from: *http://www.aln.org/alnweb/journal/jaln-vol5issue1.htm http://www.aln.org/alnweb/journal/Vol5_issue1/Curtis/curtis.htm.*

Damon, W. (1984). Peer interaction: The untapped potential. *Journal of Applied Development Psychology*, 5: 331-343.

Diaz, D., Cartnal, R. (1999). Students' learning styles in two classes: online distance learning and equivalent on-campus. *College Teaching*, 47(4), 130-135.

Dillenbourg, P. & Schneider, D. (1995). Collaborative learning and the Internet. Retrieved online August 18, 1998, from: *http://tecfa.unige.ch/tecfa/research/CMC/colla/iccai95_1.html.*

Doll, W.E. (1993). A *post-modern perspective on curriculum.* New York: Teachers College Press.

Dunn, P. & Marinetti, A. (2002). Cultural adaptation: Necessity for global eLearning. *LiNE Zine*, July. Retrieved online July, 2002 from: *http://www.linezine.com/7.2/articles/pdamca.htm.*

Espinoza, S., Whatley, S., & Cartwright, C. (1996). Online courses - The 5 W's and 2 perspectives. *The American Journal of Distance Education*. 6(3): 5-21. Retrieved online July 12, 2002 from: *http://www.coe.uh.edu/insite/elec_pub/html1996/16teless.htm#espi.*

Felder, R.M. (1996). Matters of style. *ASEE Prism*, 6(4), 18-23, December.

Copyright © 2005, Idea Group Inc. Copying or distributing in print or electronic forms without written permission of Idea Group Inc. is prohibited.

Fleming, N. D. (2001). The Active Learning Site Vark, Downloaded January 29, 2002, from the World Wide Web. Retrieved from: *http://www.active-learning-site.com/vark.*

Fleming, N.D. & Mills, C. (1992). Not Another Inventory, Rather a Catalyst for Reflection. To Improve the Academy, Vol. 11, 1992, pp. 137-149. Retrieved February 17, 2002 from: *http://www.ntlf.com/html/lib/suppmat/74fleming.htm.*

Foote, E. (1997). Collaborative learning in community colleges. Retrieved online July 12, 2002 from: *http://www.gseis.ucla.edu/ERIC/digests/dig9709.html.*

Garrison, D.R. (1989). *Understanding distance education: A framework for the future.* London: Routledge.

Gellman-Danley, B. & Fetzner, M.J. (1998). Asking the really tough questions: Policy issues for distance learning. *Online Journal of Distance Learning Administration.* 1(1). Retrieved online February 24, 1999, from: *http://www.westga.edu/~distance/danley11.html.*

Goddard, G. (2002). Icebreaker ideas submitted by DEOS Members. Retrieved online July 12, 2002 from: *http://acadweb.snhu.edu/DE/Goddard_Gretchen/icebreaker%20 activities.htm.*

Graham, C., Cagiltay, K., Lim, B., Craner, J., & Duffy, T.M. (2001). Seven principles of effective teaching: A practical lens for evaluating online courses. *The Technology Source,* March/April. Retrieved online February 10, 2003 from: *http://ts.mivu.org/ default.asp?show=article&id=839.*

Granger, D. & Benke, M. (1998). *Supporting learners at a distance from inquiry through completion.* Madison, WI: Atwood Publishing.

Hamada, T. & Scott, K. (2001). A collaborative learning model. *The Journal of Electronic Publishing.* Retrieved online July 12, 2002 from: *http://www.press.umich.edu/jep/06-01/hamada.html.*

Hanna, D., Glowacki-Dudka, M., & Conceicao-Runlee, S. (2000). *147 practical tips for teaching online groups: Essentials of web-based education.* Madison, WI: Atwood Publishing.

Harasim, L., Hiltz, S., Teles, L., & Turoff, M. (1995). *Learning networks.* Cambridge, MA: MIT Press.

Harris, J. B. (1995). Organizing and facilitating telecollaborative projects. The Computing Teacher, 22(5).

Henri, F. & Rigault, C. R. (1996). Collaborative distance learning and computer conferencing. In T. T. Liao (ed.), *Advanced educational technology: Research issues and future technologies,* pp. 45-76. Berlin: Springer-Verlag.

Hildreth, S., Masterson, N., & Wallace, G. (2000). Online activities at your (electronic) fingertips... A how-to guide for creating the best online activities that really work! Retrieved online July 12, 2002 from: *http://chabotde.clpccd.cc.ca.us/user/~astro/ edui6772.html.*

Hiltz, S.R. (1994). The virtual classroom: Learning without limits via computer networks. Human-Computer Interaction Series. Norwood, NJ: Ablex Publishing Corp.

Hsi, S. & Hoadley. C. M. (1997). Productive discussion in science: Gender equity through electronic discourse. *Journal of Science Education and Technology* 6 (1): 23-36.

Copyright © 2005, Idea Group Inc. Copying or distributing in print or electronic forms without written permission of Idea Group Inc. is prohibited.

Johnson, D.W., Johnson, R.T., & Smith, K.A. (1991). Cooperative learning: Increasing college faculty instructional productivity. *ASHE-ERIC Higher Education Report No. 4.* Washington, DC: The George Washington University School of Education and Human Development.

Jonassen, D.H., Peck, K.L., Wilson, B.G., & Pfeiffer, W.S. (1998). *Learning with technology: A constructivist approach.* Englewood Cliffs, NJ: Prentice Hall.

Kolb, D.A. (1984). *Experiential learning: Experience as the source of learning and development.* Englewood Cliffs, NJ: Prentice-Hall.

Lipman, M. (1991). *Thinking in education.* Cambridge, UK: Cambridge University Press.

Lum, G., Mebius, L., & Wijekumar, K. (1999). Welcome to the design of constructivist learning environments (CLEs). Retrieved online July 19, 2002, from: *http://tiger.coe.missouri.edu/~jonassen/courses/CLE.*

Markel, M. (1999). Distance learning and the myth of the new pedagogy. *Journal Of Business And Technical Communication,* 13(2): 99.

Martens, M. L. (1992). Inhibitors to implementing a problem-solving approach to teaching elementary science: Case study of a teacher in change. *School Science and Mathematics,* 92(3), 150-156.

Mason, R. (1998). Models of online courses. Networked Lifelong Learning: Innovative Approaches to Education and Training through the Internet. Retrieved online July 14, 2002 from: *http://www.aln.org/alnweb/magazine/vol2_issue2/Masonfind.htm.*

Maus, D.C. (1998). Walking the Line: Rectifying Institutional Goals with Student Realities. The Technology Source, February 1998. Retrieved May 14, 2003 from: *http://64.124.14.173/default.asp?show=article&id=485.*

McCormick, C. & Jones, D. (1998). *Web-based education system.* New York: John Wiley and Sons.

Miller, et al. (1998). Guiding Principles and Practices for the Design and Development of Effective Distance Education. Retrieved online July 12, 2002 from: *http://www.cde.psu.edu/DE/IDE/GP&P/GP&P.html.*

Mislevy, R. J., Steinberg, L. S., Almond, R. G., Breyer, F. J., & Johnson, L. (2001, March). *Making sense of data from complex assessments* (CSE Technical Report 538). Retrieved July 12, 2002 from: *http://www.cse.ucla.edu/CRESST/Reports/RML%20TR%20538.pdf.*

Moore, M. (1989). Three types of interaction. In M. Moore & G.C. Clark (eds.), *Readings in principles of distance education.* Pennsylvania State University, University Park, PA: American Association for Distance Education.

Oakley, W. F. & Crocker, R. K. (1977). An exploratory study of teacher interventions in elementary science laboratory groups. Paper presented at the 50th Annual Meeting of the National Association for Research in Science Teaching, March, Cincinnati, OH. (ERIC Document Reproduction Service No. ED 139 608).

Ossont, D. (1993). How I use cooperative learning. *Science Scope,* 16(8), 28-31.

Palloff, R.M. & Pratt, K. (1999). *Building learning communities in cyberspace: Effective strategies for the online classroom.* San Francisco, CA: Jossey-Bass Publishers.

Copyright © 2005, Idea Group Inc. Copying or distributing in print or electronic forms without written permission of Idea Group Inc. is prohibited.

Pressley, M. & McCormick, C.B. (1995). *Advanced educational psychology for educators, researchers, and policymakers.* New York: Harper Collins.

Recker, M., Walker, A., & Wiley, D. (2000). An interface for collaborative filtering of educational resources. In *Proceedings of the 2000 International Conference on Artificial Intelligence (IC-AI'2000),* pp. 317-323. CSREA Press. Retrieved online from: *http://rclt.usu.edu/docs/2000recker001.pdf.*

Reid, J., Forrestal, P., & Cook, J. (1989). *Small group learning in the classroom.* Portsmouth, NH: Heinemann.

Riel, M. (1993). Learning Circles: Virtual communities for elementary and secondary schools. Electronically published at URL: *http://lrs.ed.uiuc.edu/guidelines/Riel-93.html.* Retrieved August 12, 2002.

Sagan, C. (1997). The Demon-Haunted World: Science as a Candle in the Dark, Ballantine Books, New York. p. 413

Salomon, G. (1981). *Communication and education: Social and psychological interactions.* London: Sage.

Stahl, S. (2002). Bridging old ideas to new times: Learning principles of Kurt Lewin applied to distance education. 03 July 02. Retrieved online May 13, 2003 from: *http://ts.mivu.org/default.asp?show=article&id=38.*

Stirling, P. (1987). Power lines. NZ *Listener,* 13-15, June 20.

Sullivan, R. (2002). Lessons in smallness. T+D astd.org March. Retrieved online May 31, 2002 from: *http://www.astd.org/members/td_magazine/td0203/76020321.pdf.*

Tapscott, D. (1998). Growing Up Digital: The Rise of the Net Generation. New York: McGraw-Hill.

Tinker, R. (2001). E-Learning quality: The Concord model for learning from a distance. *NASSP Bulletin,* 85(628): 36-46, November. Retrieved online July 12, 2002 from: *http://www.principals.org/news/bltn_elearning1101.html.*

Tinzmann, M.B., Jones, B.F., Fennimore, T.F., Bakker, J, Fine, C., & Pierce, J. (1990). What is the collaborative classroom? [Online]. Oak brook, IL: NCREL. Retrieved online July 12, 2002 from: *http://www.ncrel.org/sdrs/areas/rpl_esys/collab.htm.*

Toh, K. & Woolnough, B.E. (1993). Middle school students' achievement in laboratory investigations: Explicit versus tacit knowledge. *Journal of Research in Science Teaching,* 30(5), 445-457.

Tuckman, B.W. (1965). Developmental sequence in small groups. *Psychological Bulletin,* 63: 384-399.

Turkel, S. (1995). *Life on the screen: Identity in the age of the Internet.* New York: Simon & Shuster.

Verdejo, M.F. (1996). Interaction and collaboration in distance learning through computer mediated technologies. In T. T. Liao (ed.), *Advanced educational technology: Research issues and future technologies,* pp. 77-88. Berlin: Springer-Verlag.

Waugh, M.L., Levin, J.A., & Smith, K. (1994). Organizing electronic network-based instructional interactions: Successful strategies and tactics, Part 1. The Computing Teacher, 21(5), 21-22. Retrieved August 12, 2002 from: *http://lrs.ed.uiuc.edu/guidelines/WLS.html.*

Wiggins, G., & McTighe, J. (1998). Understanding by design. Alexandria, VA: Association for Supervision and Curriculum Development. Retrieved August 9, 2002 from: *http://www.ers.org/ERSBulletins/0399f.htm.*

Copyright © 2005, Idea Group Inc. Copying or distributing in print or electronic forms without written permission of Idea Group Inc. is prohibited.

Chapter III

Time, Place, and Identity in Project Work on the Net

Sisse Siggaard Jensen
Roskilde University, Denmark

Simon B. Heilesen
Roskilde University, Denmark

Abstract

This chapter identifies some of the fundamental conditions and factors that affect collaborative project work on the Net. Understanding them is fundamental to developing key qualities in Net-based collaborative learning such as confidence, reliability, and trust. We argue that: (1) Collaboration and social interaction develop in continuous oscillations between abstract and meaningful frames of reference as to time and place. (2) Such oscillations condition the creation of a double identity of writer and author modes in social interaction. (3) Collaborative work creates an ever-increasing complexity of interwoven texts that we have to develop strategies for organizing. (4) One such important strategy is the negotiation of roles among the participants. Having established this theoretical framework, we discuss how to deal with these conditions in an actual Net-based learning environment, the Master of Computer-Mediated Communication program at Roskilde University, Denmark.

Copyright © 2005, Idea Group Inc. Copying or distributing in print or electronic forms without written permission of Idea Group Inc. is prohibited.

Introduction

Our reason for moving an academic program from a conventional setting to the Net was to try to solve some problems inherent in a particular kind of open-university education. Basically, these problems had to do with coordination of work in time and place. Our initial approach was a practical one, and with hindsight we may have been guilty of relying rather naively on the power of new technology to *remediate* familiar routines, i.e., to "translate" them into the new medium, expecting them to work in much the familiar way, but hopefully better (Bolter & Grusin, 1999). Complex tasks do not lend themselves easily to an unproblematic remediation, and in moving from the physical world to the Net, we have come to realize that problems with time and place remain, but that they become of a different order.

Reflecting on our practice, we will discuss different time and place interrelations as the key fundamental factors affecting the Net-mediated learning environment. Further, we will examine how they influence identity and trust, both of which in our experience are essential for successful collaboration in a Net environment. In collaborative learning, there is every indication that social relations based on distinct identity and the building of trust encourage interaction, dialogue, and reflective practices. The question therefore is how the process of remediating learning activity from a located and shared context such as a lecture room to a displaced networked environment influences factors significant for social relations such as identity and trust. In this examination, we will draw on social theory and philosophy on time and place (Adam, 1990, 1995, 1998; Latour, 1988a, 1988b, 1998; Mead, 1929,1932; Mead, Morris, Brewster, Dunham & Miller, 1938; Rämö, 1999) and on social interaction, discourse, and intertextuality (Fairclough, 1992, 2003; Jensen, 2001a, 2001b; Mead, 1934/1970, Mead et al., 1938; Stacey, 2001). A brief guide to this theoretical framework may not be amiss.

Barbara Adam's (1990, 1995) integrating theories from the natural and social sciences with the philosophy of time has shown how essential and useful the understanding of time is to social theory and with that the basic understanding of any human activity and social interaction. George Herbert Mead's (1932) philosophical analysis of time based on a phenomenological approach influenced Martin Heidegger in his philosophy on "Sein und Zeit," and it is still a classic when it comes to understanding time as a human and social phenomenon. Mead (1934), Mead et al.(1938) through his symbolic interactionism also provided a new way of understanding human interaction and especially cooperation. Hans Rämö (1999) has applied a philosophical and Aristotelian conception of time and place to virtual organizations. Thus, he exemplifies the Aristotelian distinction between abstract and meaningful time and place, which also has influenced 20th century philosophy. Norman Fairclough (1992, 2003) introduced the concept of intertextuality to discourse analysis in order to emphasize the social nature of any text. The concept of intertextuality is useful because social interaction on the Net is very often expressed in patterns of interrelated text. Finally, Bruno Latour's (1988b) thoughts on time and identity are introduced, particularly his distinction between a writer and an author mode or identity, which is relative to different relations of time and place (1988b).

Copyright © 2005, Idea Group Inc. Copying or distributing in print or electronic forms without written permission of Idea Group Inc. is prohibited.

Net-Based Learning at Roskilde University

Before embarking on the discussion of time and place, we will describe briefly the setting as well as the academic programs that provide our empirical material.

In its three decades of existence, Roskilde University, Denmark, has developed and refined a pedagogy focusing on interdisciplinary problem-oriented project work done by students working in groups (for a short description, see Cheesman & Heilesen, 1999). In its conventional face-to-face form, studies of this nature require students to work together quite intensively on-campus in buildings actually designed for group work. The challenge in introducing Net-based education has been to transform and remediate a well-functioning university culture to the computer medium. The possible reward has been to make academic programs available to students unable to attend classes and group meetings at set times and at specific locations.

In the mid-1990s, the Institute of Communication Studies, Journalism, and Computer Science began developing a Net-based, open-university, part-time program in communication studies. It was launched in 1996 as InterKomm+ (as of 2003: Master of Professional Communication, http://www.mpk.ruc.dk/), and in 2000 it was supplemented with a master's degree program in Computer-Mediated Communication (http://www.mcc.ruc.dk/, for a short description, see Cheesman & Heilesen, 2001).

Both programs combine face-to-face meetings at six or seven annual weekend seminars with "Net seminars" involving assignments to be done as group work (four to five weeks each) and a one-semester self-defined project usually also done as group work. In the Master of Computer-Mediated Communication program, which we will be focusing on in this chapter, a class consists of 16 – 24 students who work together in groups of three to five for short assignments and two to six for large projects. An assignment consists in producing a paper, a website or a PowerPoint presentation on a subject defined by the teacher. A project involves writing a paper (30-60 pages depending on group size) on a subject approved by the teacher and also in developing a product (e.g., a website or multimedia application).

As technical platform, we use Basic Support for Cooperative Work (BSCW, http://www.bscw.de), which is a general-purpose system for Computer-Supported Collaborative Work (CSCW). Apart from being particularly suitable for supporting project work (Sikkel, Gommer, & van der Veen, 2001), BSCW also imposes no implicit or explicit pedagogical restraints and offers an extremely configurable user interface – both somewhat unusual but useful qualities in e-learning software.

Extensive evaluations of our Master's program have shown that most students will supplement the use of BSCW with other means of communication, either electronic (email, chat) or conventional (telephone, face-to-face meetings). Thus the program should perhaps be characterized as a Net-supported rather than a purely Net-based program. Not being on the Net exclusively, but only when it serves a purpose, does not suggest failure to us, but is rather evidence of a maturing use of new media. New and interesting hybrid forms of education are evolving. In this chapter, however, we will focus only on the Net-based communication.

Copyright © 2005, Idea Group Inc. Copying or distributing in print or electronic forms without written permission of Idea Group Inc. is prohibited.

Time, Place, and Identity

Time and place are basic conditions in any human activity (Adam, 1990, 1998; Mead, 1932, 1934/1970, Mead et al. 1938). Visions about the information society tend to claim that the networked learning environment offers a great deal of freedom and flexibility, if not actually independence of time and place. As we have described above, such ideas were not far from mind when we first introduced Net-based open education.

Certainly, when physical presence is no longer required, changes occur in the controlling institutional structures and the synchronization of activities (both characteristics of the industrial age learning environments), and they are supplemented with a notion of an abstract time and space relation in a cyberspace or virtual space. Understanding the changes as well as the new time and space interrelations in the networked environment is fundamental to understanding how social interaction unfolds in such an environment and thus, by extension, how we can improve on the "habitability" of the Net environment.

Problems of Time and Place

In Net-based learning, students and teachers are not liberated from the constraints of time and place. On the contrary, they may become pressing problems. This was demonstrated already in the early years of Net-based learning in interviews with pioneers in the field (Jensen, 1991, 1993). In those days, technical problems loomed quite large, but then as now the truly time-consuming tasks were to understand and practice new ways of communicating, coordinating, and cooperating in an environment characterized by urgency. Net-based communication allows for prompt exchanges, and usually the participants expect nothing less. In order to deal with a pressure that easily becomes constant, strategies have to be developed for scheduling response and moderating expectations. As illustrated by the common experience that even minor deviations from the rules cause frustrations, networked learning makes heavy demands on the predictability of communication—and with that the reliability that inspires confidence and trust.

Subsequent qualitative video analysis of networked learning has reinforced the impression that the learning activities are far from being independent of time and place (Jensen, 1994, 2001a). Of course the place is different. A computer screen has replaced the physical room with its conventional educational technologies. But behind the screen, so to speak, are not only learning materials and tools, but also the teacher and one's fellow students. They too inhabit the space and emerge face-to-interface, although now in the shape of complex patterns of intertextuality (see Figure 2). Furthermore, the complexity of these patterns is constantly growing as the collaboration and social interaction of Net-based project work develops. Such activities naturally require some kind of synchronization, and we have thus observed that in the face-to-interface situation there is a strong need for dealing with time-related questions, a need that may in fact be greater than in a conventional classroom setting.

Copyright © 2005, Idea Group Inc. Copying or distributing in print or electronic forms without written permission of Idea Group Inc. is prohibited.

A Matrix of Time and Place

As suggested above, a new relation to time is brought about by the transformation of place from a physical room to a computer screen that is the interface to a virtual space, which is in fact not altogether abstract. The actual architecture of virtual space is not central to the discussion of time/space interrelations. But being important both for the design and implementation of Net-based learning, we will offer some reflections on metaphorical space in an excursus (see p. 65).

To further discuss some of the empirical observations of time and place we shall introduce a conceptual framework of time and place inspired by premodernist, preindustrial societies. The framework is based upon an Aristotelian conception, but it is a modern version that is modeled as a matrix of time and place by means of which we can analyze and discuss new forms of organization such as virtual environments (Rämö, 1999). We shall use this conceptual framework for a discussion of some of the empirical observations.

The familiar ancient Greek concepts of time and place are "chronos" and "topos," whereas the two related concepts, "kairos" and "chora," have survived only in specialized terms. Chronos and kairos are both concepts of time, chronos being the abstract *chronos time,* kairos the meaningful *kairos time* (Kairos was god of the favorable moment). In the Aristotelian sense of the word, kairos is closely linked to the notion of "phronesis," which means wisdom and judgment. The "kairic" feeling for the right moment is always connected with the wisdom and judgment obtained by acting in concrete and meaningful practices. In the pair of concepts, "chora" and "topos," there is a similar distinction between the abstract and the concrete and meaningful. Chora is a notion of abstract space, whereas topos refers to a concrete place.

In other words, time and place configurations may take the form of the following four conjunctions of the abstract and the meaningful: "chronochora," "chronotopos," "kairochora," and "kairotopos," as summarized in Figure 1.

Figure 1: A matrix of time and space manifold (based upon Rämö, 1999).

	Abstract Space	**Meaningful Place**
Abstract Time	*chronochora,* abstract time and abstract space. Purely abstract and generally applicable conjunction.	*chronotopos,* abstract time and meaningful place precisely regulated by the clock.
Meaningful Time	*kairochora,* meaningful time and abstract space. Implies the ability to communicate and act on human "right moments" in an abstract space.	*kairotopos* meaningful time and meaningful place. A place world with implaced activity regulated by the ability to communicate and act on human right moments.

Copyright © 2005, Idea Group Inc. Copying or distributing in print or electronic forms without written permission of Idea Group Inc. is prohibited.

Negotiating Time and Place

In relating these concepts to Net-based learning, we may start by observing that trust is one of the most important "connectors" when physical presence no longer enables and commits participants. If people meet only on rare occasions, connections are best sustained if based on trust. Along with trust comes the phenomenon of timing or "the right moment" in the kairic sense of the concept (Kirkeby, 2001). Trust emerges if action and reflection are enacted and communicated at the right time. The time dimension of networked learning therefore is dominated by kairos time rather than the time-scheduled chronos time of conventional learning environments. When it comes to space or place, it is common to emphasize the chora quality (abstract space) of Net-based learning environments and organization. If perceived from such a perspective, the networking activity is not conceived of as being located and situated. Thus, on the face of it, the networked environment should be characterized as *kairochora* according to the matrix of time and space manifold (Figure 1).

Our empirical observation, however, diverges from this accentuation of abstract spacing in the kairochora neologism. When observing networked learning activities from the participants' point of view, there is nothing abstract about the location of learning activity. On the contrary, participants are implaced and located by the interface be it at their writing table, in the kitchen, or on the train. They are involved in meaningful activity in what can be described as the kairotopos of the matrix (Figure 1). Occasionally they may meet physically, but here-and-now learning activities are physically present in the wide range of different symbolic and mediated inscriptions which form patterns of intertextuality on the computer screen (Fairclough, 1992, 2003; Latour, 1998). Students locate their learning activity face-to-interface—on websites, in computer conferencing, and when using collaborative tools including a whole range of functionalities facilitating coordination and cooperation.

So, in a sense, networked learning is implaced and situated although located differently than traditional education. Time is neither scheduled nor floating, both of which would give rise to frustration, whereas institutional time schedules are transformed into concurrently negotiated time and place relations accompanied by new rhythms of communication. Social relations and interaction are thus based on the concurrent negotiation of the right moments, and that is the timing of activities and cooperation. This kairic feeling for the right moment, or timing, is best developed in concrete and meaningful practices, if we are to believe Aristotle. In this view the coordination and synchronization of collaboration and learning on the Net is best enacted if based on concrete and meaningful practices rather than on abstract pre-scheduled and administrative procedures. This suggests that both teachers and students should learn how to negotiate the time and place related issues of networked learning.

Meaningful and Abstract Time and Place

Above we argued that negotiating time and place in networked learning is accompanied by new rhythms of communication. Let us use an example of text-based collaborative

Copyright © 2005, Idea Group Inc. Copying or distributing in print or electronic forms without written permission of Idea Group Inc. is prohibited.

Figure 2: Left: Involved learning activity located face-to-interface. This is the writer mode of identity. Right: Patterns of intertextuality from dialogues in Net seminars. This is the author mode of identity

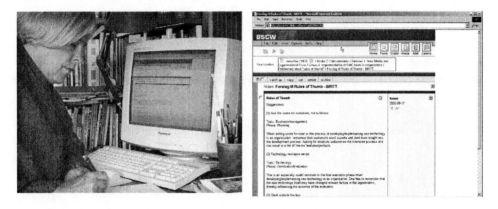

work on the Net in order to illustrate these new rhythms, which are based on the continuous oscillation between meaningful and abstract time and place/space relations, between kairotopos and chronochora.

When preparing a contribution to the learning environment, the participant (student or teacher) is located face-to-interface in a concrete and physical environment (at home, in the office, on a train, etc.). No matter where she is located, as the writer of a text, she continuously has to identify with her writings and, in this process of identification, the learning activity is implaced and involved as the participant is engaged in meaningful activity (kairotopos).

Next, in the collaborative learning environment, social interaction takes place as the text written by the participant is submitted and circulated to the other participants so as to make response possible. To put it in a more formal way: participation means writing a text (on a computer) in an interrelated process of identification and expression in the reflective practices of learning (Dewey, 1933), the text subsequently becoming part of a social activity of gesture and response activities (Mead, 1934/1970) in the form of patterns of intertextuality (the instantiation of which depends upon the system in use).

Along with this transformation, the identity of the participant changes from being a *writer* implaced and located face-to-interface to becoming the *author* of a text represented in an abstracted environment of computer system time and space (chronochora). In the Net environment, the participant takes part in the social interaction as an author whose identity is concurrently constructed in the process of social interaction – in even more abstract terms, in the intertextuality of computer-mediated texts. In this abstracted system of time and space, the identity of the participant when perceived as an author is open to design and decision and thereby also to a diversity of expressions in shifting roles and role plays.

Finally, there is again oscillation from the abstract system time and space to the meaningful time and place interrelations of kairotopos, as the other participants settle down to read, reflect on, and answer the posting of the first participant.

Copyright © 2005, Idea Group Inc. Copying or distributing in print or electronic forms without written permission of Idea Group Inc. is prohibited.

The Double Identity of Writer and Author

The split of identity into *writer* and *author* modes or rather the creation of a *double identity* in the oscillation between different frames of reference in time and place interrelations (Latour, 1988b) is a characteristic of communication and collaboration in networked learning. Modes of identity or self-consciousness in social interaction exist also in face-to-face communication. They are well known from the analysis of the simultaneous interrelations of the "I" and the "Me," which are the two aspects of our social identity that form the basis of any social interaction (Mead, 1934/1970; Latour, 1988b; Stacey, 2001).

What is different is the extension and prolongation of the double identity in rhythms of oscillation between different frames of reference in time and place interrelations in social interaction. These extensions and rhythms make possible new forms of interaction based on designed and decided shifts of roles and advanced role plays. This is because the interrelations of the "I" (writer mode) and the "Me" (author mode) in social activity are *clearly marked* and easily differentiated in networked learning and thus open to conscious design and decision.

The participant in Net-based collaborative work and the gestures (digital text) by which she participates and creates her identity and self-consciousness in the act of acting are

Figure 3: A cycle of gesture~response patterns in social interaction and the double identities of writer~author and reader~writer when oscillating between meaningful and abstract time and place.

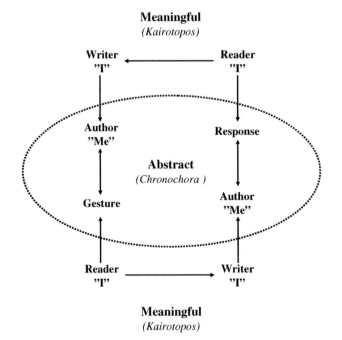

Copyright © 2005, Idea Group Inc. Copying or distributing in print or electronic forms without written permission of Idea Group Inc. is prohibited.

distinctively differentiated from the way in which she perceives the other participants, from their perception of her participation as an author of a text (in computer-mediated intertextuality), and thus also from their responses (social interaction). In face-to-face communication and interaction, on the other hand, there is a concurrent and simultaneous interrelation between these processes, and meaning occurs in the ongoing and immediate interpretations of the social activity when located within a shared frame of reference as to time and place interrelations.

To illustrate these points, Figure 3 visualizes a cycle of gesture~response patterns of social interaction across abstract and meaningful time and place interrelations. In order to take part in Net-based collaborative work, a participant undergoes the transformation from the identity of *writer* in implaced and meaningful time and place (kairotopos) to become an *author* in the shape of digital texts that are parts of the patterns of intertextuality in the abstract time and space of the computer system (chronochora). In computer-mediated communication, such a transformation is the basic gesture of social interaction. It only turns into a gesture, however, if the author is perceived as author by a reader who acts in response to the gesture, and thus transforms from the identity of being a reader of the gesture to becoming a writer of the response (kairotopos). In such patterns of social interaction an integral part is concurrent shifts of role between writer, author, and reader across different frames of reference in time and place interrelations. In our view, taking a conscious, exploratory, and experiential attitude towards these new possibilities of defining and designing different roles as part of learning activity is indeed one of the most inspiring challenges we currently face in Net-based collaborative project work.

New Rhythms of Interaction on the Net

As we have seen, successful acting and reflection in Net-based collaborative work requires the ability to deal with the oscillation between meaningful/abstract time and place, as well as some talent for juggling with double identities that may give rise to unexpected and conflicting or contradictory interpretations. In face-to-face conversation, different rhythms underlie patterns of gesture~response or turn taking, and they are of vital importance to our interpretation of the situation. If, for example, a pause becomes painful or if something is enacted at the wrong time in the wrong place, then we can almost feel it physically as it influences our bodily rhythms. This of course cannot be so in Net-based collaborative work. We are as yet unfamiliar with the rhythms of communication that underlie computer-mediated communication, and hence the interpretation of interaction patterns is difficult. In interaction patterns, what is *not* said or done is as important as that which is. In conversation, pauses in and between turn takings are as important as the words of the conversation. Pauses are pregnant with meaning, but only if the participants are able to interpret that meaning, being familiar with the rhythm of the turn taking. In Net-based communication, on the other hand, it is difficult to interpret the meaning of pauses, as there is no familiarity with the rhythm of the gesture~response patterns of intertextuality. Trust arises from mutual familiarity within shared rhythmicity, as it is the prerequisite for acting judiciously, wisely, and at the right moment. In Net-based collaboration and project work – as shown above – "kairic

Copyright © 2005, Idea Group Inc. Copying or distributing in print or electronic forms without written permission of Idea Group Inc. is prohibited.

rhythmicity," and with that negotiated timing, grounds social interaction based on trust. Therefore, we need to know more about the right moments and the rhythms of communication that underlie interaction patterns when identity and roles are concurrently changed across and between different time and place conjunctions.

Organizing Net Based Project Work

Having outlined a theoretical understanding of the fundamental conditions of time, space/place and identity in Net-based communication, we shall now discuss how to deal with these conditions in an actual Net-based learning environment, the Master of Computer-Mediated Communication program at Roskilde University. We do so by focusing on some key issues: organization and synchronization, decision making, and identity and roles.

Coordinating and Synchronizing Work

Coordination and synchronization of Net-based collaborative work take place in the chronochora, which is perfect for dispassionate scheduling. However, every individual involved in the process is located in kairotopos at different locations, under different circumstances, each with his or her own obligations and a schedule of everyday tasks. This provides for a more complex situation than in a conventional learning environment.

Two types of synchronization can be distinguished, characterized respectively by asymmetrical and symmetrical relations. Asymmetrical relations exist between teacher and students. The teacher is in charge of overall planning and defines phases of work and deadlines that the students must meet. To some extent this is a purely abstract exercise. After all, the academic program consists of a number of modules (seminars) with terminal dates that the students will just have to accept. It should be noted, however, that it is important to offer some latitude. If the granularity becomes too fine, it is likely to have adverse effects on the kairotopos reality of the students. A case in point: Initially in our master's program we introduced a weekly deadline for a progress report, indicating also that certain milestones had to be reached. As a result, the groups of students started making their own frequent deadlines just to make sure that they could meet the weekly one, the worst cases having a deadline every other day. This of course makes a complete farce out of the claim that Net-based studies are independent of time. The students had never been more stressed in their lives. Later we reduced the number of deadlines to one for short projects and three for long projects, and that proved sufficient to ensure progress.

Ideally, symmetry characterizes the relations between students participating in project-oriented group work. The group members are supposed to be peers and, within the boundaries set for the project, they will have to negotiate a "project time." Project time is meaningful time in relation to the tasks required and the actors involved in doing them. Unlike the abstract time of computer system and course planning, project time is "owned"

Copyright © 2005, Idea Group Inc. Copying or distributing in print or electronic forms without written permission of Idea Group Inc. is prohibited.

by the members of the project group and to some extent, it is a reflection of the kairotopos of each individual.

Decision Making

Negotiating project time is just one of the tasks in project work requiring decision-making. Making decisions is an extremely difficult exercise in a Net-based environment, one that illustrates well the kairotopos/chonochora dichotomy. Sitting in front of the screen, it is not easy to decide when a discussion in the abstract time/space is at an end. How do we know that all arguments have been heard, that we all agree, and that we have in fact reached a common understanding? Who should conclude the discussion and how? If such questions cannot easily be answered, the discussion is likely to drag on.

A particularly apt example from an online course at Roskilde University is when we gave a group of a dozen students in an asynchronous discussion forum a choice of two subjects to discuss. The meta-discussion of what subject to choose dragged on for five days, leaving little time for substance in a one-week assignment (Heilesen, Thomsen, & Cheesman, 2002). If the students had met face-to-face, the decision probably could have been made in half an hour at most.

Having students meet face-to-face saves time and prevents misunderstandings. In our master's program, all work on the net starts with a physical meeting where the students discuss their understanding of the problem, prepare a plan of activities for the project work, and decide on a set of rules. If physical meetings are not possible, the second-best solution is to use a synchronous means of communication.

When moving on to the asynchronous work form on the Net, completely symmetrical relations between participants tend to become a problem. Of course the students are equals when it comes to contributing to the learning environment, but even a small measure of asymmetry in relations goes a long way to make project work effective. Quite often someone spontaneously takes charge to get things moving. But it may be better to deliberately introduce a measure of asymmetry in the roles that the students should assume as an integral part of project work. In this case, it involves designating a coordinator who, as "master of the chronochora," will be responsible for organizing the workspace and the time relations of the entire group. This does not rule out that consensus should be aimed at in important matters, Only that someone has to establish the fact that a decision has been reached and make sure that it is acted upon.

Identity and Roles

In the theoretical framework, we have distinguished between the "I" of kairotopos and the "Me" of chronochora. As mentioned above, in the Master of Computer-Mediated Communication program, so far we have insisted that the students should meet face-to-face at the beginning of each Net seminar. Having met in meaningful time and place to discuss what is going to happen online, the students have formed an impression of the personal and professional qualities of their fellow students far beyond what is possible in a pure online environment. They will also have agreed on how to define roles of all

Copyright © 2005, Idea Group Inc. Copying or distributing in print or electronic forms without written permission of Idea Group Inc. is prohibited.

individuals involved in the project. This experience is invaluable in furthering collaboration and building identity and trust on the Net. Still, when meeting in chronochora, for better or for worse, the "Me" of the authors may differ considerably from the face-to-face experience, and roles too may be misunderstood or gradually displaced. Thus, even when founded on face-to-face acquaintance, the social relationships on the Net certainly are complex and even quite vulnerable.

Awareness

Awareness is a Computer Support for Cooperative Work (CSCW) concept with many different meanings (Schmidt, 2002). In the present context, it is seen as an important means of strengthening the kairotopos aspect, meaningful time and place. Awareness of the activities of others and awareness of others being aware of you are essential in Net-based collaborative work. A posting is but archived text until somebody reads it and perceives you as an author. Once you receive a response, you have been established as a participant with an identity – a role that has to be asserted continuously, because not just quality but also the number and frequency of your postings help define your online identity.

In an ideal collaborative Net-learning environment, participants are considerate and respond quickly, if nothing else than with a polite show of "appropriate obtrusiveness"

Figure 4: BSCW has two history functions that help create awareness: One showing when and by whom a file has been modified, and one showing when and by whom it has been read.

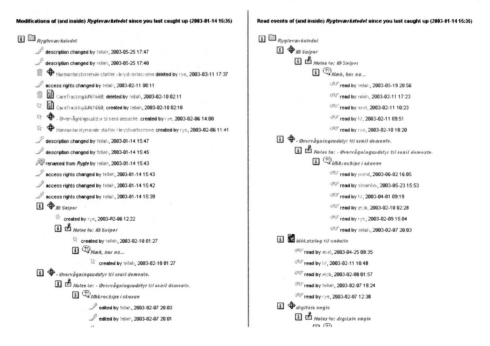

Copyright © 2005, Idea Group Inc. Copying or distributing in print or electronic forms without written permission of Idea Group Inc. is prohibited.

(Schmidt, 2002) in the form of an encouraging note. Often, however, the project workspace can be a very lonely place when you have made a contribution and no one is seen to react. Silence may mean acceptance, indifference, or just inattention rather than rejection. With the absence of established cultural codes in computer-mediated communication, a pause can be difficult to interpret, and you are more or less stuck until someone responds. The situation is aggravated if you have no idea whether or not your contribution has been read at all or, indeed, if there is or recently has been anyone present in the Net environment. Some Computer Support Cooperative Learning/Work (CSCL/W) systems offer technical remedies in the form of more or less developed "history functions" that record all events happening to an object in the system (Figure 4 shows two such BSCW records). By indicating some, however potential, progress in the gesture~response cycle from abstract towards meaningful, even a simple device of this kind does establish the author as a participant in the social interaction and helps boost morale.

Student Roles and Teacher Roles

Above, in several contexts, we have mentioned that deliberately assuming and acting out roles is important in collaborative project work on the Net. It is a different and more complex kind of role than the activity-dependent roles played by the participant, as he or she constantly shifts between being a writer, an author, and a reader. They are not totally unrelated, however, because the dichotomy of the "I" and the "Me" probably facilitates assuming an online identity.

Student roles may be established on the basis of personality and inclination or from deliberate choice. As to the first, it is a common phenomenon in Net-based communities to find all kinds of personalities ranging from helpful to aloof, from constructive to sarcastic, from hyperactive to passive. In Net-based project work, the students are interdependent, and thus truly deviant behavior threatening social interaction in the group is likely to be censured. Still, in any group, there are different dispositions and different qualifications, and this tends to materialize in a division of labor. However, chosen roles should not be permanently tied to an individual's professional qualifications. Rotation from project to project or at regular intervals in large projects helps each individual acquire new qualifications and experience and contribute to project work from different points of view to project work.

There is a touch of game playing to choosing and defining a role as coordinator, programmer, copywriter, editor, designer, librarian, etc., and bringing it to the Net to develop it further in the continuing negotiations with the other participants in the project. But apart from being perhaps motivating, roles are also extremely useful in the Net environment, where they contribute heavily to constructing identity and to stimulating the work process, as each student understands what contribution is expected of her, and the other group members know what to expect from her and welcome her contribution.

The teacher also has a variety of roles to choose from – or, rather, live up to: advisor, administrator, agent provocateur, coach, moderator, observer, and more. These are roles to be played as needed in the context of project work. Sometimes one of them may even be delegated to the students when, for example, reviewing the work of a project. No matter

Copyright © 2005, Idea Group Inc. Copying or distributing in print or electronic forms without written permission of Idea Group Inc. is prohibited.

what role the teacher constructs in the context of a project, it is important that she signals a strong identity on the Net, i.e., that she is constantly monitoring events and can be reached easily.

Primarily each teacher has to establish an identity, partly by interacting with the students and partly just by diligent use of the system (all activity likely to be keenly monitored by the students). In addition, we try to reinforce teacher identity by the simple means of placing a subfolder with the teacher's name on it in each project folder. This folder serves as a mailbox for submitting reports, providing responses, and exchanging messages in general. As a rule, messages and files posted to the mailbox will be answered or at least acknowledged within a day. By adding a measure of urgency and high predictability to the exchanges, we hope to encourage a rhythm of communication emphasizing kairotopos, the meaningful time and place in which reflection and creative work takes place.

Another effect of the teacher's folder is that it naturally becomes a focus of the synchronization with respect to workflow and syllabus, adding a touch of personalization and with that possibly a sense of obligation to this otherwise impersonal scheduling in abstract time. Finally, the teacher's folder also is central in defining the rules for communication with the teacher and for regulating expectations of her involvement in the project work in progress.

Conclusion

Reflecting on our practice in developing a Net-based academic program based on the Roskilde University style of project work, we have identified some fundamental conditions and a number of factors affecting the Net environment, most importantly:

- the different frames of reference conditioned by the interrelations of time and place;

- the creation of a double identity of writer and author in a social interaction oscillating between different frames of reference conditioned by time/place inter-relations;

- the development of strategies for organizing and reorganizing the complexity of project work in the Net environment; and

- the need for organizing activities on the basis of negotiated roles in the context of the demands of a given project.

The brief discussion of these factors on the previous pages is far from exhaustive. There is a need for further work, both of a practical and a theoretical nature. Here we will suggest some directions for such future research.

Interpreting time and place interrelations as fluctuations rather than permanent states seems helpful in developing a better understanding of the dynamics of Net-based collaborative work. Building on Rämö's (1999) systematization of Aristotelian concepts

Copyright © 2005, Idea Group Inc. Copying or distributing in print or electronic forms without written permission of Idea Group Inc. is prohibited.

of time and place, we have developed an understanding of complex time/place interrelation that—combined with theories of social interaction and self-consciousness—can lead to a new understanding of what determines the construction of such key qualities in Net-based learning as identity and trust. Further analysis of empirical evidence is needed to test the explanatory power of these theories.

Above, we have suggested that complexity is a characteristic of Net-based collaborative work. We can observe the phenomenon but have yet to pinpoint the significance, singly or combined, of the various obvious parameters such as group size, duration and nature of the work, form and intensity of the communication, qualifications and work patterns of the students and teachers participating. An improved understanding of the processes will not only help develop more effective strategies for handling complexity, it is also likely to facet our understanding of time and place in emphasizing the significance of negotiating meaningful time/place relations.

Two kinds of roles have been mentioned. We have already indicated the importance of further clarifying the meaning of double identity and the concurrent shifts of roles in time/place oscillations in Net-based collaboration. But deliberate role taking is an equally promising area for further study. The idea of organizing communication, coordination, and collaboration on the basis of a diversity of roles negotiated and constructed by the participants in collaborative work is an insight that grew out of the theoretical analysis of the way social interaction is influenced by the new rhythms of communication.

A prerequisite for observing and later analyzing social interaction in Net-based collaborative work is the availability of suitable software. The study of collaborative work on the Net should go hand-in-hand with an effort to promote the design of CSCL/W systems featuring advanced facilities for visualizing time/space patterns of interaction, an awareness of which most likely will provide for a more informed way of interacting, and with that creating the mutual trust that is indeed the heart of Net-based communication.

Excursus: Metaphorical Space

Most CSCL/W software makes use of spatial metaphors in an attempt to make the chronochora dimension less abstract and thus more intelligible and appealing. The importance of such metaphors may perhaps be overrated (Harrison & Dourish, 1996). Certainly, not many users will actually feel immersed in a metaphorical space defined by remediated 2- or 3-dimensional familiar units such as rooms and objects (classroom, library, blackboard, assignment box, etc.). Such metaphorical spaces and objects do however provide an organizing principle that is easy to understand. Developing strategies for handling complexity in cyberspace is what really matters.

Spatial metaphors may feel comfortable, particularly to novices, but they do not by any means have to be elaborate. Project work in our master's program takes place in a CSCW system, the basic metaphor of which is a file archive that is rather similar to that on a PC (e.g., Windows Pathfinder, Figure 5).

A hierarchy of folders may not seem appealing as a spatial setting for social interaction, but it is quite flexible in providing for organization on two levels. On a basic level, it compartmentalizes abstract space exhaustively, suggesting that all elements can be

Copyright © 2005, Idea Group Inc. Copying or distributing in print or electronic forms without written permission of Idea Group Inc. is prohibited.

Figure 5: BSCW hierarchy, from program level to student project level.

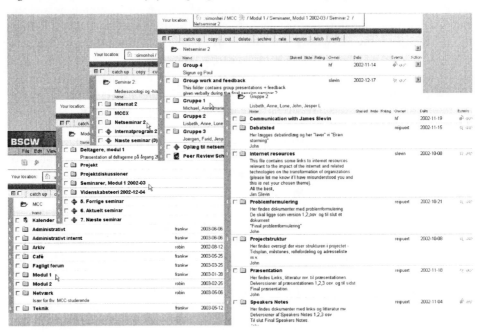

arranged according to simple and readily understood spatial relations such as "outside – inside" containers (folders) and "higher – lower" in the hierarchy of folders. On the higher level of providing organization, the hierarchy of folders allows teachers and students freely and constantly to organize and to reorganize structure and contents. Empowering the participants to control the environment is important in Net-based learning in general and in project work in particular. Working as a group on a project in real life or on the Net will nearly always involve dynamically adapting an organization to changing requirements. In doing so, the available "space" is turned into meaningful "place" – to use the catchphrase of Harrison & Dourish (1996): "Space is the opportunity, and place is the understood reality."

In our master's program, we try to impose a minimum of structure in order to encourage the students to take charge of organizing their Net environment. From the outset, and as the academic year unfolds, we create a number of top-level folders for "administration," "news," "technical matters," and the like, as well as a shallow hierarchy of folders for "net seminars" and "project work" (subdivided into groups). Within this rather prosaic metaphorical space that identifies the building blocks of the program, the groups of students are free to create their own hierarchies of folders, adding, if they wish, their own metaphorical space.

Copyright © 2005, Idea Group Inc. Copying or distributing in print or electronic forms without written permission of Idea Group Inc. is prohibited.

Complexity

Computers handle complexity – and also foster complexity. In the unrestricted metaphorical project workspace on the Net, complexity increases rapidly as the students start working on their projects. Social interaction by means of patterns of intertextuality is likely to remain meaningful and useful only if the workspace is supervised and reorganized constantly. An analogy would be the selection, ordering, and editing necessary to produce a written, concise and generally intelligible summary of a discussion or a meeting. Moving texts and reorganizing established folder structures may add somewhat to the cognitive load of the students, but, as mentioned, being able to reorganize space into a meaningful place is essential in project work, and if it is done as part of a well-understood social process, it should not cause confusion.

Organizing strategies can emphasize divisions according to theme (tasks, archives, functions) or chronology (phases, time-spans, deadlines, history). Usually a project involves both organizing principles, either of which may be governing. A project phase model reflecting dependencies is universally recognized as being clear and efficient. But in a small project a hierarchy of subject folders and project diary may suffice.

History is a form of chronological organization that is burdensome in conventional project work, but easy to implement in a Net environment where it is a simple task to store and retrieve masses of information. It may have the form of (automatic) file versioning or of manual reorganization into current and dated material. In either case, it offers not only documentation of the work process (and a chance to reflect on it), but also reversibility, a feature that is likely to stimulate experimentation and to make interdependence easier to cope with for the students.

Moving files and reorganizing folder structure in the abstract space and time of the computer system inevitably affects the interpretation of the social interaction, and therefore it is likely to cause major confusion among the group members if it is done unexpectedly. The group therefore has to agree on rules on when, how, and by whom structural changes should be made. This is one of the many situations in Net-based project work where it is important to establish and continuously negotiate roles for the various members of the group.

References

Adam, B. (1990). *Time and social theory*. Oxford, UK: Polity Press & Blackwell Publishers.

Adam, B. (1995). *Timewatch: The social analysis of time*. Oxford, UK: Polity Press.

Adam, B., Geissler, K.A., & Held, M. (eds.) (1998). *Die nonstop-gesellschaft und ihr preis*. Stuttgart & Leipzig, Germany: S. Hirzel Verlag.

Bolter, J.D. & Grusin, R. (1999). *Remediation: Understanding new media*. Cambridge, MA: MIT Press.

Copyright © 2005, Idea Group Inc. Copying or distributing in print or electronic forms without written permission of Idea Group Inc. is prohibited.

Cheesman, R. & Heilesen, S. (1999). Supporting problem-based learning in groups in a Net environment. Paper presented at the *Computer Support for Collaborative Learning (CSCL 1999)*, December 12-15). Stanford University, Menlo Park, CA. Retrieved online 15 November 2002 from: *http://kn.cilt.org/cscl99/A27/A27.HTM*.

Cheesman, R. & Heilesen, S. (2001). Using CSCW for problem-oriented teaching and learning in a Net environment. Paper presented at the *European Perspectives on Computer-Supported Collaborative Learning*, Maastricht, Netherlands. Retrieved online 15 November 2002 from: *http://www.ruc.dk/~simonhei/docs/papers/cscl2001.pdf*.

Dewey, J. (1933). *How we think: A restatement of reflective thinking to the educative process*. Boston, MA: D.C. Heath and Company.

Fairclough, N. (1992). *Discourse and social change*. Oxford, UK: Polity Press & Blackwell Publishers.

Fairclough, N. (2003). *Analysing discourse: Textual analysis for social research*. New York: Routledge.

Harrison, S.& Dourish, P. (1996). Re-place-ing space: The roles of place and space in collaborative systems. In *Proceedings of the ACM Conference on Computer-Supported Cooperative Work CSCW'96* (Boston, MA), p. 67-76. New York: ACM.

Heilesen, S.B., Thomsen, M.C., & Cheesman, R. (2002). *Distributed CSCL/T in a groupware environment*. Paper presented at the *Computer Support for Collaborative Learning: Foundations for a CSCL Community (CSCL 2002)*, January 7–11, Denver, Colorado. Retrieved online 15 November 2002 from: *http://newmedia.colorado.edu/cscl/166.html*.

Jensen, S.S. (1991). Lærervirksomhed og lærerkvalifikationer i computer-mediated communication. Paper presented at *Nordens Elektroniske Kunnskapsnettværk*, Oslo, Norway.

Jensen, S.S. (1993). Teachers and tutors in computer-mediated communication—Global online interviews. *Rapport fra Udviklingsprojektet Blaagaard Online*: Blaagaard Statsseminarium.

Jensen, S.S. (1994). *CMC situeret lærervirksomhed – Interface og designprocesser. Rapport fra Udviklingsprojektet Blaagaard Online*: Blaagaard Statsseminarium.

Jensen, S.S. (2001a). *De digitale delegater: Tekst og tanke i netuddannelse. En afhandling om hyperlinks i refleksiv praksis, der er face-to-interface*. København, Denmark: Multivers. Retrieved online 15 November 2002 from: *http://www.afhandling.dk/*.

Jensen, S.S. (2001b). Kairic rhythmicity in the *Turing-Galaxy*. Paper presented at the *International conference on spacing and timing: Rethinking globalization and standardization*, November 1-3, Palermo, Italy.

Kirkeby, O.F. (2001). Aspekter af tillidens fænomenologi. In A. Bordum. & S.Wenneberg. (eds.): *Det handler om tillid*. København, Denmark: Samfundslitteratur.

Latour, B. (1988a). The politics of explanation. In S. Woolgar (ed.), *Knowledge and reflexivity: New frontiers in the sociology of knowledge*, pp. 155-176. London: Sage.

Copyright © 2005, Idea Group Inc. Copying or distributing in print or electronic forms without written permission of Idea Group Inc. is prohibited.

Latour, B. (1988b). A relativistic account of Einstein's relativity. *Social Studies of Science, 18*(1): 3-44.

Latour, B. (1998). *Artefaktens återkomst: Ett möte mellan organisationsteori och tingens sociologi.* Stockholm, Sweden: Nerenius & Santérus.

Mead, G.H. (1929). The nature of the past. In *Essays in honor of John Dewey, on the occasion of his seventieth birthday, October 20, 1929.* New York: H. Holt and Company.

Mead, G.H. (1932). *The philosophy of the present.* La Salle, IL: Open Court Publishing Company.

Mead, G.H. (1934 repro. 1970) (Ed. by C.W. Morris). *Mind, self & society/George Herbert Mead.* Chicago, IL: University of Chicago Press.

Mead, G.H., Morris, C.W., Brewster, J.M., Dunham, A.M., & Miller, D.L. (1938). *The philosophy of the act.* Chicago, IL: The University of Chicago Press.

Rämö, H. (1999). An Aristotelian human time-space manifold. From chronochora to kairotopos. *Time & Society, 8*(2): 309-328.

Schmidt, K. (2002). The problem with "awareness." Introductory Remarks on "Awareness in CSCW." *Computer Supported Cooperative Work, 11*(3-4): 285-298.

Sikkel, K., Gommer, L., & van der Veen, J. (2001). A cross-case comparison of BSCW in different educational settings. Paper presented at the *European Perspectives on Computer-Supported Collaborative Learning,* Maastricht, Netherlands. Retrieved online 15 November 2002 from: *http://www.mmi.unimaas.nl/euro-cscl/papers/146.doc.*

Stacey, R.D. (2001). *Complex responsive processes in organizations: Learning and knowledge creation.* London: Routledge.

Copyright © 2005, Idea Group Inc. Copying or distributing in print or electronic forms without written permission of Idea Group Inc. is prohibited.

Chapter IV

The Collective Building of Knowledge in Collaborative Learning Environments

Alexandra Lilaváti Pereira Okada
Pontifical Catholic University and Dante Alighieri School, Brazil

Abstract

The intention of this chapter is to investigate how collaborative learning environments (CLEs) can be used to elicit the collective building of knowledge. This work discusses CLEs as lively cognitive systems and looks at some strategies that might contribute to the improvement of significant pedagogical practices. The study is supported by rhizome principles, whose characteristics allow us to understand the process of selecting and connecting what is relevant and meaningful for the collective building of knowledge. A brief theoretical and conceptual approach is presented and major contributions and difficulties about collaborative learning environments are discussed. New questions and future trends about the collective building of knowledge are suggested.

Copyright © 2005, Idea Group Inc. Copying or distributing in print or electronic forms without written permission of Idea Group Inc. is prohibited.

Introduction

A few years ago, in the beginning of a discipline at a large university in Sao Paulo, a professor asked if his students would like to use a virtual learning environment. His intention was to promote discussion, group learning, and more interaction towards collaborative learning. Then, one of them said, *"I prefer just face-to-face classes. Interacting in virtual environment means spending more time."* Although one of the great advantages in virtual learning environments is communication anytime from anywhere, some participants revealed that such flexibility provokes intensive interactions, information overflow, difficulty in organizing what is relevant, and consequently, time becomes a great problem.

Due to the rapid growth of online learning and the incredible increase of information on the Web, developing methodologies to build knowledge collectively, articulating what is meaningful, has been quite essential to eliciting better collaborative strategies in online courses. For that, I have been investigating how to manage information overflow and to incentivize collective building of knowledge through virtual learning environments using the software Nestor Web Cartographer (www.projeto.org.br/nestor/) and other freeware resources available on the Internet (Okada, 2001, 2002).

The purpose of this study is to develop strategies for designing and mediating collaborative learning environments from a net of knowledge perspective. This network perspective is supported by some characteristics of cyberspace, which not only highlight the intersection of oral and written language with memory, but also facilitate the process of weaving the meanings offered by subjects into a collective building of knowledge.

As a theoretical basis for the comprehension of collective building of knowledge, some authors have been selected, such as: Humberto Maturana and Francisco Varela, who consider knowledge a biological phenomenon of which knowing, being, and living are inseparable dimensions; Paulo Freire, who defines knowledge as conscious reading and rewriting of the world by the subjects themselves; and Pierre Lévy, for whom knowledge is a complex net where technical, biological, and human actors interact all the time.

In the light of the above theories, we have tried to unveil the practice behind virtual environments created during a workshop about the software Nestor Web Cartographer, a subject of the Education post-graduation course at Pontifical University of São Paulo. The participants of the workshop were two professors, doctors in Education, and twelve post-graduate research students from Education and other areas such as Administration, Computer Science, Communication, and Semiotics. One of the students was invited by the professors to organize the workshop. The data collected on the six environments were mapped and analyzed (14 descriptions about the participants, 130 forum messages, 173 emails, 15 webmaps, 10 websites, 19 papers, and all feedback comments relating to the group tasks and learning environments).

The methodology used to develop this investigation was based mainly on qualitative research. It involves description and interpretation of data obtained during the workshop from interactions and reflections of all researchers (teachers and students). Not only were the results analyzed, but the processes were also investigated to reveal how virtual learning environments can elicit the collective building of knowledge. Thus, interrela-

Copyright © 2005, Idea Group Inc. Copying or distributing in print or electronic forms without written permission of Idea Group Inc. is prohibited.

tions between subjects or between subjects and objects in their multiple interfaces could be better understood.

At the end of our study, some important findings that emerged from the Virtual Learning Environment (VLE) Workshop are presented that make collaborative learning environments contribute to the collective building of knowledge: a clear and common purpose, self-organization, collective building, contextualization, argumentative consensual dialogue, pleasure, and well-being.

Overview of the Course

The Nestor Web Cartographer Workshop was a course offered as part of the subject Epistemology and Computer Uses in Education of the post-graduation program, Education - Syllabus at PUC University of São Paulo.

Nestor Web Cartographer, developed in France by Romain Zeiliger, is a graphic web browser-an editor of HTML pages and a cartographer with synchronous and asynchronous resources that supports collaborative learning. This software was developed to promote the construction of a personalized web space. For that purpose, it dynamically builds a flexible and navigable overview map of the hyperspace when users interact with it. In turn, users can rearrange the map creating new objects: documents, links, annotations, sub-maps, tours, search keywords, and conceptual areas. Consequently, it allows users to solve their own navigation problems: identifying documents, delineating pertinent materials, organizing links into categories, and selecting information through contextual navigational (Okada & Zeiliger, 2003).

Figure 1: Nestor Web Cartographer. Free download: http://www.gate.cnrs.fr/~zeiliger/ nestor.htm.

Copyright © 2005, Idea Group Inc. Copying or distributing in print or electronic forms without written permission of Idea Group Inc. is prohibited.

This software encourages users to reflect on their interactions with an information space, to discuss those interactions with annotations, to collaborate with others through the sharing of tours and annotated maps, and to apply their own methodologies to build knowledge-based structures. Zeiliger, Belisle, and Cerrato (1999) emphasize six important issues about this software:

1. Representing Self-Navigational Experience: Every visited document is represented as a symbol (icon). Users can rearrange the layout deleting the non-significant web pages, changing the icons, grouping them in conceptual areas, and creating arrows to connect information.

2. Constructing a Personal Web Space: Users can create web pages using Nestor Editor, insert the converted maps, and weave them with the public network. They can build thematic maps and develop personal hypertexts about what is relevant and meaningful.

3. Note-taking: Users can attach annotations to every visited public or personal document. When an annotated document is visited, the corresponding annotation is displayed in a separate window—"the bag"; a visible clipboard can be used also to select and gather important information during the navigation process.

4. Creating Keywords Objects: Users can also insert keywords, areas, and sub-guides in maps. The created keywords are automatically searched in the visited document's text and highlighted when found (both on the map and in the document). This is especially useful when users want to seek relevant information.

5. Creating and Saving Navigational Objects: All objects created by users (maps, keywords, conceptual areas, annotations, and routes) can be saved to an HTML file, retrieved, and published. Those objects are considered as "navigational objects" because they can serve to initiate new navigational operations.

6. Sharing Maps: NESTOR allows users to build maps collectively using synchronous and asynchronous resources and also to share objects published in the cyberspace. Nestor users can construct meaningful information through computer-mediated communications and collaborative navigation.

The aim of the workshop was not only to demonstrate the software Nestor Web Cartographer in order to develop maps of investigation, but also to go deeper into some theories following participants' expectations through our own collaborative environment that we created by using just freeware resources available on the Internet.

All students and teachers were encouraged to install this free software and to participate in the workshop outside of class time. The student responsible for the workshop developed the environment about Nestor Web Cartographer using the same software. The professors created the subject's environment to discuss theory and practices and the twelve researchers, organized in four groups according to their interests, developed four environments about autonomy, collaboration, pedagogical mediation, and interactivity. The six environments were connected with each other and could be accessed by everybody.

Copyright © 2005, Idea Group Inc. Copying or distributing in print or electronic forms without written permission of Idea Group Inc. is prohibited.

Figure 2: Epistemology and computer uses in education subject organized by professors Maria Elizabeth B. Almeida and Maria Cândida Moraes PUC-SP 2001.

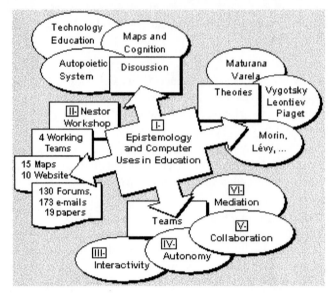

Background for Collaborative Learning Environments

What is a Virtual Learning Environment? What attributes constitute a "VLE"? Many authors, for example, Colin (1999), and Britain and Liber (1999), define virtual learning environments as software packages installed in a server to administer the learning process (interactions, information access, discussion, support. etc.) into an integrated environment. The purpose is not to reproduce the same environment of the classroom, but to offer technological resources to facilitate the apprenticeship.

Maturana and Varela (1980), both biology researchers in the autopoietic theory, consider environment as a life organization. Organisms are adapted to their environments, and their organization represents the environment in which they live. To those authors, living systems are units of interactions that exist in an ambience and are essential for its maintenance as a unit. And considering the biological point of view, it is impossible to understand those units independently or outside the ambience with which they interact.

Dodge and Kitchen (2001, 2002), both cybergeographers, define environment as a space of interactions, places of production and consumption that are recognized by their own relations inside and outside. They emphasize that our lives are rooted and given context by places we live in, the communities we inhabit, the sites of our homes, work, and leisure, and are shaped by complex socio-space processes that operate across many scales, from local to global. In turn, spaces are produced and given meaning through social practices creating places.

Copyright © 2005, Idea Group Inc. Copying or distributing in print or electronic forms without written permission of Idea Group Inc. is prohibited.

Figure 3: A schematic of a prototypical VLE by Britain and Liber (1999).

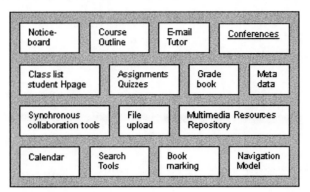

Using the conceptual approach by those authors, we can define virtual learning environments as a network of technological, human, and biological components and their interactions. Thus, it is important to consider virtual learning environments not only as technological resources (computers, modem, connectors, web servers, software, web services, synchronous and asynchronous interfaces), but also all participants (teachers, students, guests, technicians, specialists, and apprentices, including their interactions), the traffic of text, documents, images, sounds, the sharing of messages, the forum of discussion, the registering of databank and forms, the access of websites, and all information.

This information flow describes an interactive learning process and could not have been completed in isolation. Virtual learning environments begin to reveal the development of a new paradigm of education: the transformative nature of the learning process where students and teacher can learn and contribute to each other. Consequently, a network of interactions and collaborative attitudes between all participants can be formed, through which the process of knowledge building is collaboratively created.

Maturana and Varela (1980) consider living systems as emergent from or constituted by the interactivity of beings, not as a priori abstract units. The authors define social systems as a bundle of specific interactions among its participants realized primarily in linguistic consensual domains. Those interactions (e.g., regarding frequency, connectivity, membership) define the character of a social system. To Maturana and Varela, the social system exerts influence upon individual participants through affordances for and regularities in their interactivity, and this influence is recursively exercised upon the emergent social system through the participants' ongoing interactions.

About social systems, Dodge and Kitchen (2001) emphasize that information and communication technologies (ICTs) allow the reconfiguration of space-time relations and radically restructure the materiality and spatiality of space and the relationship between people and place. It is possible to interact anywhere, any time, changing any kind of information quickly and cheaply, and everybody can be emissor and receptor at the same time. It means a new way to build knowledge, interlacing thought in diverse facets, collectively and with autonomy.

Copyright © 2005, Idea Group Inc. Copying or distributing in print or electronic forms without written permission of Idea Group Inc. is prohibited.

Figure 4: A schematic of a prototypical VLE as a living system.

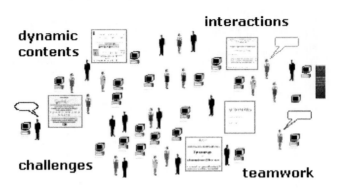

Considering VLE as a social system whose characters are defined from its interactions, and based on the studies of Mason (1998), we can realize three models of environments:

- Instructional. The level of online interaction is low, concentrated between the student and the material, or the student and the teacher. The core of the environment is the contents, which are already produced. The methodology is based on tutorials; this model reflects the traditional teaching environment. The knowledge is built through readings and memorizing by individuals.

- Interactive. The level of online interaction is high among all participants. The environment contents are produced during the process that involves activities and online discussions. The groups build knowledge through consensual dialogue.

- Collaborative. The level of online interaction is very high and centered around collective activities and common purpose. The environment contents are dynamic and are determined largely by individual and group needs. The knowledge is the result of collaborative activities, discussions, consensual dialogue, joint assignments, and common challenges by teamwork.

Based on the Maturana and Varela's (1980) theory about cognition as a biological phenomenon, it is possible to associate collaborative learning environments as a cognitive system whose organization realizes and produces the network of interactions in which it can act with relevance in order to keep its existence. "Living systems are cognitive systems and living as a process is a process of cognition" (Maturana & Varela).

According to autopoiesis theory, a cognitive system needs to manage its complex context to maintain its existence. The world that any organization inhabits is much more complex than the organization itself, and the variety of organization is much larger than variety of organism. Therefore, in order to keep itself lively, a cognitive system can amplify or attenuate its variety, enlarge or reduce its domain of interactions, by making its internal states modifiable in a relevant manner.

Copyright © 2005, Idea Group Inc. Copying or distributing in print or electronic forms without written permission of Idea Group Inc. is prohibited.

Figure 5: Collaborative learning environments as a living system.

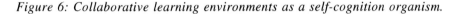

Figure 6: Collaborative learning environments as a self-cognition organism.

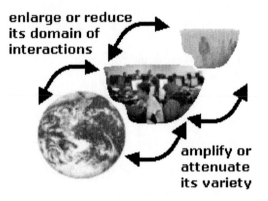

Taking the virtual learning environment as a cognitive system, its characteristics (members' roles, purposes, context, common interests, etc.) define its identity and its initial structure (interfaces, communication channels, design, contents, etc.) to keep the environment's existence. Participants also have their own characteristics (personality, skills, difficulties) and initial states (time, motivation, expectations, intentions, interests).

The key problem for a collaborative learning environment in maintaining itself as a lively cognitive system is to know how to manage its complexity and the context in which it finds itself.

Complexity, according to Morin's studies, has provoked an important discussion about the relationship between order and disorder and new ways to deal with this. Some aspects, such as the unforeseen, uncertainty, ambiguity, and subjectivity, are increasingly being studied in the social and natural sciences. Ordered and linear conceptions of universe, nature, and human civilization have been dismantled (Demo 2002; Morin & Kern, 1999). The sciences of chaos and complexity show us the profound role of disorder and the importance of knowing how to create new alternatives, to innovate, improvise, organize, and self-organize, to disorganize and reorganize, as a constant dynamic and non-linear process.

After reflecting about learning environments as a cognitive living system, it is important to discuss how knowledge can be built collectively. What does net of knowledge mean?

Copyright © 2005, Idea Group Inc. Copying or distributing in print or electronic forms without written permission of Idea Group Inc. is prohibited.

Theoretical Issues about Knowledge as a Network

Web of knowledge and knowledge in network are constructs that result from the flexibility, plasticity, interactivity, adaptability, cooperation, sharing, support, and self-organization that characterize the knowledge-building process (Moraes, 1999). The net metaphor seems to be the key to the emergence of knowledge as a new interdisciplinary work. To understand is to apprehend the meaning by seeing the relations among things. The more relations can be established between one topic and other areas of knowledge, the closer that topic will be to its thorough meaning, to its "completeness." Such relations connect different topics in a non-linear way. In other words, the meaning of a topic "X" can be apprehended through multiple relations established between "X" and other topics, "A", "B", "Y", "M", and "G", those being or not being the references in the topic that is studied (Machado, 2000).

There are six important principles about the concept of net and rhizome presented by Deleuze and Guattari (1987) and Pierre Levy (1994):

- Metamorphosis - there is the need for a constant change.

- Multiplicity - the components and interconnection have multiple scales.

- Heterogeneity - the structure is always different.

- Exteriority - the feeding information should come from outside.

- Acentrism - there is no beginning, no end, and not one center but mobility of the centers.

- Proximity - the interaction allows association of components.

The metaphor of network and rhizome allows to associate three theories and to conceive the building of knowledge as the result of biological, social, and technological process.

Maturana and Varela (1980) consider knowledge a biological phenomenon, of which knowing, being, and living are inseparable dimensions. The living being can develop knowledge:

- through the dynamic and flexible changes of the components (metamorphosis);

- as a process where components produce multiple dynamics of production (multiplicity);

- from the operation of different components (heterogeneity);

- as resulting of internal and external interactions to keep the structural congruence (exteriority);

Copyright © 2005, Idea Group Inc. Copying or distributing in print or electronic forms without written permission of Idea Group Inc. is prohibited.

- by configuring enterprise (re-)engineering practices for mutual orientation and self-organization (acentrism); and
- through the interaction that allows association of components (proximity).

Paulo Freire (1987) defines knowledge as a social process of conscious reading and rewriting of the world by the subjects themselves. People can develop knowledge:

- by transforming the reality for an equal and just world (metamorphosis);
- through decodification as a multi-dimensional step where there is breaking down of the knowable object for critical analysis and future action upon this reflection (multiplicity);
- as a way to achieve critical consciousness through the consensual dialogue considering different opinions and points of view (heterogeneity);
- through dialectical movement of reflection and action managing the pluralities within, across, and outside communities with different interests (exteriority);
- by being co-learners—both teacher and students must participate in and be responsible for their learning process as social-historic subjects (acentrism); and
- by coming closer to the object and to each other. Humans are the only beings capable of being both objects and subjects of the relationships woven with others and with the history that we make and that makes and remakes us (proximity).

Pierre Lévy (1994) emphasizes that knowledge is a complex net where technical, biological, and human actors interact all the time. Web users can weave knowledge:

- in a continuously space of changing: the cyberspace;
- by building a network of information in multiple scales (multiplicity);
- by interconnecting different components: sounds, images, text, ideas, thoughts, etc. (heterogeneity);
- by feeding information even outside the web as experiences, practices, other examples lived (exteriority);
- by navigating and building diverse hypertext and journeys in the cyberspace without a specific beginning or end (acentrism); and
- by logging on and interacting with anybody, anywhere, and anytime (proximity).

In fact, those principles can be considered as the essence of Internet, non-linear access of information and non-linear building of knowledge. It also allows the association of an unimaginable amount of information routes. All those characteristics allow understanding of how the interaction can occur and how the environment can maintain itself.

Copyright © 2005, Idea Group Inc. Copying or distributing in print or electronic forms without written permission of Idea Group Inc. is prohibited.

Figure 7: Knowledge as a contemporary network.

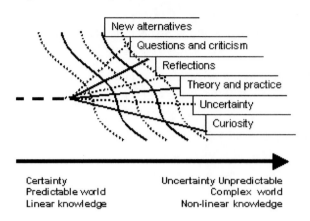

This networking approach, based on a biological, social, and technical notion of knowledge, can offer a useful framework for conceptualizing the pluralistic and dynamic nature of cultural knowledge. Considering this aspect, it is necessary to recognize that knowledge can develop better within an unpredictable and complex world.

The knowledge must be built to reduce and to overcome uncertainty. Consequently, theories must be questioned and criticized in order to be substituted for a better one, and then, knowledge can be improved into a complex and more advanced level while it is deconstructed.

Moreover, questioning is not only to show emptiness and inconsistencies, it is also meant to see through other angles, points of view, different contexts, and multiple levels in order to create new alternatives.

In this way, it is essential to investigate how to engage co-learners into weaving a net of meaning collectively. How can environments elicit collective knowledge building as a network of meanings? What are the strategies for designing and mediating collaborative learning environments from a net of knowledge perspective? For that purpose, all data collected in the virtual learning environments were mapped and analyzed along with all the feedback comments.

Finding Results about Collective Building of Knowledge in VLE

The six environments created by students and teachers during the workshop were developed using just free software available on the Internet. Web pages and web maps were built using Nestor Web Cartographer; the discussion, using ForumNow; the information exchange, using Yahoogroups.

Copyright © 2005, Idea Group Inc. Copying or distributing in print or electronic forms without written permission of Idea Group Inc. is prohibited.

About collected data, it was found that:

- In the first weeks, teachers and learners whose role was to mediate the environment interacted much more than others. About 173 emails were sent:
    - 33% were sent by students: 9% activities doubts, 9% technical support, 8% suggestions and new ideas, 7% reflection about environment.
    - 67% were sent by teachers: 20% reflections and questions about the environments, 18% feedback (support and answers), 14% reflections and questions about activities, 13% incentive.
- After the first month, messages in forum were more frequent than e-mails, and students interacted much more. About 130 forum messages:
    - 28% were sent by teachers: 11% questions, 9% feedback and reflections, 8% incentive.
    - 72% were sent by students: 23% Theories (questions and reflections), 17% Maps and papers, 12% Nestor software, 10% others (incentive, absent reasons, etc.).
- Of the theories presented in the 19 papers and 15 maps: 27% references expected, 63% new references.

To analyze the content of these outcomes, a qualitative research approach was adopted, and investigation methods consisted of document analysis, interaction observations, and description and interpretation of the co-construction process. Over four months, data was collected during the workshop from six environments. The results were analyzed and the process was investigated. This allowed interrelations between subjects or between subjects and objects, in their multiple interfaces, to be better understood.

The focus of the study is on how virtual learning environments can be used to elicit collective building of knowledge. Six important issues could be observed:

- common and clear purpose;
- contextualizing;
- self-organization;
- argumentative dialogue;
- co-construction; and
- pleasure / well-being.

Copyright © 2005, Idea Group Inc. Copying or distributing in print or electronic forms without written permission of Idea Group Inc. is prohibited.

Common and Clear Purpose

In the beginning, teachers intended to find a way to weave theory and practice about Epistemology and Education. The students wanted to go deeper into some theories in order to improve their research.

The first week of the workshop was very difficult, because it was not the students´ spontaneous option, and the environment was new and unknown to them. Interaction was very poor. Although the aim of the workshop was to facilitate the process of researching and to allow a theoretical and practical approach through the software Nestor Web Cartographer, a common purpose among all participants had not yet appeared. It could be realized that just the workshop intention and authentic activities were not enough to guarantee the participants involvement.

For learners to become actively engaged with each other, they were invited to introduce themselves, to write about their interests, expectations, experiences, and preferences. When the students started to discuss and to build maps about their preferred theories, common interests were identified so they started to exchange maps and to share bibliographic references. Consequently, they became more involved, not only in the workshop but also in collaborative action.

This process allowed four themes of interests to emerge: interactivity, collaboration, autonomy, and pedagogical mediation. Then, they organized four working teams and started to build their learning environments using the same free resources.

After clearing up the purpose in their environments, the participants started to interact not only in their own working team, but also in others, bringing related issues. It could be realized that common purpose contributed to increased trust, to communicate with confidence. and to develop authentic presence. Expressing it clearly in their own environment allowed developing initiative, collaborative action, and continual learning.

Contextualizing

Environments exist in their own particular context. Cognitive process occurs within the context of an environment. In order to understand the collective building of knowledge within the collaborative learning environments, it is important to know their contexts and the motivational aspects of their interactivity. It is through interaction that theory and practice, identity and meaning, collaborative and continuous learning can emerge and evolve-all of which interactively constitute context.

About virtual learning environments developed in the workshop, the participants wrote about themselves, inserted their pictures, described when, where, how, and why they had discussed and developed concepts. The more learners can relate their life experiences and what they already know about the context, the more meaningful will be what they will learn.

Contextualizing is a process to express or to make meaning from the context itself. Through a contextual learning environment, meaning can be developed and understood. Context in the environment allows not only production of meanings about the communal

Copyright © 2005, Idea Group Inc. Copying or distributing in print or electronic forms without written permission of Idea Group Inc. is prohibited.

Figure 8: Interactivity working team Website.

The class | *The team* | *Partial results* | *Readings* | *Authors and theories*
Site of Subject | *Our Forum* | *Workshop* | *Autonomy* | *Cooperation* | *Mediation* |
All rights reserved © Glak – 2001

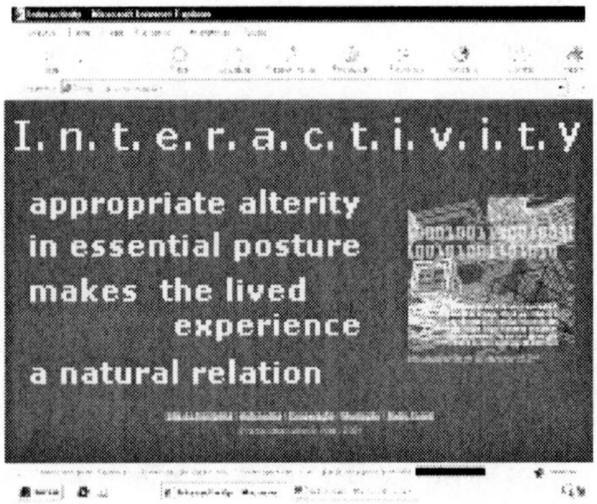

world, but also formation of identities that help participants to discover their similar interests.

For instance, the working team organized by Adhara, Graffix, Krugger and Luyten about interactivity described how they could co-elaborate their own approach about this concept.

"Why did we discuss interactivity? When we met at PUC cafeteria, on Tuesday, 27th March 2001, we started to discuss the interactivity as Luyten had suggested. The main purpose was to find a general concept, since everybody had already read about some theories and had written their opinion. One of our conclusions was: who thinks about interactivity, thinks about multiple levels, because it is a broad concept and can be selected depending on who uses it. Then, we tried to summarize our view in one sentence:

- *Appropriate alterity is a concept developed by Graffix that expresses the capacity to see, to think, or to feel things in such a way that one feels almost as being the other,*

- *in an essential posture is what the Lyten in the group discussion introduced in order to relate the concept with the educational question and the media,*

- *makes the lived experience, underlined by Adhara, the only way for interactivity to occur,*

- *a natural relation is the mathematical view of Krugger´s of interactivity as relation that only exists when there are two elements in action."*

Practices and meanings are only fully contextualized within the context of their authentic use. The mutual relationships between context and content, individuals and environment, knowing and doing could be developed continuously from that which is known.

Copyright © 2005, Idea Group Inc. Copying or distributing in print or electronic forms without written permission of Idea Group Inc. is prohibited.

Self-Organization

There are seven important categories that describe the self-organization process, according to Whitaker (1995). The Workshop environments were analyzed from this point of view. The categories are:

- self-creation - the capacity that a VLE has to be originated by circumstances in which it occurs. Specific circumstances and attitudes such as: encouraging innovation, stimulating initiative, and supporting doubts allowed the participants to create collectively their own environments, maps and papers;

- self-configuration - the ability that a VLE has to actively define the arrangement of its constituent parts. Freeware resources such as: Nestor Web Cartographer, ForumNow, and Yahoogroups facilitated the students´ and teachers´ participation in the workshop and also the configuration of their environments;

- self-regulation - the ability that a VLE has to control the course of its internal transformations, typically with respect to one or more parameters. Each team could define its own interactions process, purpose, and tasks;

- self-steering - the ability that a VLE has to actively control its course of activity within some external environment or a general set of possible states. All participants could navigate in their environments through links, hypertexts, and maps selected and created by themselves;

- self-maintenance - the ability that a VLE has to actively preserve itself, its form, and/or its functional status over time. However, learners´ participation rhythm, number, and frequency of access were very different from one another. Roles were defined according to their interests by the learners themselves. Some participants became responsible for technical aspects of the environments, some for pedagogical mediation, and others for motivation of the group;

- self-(re-)production - the ability that a VLE has to generate itself anew or produce other systems identical to itself. Two environments (Epistemology and Computers Uses in Education, and Nestor Workshop) could give rise to another four VLEs (Autonomy, Cooperation, Pedagogical Mediation, and Interactivity).

- self-reference - the ability that a VLE has to value its essence, to make its character or its behavior meaningful to itself. All teams had autonomy to make decisions and agreements, to express their opinions, and to be a source of information and reference to themselves and to the others.

All these concepts are not mutually exclusive. Any approach to treating virtual learning environments as self-organizing entities should, therefore, consider which (or how many) of these connotations to include.

Copyright © 2005, Idea Group Inc. Copying or distributing in print or electronic forms without written permission of Idea Group Inc. is prohibited.

Argumentative Dialogue

Another important aspect of the interactions was the argumentative dialogue among learners. Interactions involve the attempt to resolve expressed conflicts of opinions with respect to proposals, theories, opinions, and justifications. Some special circumstances are required in order for argumentative interactions to be produced by learners.

Such circumstances mean encouraging students to express their ideas into a linguistic form as a preparation for debating, developing individualized texts, describing the verbal conflict situation, and individual reconstruction of the agreed conclusion and justification.

Learners are not naturally likely to argue spontaneously with each other, at least with respect to the subjects that they have not been in contact with yet. And sometimes, interpersonal conflicts or individual contradictions are not sufficient to provoke the incidence of argumentation.

It could be noticed that in the working team environments spontaneous argumentative dialogues resulted from common shared ground (theories read, papers written, maps built) related to the topic discussed. A conflict of opinions was openly declared and understood: participants knew their own arguments in the discussion. Participants had enough arguments at their disposal, and committed themselves to the debate: they have *something to argue about.*

The emergence of a critical discussion was predicted as soon as the appropriate dialogical attitudes ("pro"or "con") had been expressed and the communication between participants had been established. This implied that points of view had already been constituted, so students could discuss together, in pairs or in teams.

Argumentative interactions are an essential condition for development of a consensual and critical knowledge.

Figure 9: Argumentative interactions.

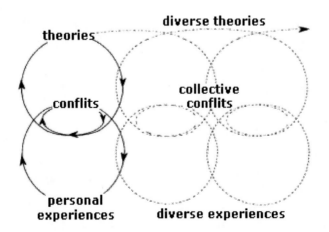

Copyright © 2005, Idea Group Inc. Copying or distributing in print or electronic forms without written permission of Idea Group Inc. is prohibited.

Co-Construction

Teachers and students as co-learners become partners in collaborative learning. When they have a common and clear purpose, they become co-investigators. They can share experiences and pursue a search for knowledge together.

Co-learners invite everybody to participate in the construction of knowledge and the creation of the learning environment. However, they do not only exchange certainties about the subjects, but also questions and unresolved issues that provoke a real opportunity for everybody to learn with each other.

Thus, they are never sure about what the direction the discussion will take. Surprises are more likely. Issues they have not thought about are more likely to arise. Such unpredictable ideas lead them or set them free to think freshly about the subject.

Then, co-learners can feel that they have thoroughly explored and tracked something together. They feel freer to share their thoughts and ideas and consider the environment as their own space where everybody can be respected as a thinker and a learner.

Table 1: Two types of knowledge by Nonaka and Takeushi (1995).

Tacit Knowledge (Subjective)	Explicit knowledge (Objective)
Knowledge of experience (body)	Knowledge of rationality (mind)
Simultaneous knowledge (here and now)	Sequential knowledge (there and then)
Analog knowledge (practice)	Digital knowledge (theory)

Table 2: Four modes of knowledge conversion by Nonaka and Takeushi (1995).

to

		Tacit knowledge	Explicit knowledge
f r o m	**Tacit knowledge**	Socialization	Externalization
	Explicit knowledge	Internalization	Combination

Copyright © 2005, Idea Group Inc. Copying or distributing in print or electronic forms without written permission of Idea Group Inc. is prohibited.

When co-learners share their production in an opened learning environment, they can socialize with anybody outside the environment who is interested. New interactions can broaden the knowledge of individuals, groups, and environments. This process can be better understood from the spiral of knowledge theory developed by two Japanese researchers.

Nonaka and Takeushi (1995), both consultants and professors, stressed that organizational knowledge requires human/individual knowledge and starts with it. Knowledge is the basic unit of analysis to explain firm behavior. Organizations do not merely process knowledge – they also create it.

The authors explained that human knowledge can be classified into two kinds:

1.   Explicit knowledge, which can be articulated in formal language including grammatical statements, mathematical expressions, specifications, manuals and so forth; and.

2.   Tacit knowledge, which is hard to articulate with formal language, is personal knowledge hard to transfer.

The interaction between explicit and tacit knowledge is the key dynamic of knowledge creation by the individual, group, and organization. The two main dimensions of knowledge creation are: epistemological and ontological. And there are four major processes of knowledge conversion: Tacit–explicit, Explicit–explicit, Explicit–tacit, Tacit–tacit.

The most precious knowledge can neither be taught nor passed on. Tacit knowledge embraces values, ideals, and emotions, as well as images and symbols. The most powerful learning comes from direct experience. It is essential to learn with the body, not only with the mind. Children learn through trial and error. Tacit knowledge involves two concepts:

1.   Know how: Technical dimension that encompasses the kind of informal, hard-to-pin-down skills or crafts and "fingertips" feelings; and

2.   The "cognitive" dimension: Schemata, mental models, beliefs, and perceptions that reflect our image of reality ("what is") and our vision for the future ("what ought to be").

Both info and knowledge can be developed in a specific and relational context in that they depend on the situation itself and are created dynamically in social interaction among people.

Observing the discussion in the environments´ forum, Yahoogroups, and emails, it was possible to identify those four important moments (as seen in Table 2). First, participants *socialized*, exchanging previous opinions about the subject resulting from experience and previous knowledge (tacit knowledge). Second, conceptual issues related to the subject could be *externalized* through maps, texts, papers, and bibliographical references (explicit knowledge). Third, theory and practice could be woven; tacit and explicit knowledge could be connected, discussed, and *combined* through critical and consen-

Copyright © 2005, Idea Group Inc. Copying or distributing in print or electronic forms without written permission of Idea Group Inc. is prohibited.

Figure 10: Spiral of organizational knowledge based on Nonaka and Takeushi (1995) Theory.

Figure 11: Spiral of collective building of knowledge based on Nonaka and Takeushi (1995) Theory.

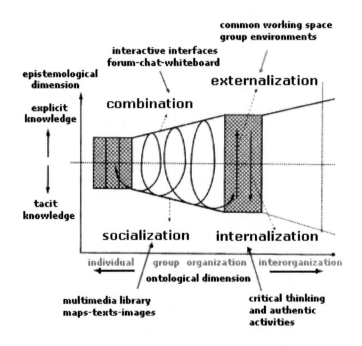

Copyright © 2005, Idea Group Inc. Copying or distributing in print or electronic forms without written permission of Idea Group Inc. is prohibited.

sual conclusion. Fourth, the theory and practice combined could be *internalized* (explicit knowledge became tacit knowledge).

Concerning the spiral of knowledge, Nonaka and Takeushi (1995) developed a theoretical framework by pointing out the two dimensions — epistemological and ontological — of organizational knowledge creation. As depicted in Figure 10, the epistemological dimension, graphically represented on the vertical axis, is where knowledge conversion takes place between tacit knowledge and explicit knowledge. And the ontological dimension, on the horizontal axis, is where knowledge created by individuals is transformed into knowledge at the group and organizational levels.

These four models allow us to understand the conversions between tacit and explicit knowledge: socialization, externalization, combination, and internalization. These four processes are not independent of each other, but their interactions produce a knowledge spiral when time is introduced as the third dimension.

Another spiral takes place at the ontological dimension, when knowledge is developed; for example, the project-team level is transformed into knowledge at the divisional level, and possibly at the corporate or inter-organizational level. Again, the authors introduced time as the third dimension to develop the five-phase process of organizational knowledge creation: sharing tacit knowledge, creating concepts, justifying concepts, building an archetype, and cross-leveling knowledge.

The five enabling conditions promote the entire process and facilitate the spiral.

The transformation process within these two knowledge spirals is the key to understanding their theory. Innovation emerges out of these spirals.

The cyclical movement and organizational spiral can be observed in the environments since working teams started their production in an opened access site. Everything was shared and socialized on the Internet. Interactions occurred not only among researchers

Figure 12: Pedagogical mediation role.

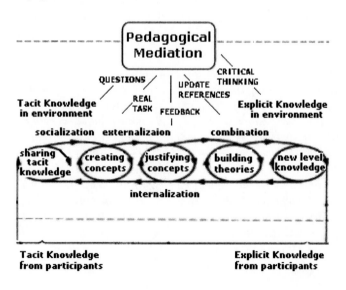

Copyright © 2005, Idea Group Inc. Copying or distributing in print or electronic forms without written permission of Idea Group Inc. is prohibited.

and teams, but also with five other persons, researchers from Brazil, France, and the U.S. interested in the same subject. They accessed the environments site, wrote new information, and contributed to the discussion.

To guarantee this process, it could be observed that pedagogical mediation was essential to provoke reflection through questions, comments, articulations between personal experiences, theories, and new meanings . A friendly environment was important to establish a comfortable and motivating atmosphere to exchange ideas, uncertainties, doubts, new concepts, reflections, and criticism. Authentic activities based on common and clear purposes facilitate collaboration learning. Other important aspects are the quality of interactions, such as exchanging opinions, reorganizing and synthesizing comments collectively, reconstructing new concepts together, criticizing, and deconstructing approaches in groups in order to go deep into theories. It seems to be essential that the interactions are not only intensive but also meaningful to promote the collective building of knowledge.

Pleasure/Well-Being

One of the most important and necessary aspects noticed in the environments is pleasure and well-being. The meaning of "university," based on the medieval Latin word "*universitas*" describes a group of people getting together to learn for pleasure. Those who learn and those who teach should encourage the intellectual, cultural, and creative abilities of each other as a spontaneous and pleasant process. Thereby, co-learners can enjoy learning for pleasure. It contributes to self-esteem and self-knowledge.

About the environments developed in the workshop, it could be noticed that co-learners felt gladness and gratification in being authors of maps, papers, and their own environments. They experienced the excitement of freely discussing and debating ideas at nearly level ground with persons who became not only colleagues but friends.

The possibility of being subjects of their own knowledge, of creating and innovating, of leading discussions, and of being pedagogical mediators provides the impetus for such preparation.

This discipline Epistemology and Education comprises innovation and daring: practicing theories, creating an environment of responsibility among students. The opportunity given to share proposals, actions, is different from the usual learning. I am learning things from many angles...I could not evaluate everything that has been happening with me yet. Clearly, we are here the subject of researching experiment. I feel in this discipline, the chance of "looking within" and the invitation to "looking outside." I do not know if I am being very confusing, but I felt as student that you, professors, have awakened this reflective view. But, it is different when you provoke the reflection and when you allow interference. Is it a practice of detachment? Is it the change of paradigm, isn't it? I do not know, but it seems fantastic the way we are taking. Between the perplexity and the ecstasy, it becomes almost another research.... hehehehe)? :)))). (Ross 28/03)

Copyright © 2005, Idea Group Inc. Copying or distributing in print or electronic forms without written permission of Idea Group Inc. is prohibited.

Some Problems

Managing Time

Although one of the great advantages in virtual learning environments is communication at different times from different places, some participants revealed that such flexibility provokes intensive interactions and, consequently, time became a great problem. The challenge was managing time: feeding the environment, being involved with technical aspects, and weaving theory and practice in order to develop new concepts.

One way to minimize this problem is to invite the participants to an explicit conversation to create strategies for managing their time

Evaluating the Environment

Another difficult issue presented by participants was how to evaluate the environment, how could they know the quality of the productions and interactions. Some different kinds of feedback are necessary to help them calibrate their participation with their expectations. It could be noticed that talking about the quality of their communication was very important. The teacher can provide some feedback but it is even better if the teacher can encourage participants to develop a norm of providing feedback to each other about communication style, quantity, frequency, clarity, etc. Teachers can help team participants access more of their own feelings and reactions to messages in different media. This kind of self-organization is an important skill.

TechnoStress

Besides managing time and evaluating the environments, participants talked about TechnoStress.

There is technostress in the environment due to many interactions to take place through the computers. Very often I stay in front of computer instead of staying with my family. The flexible virtual class time frequently overcame my leisure time with my family, reading books by myself, or discussing face to face with colleagues. (Krugger 11/3 14h)

For Weil and Rosen (2001),

TechnoStress is our reaction to technology and how we are changing due to its influence. Over the past 15 years, as technology has become an increasingly prevalent part of our lives, we have watched TechnoStress develop and impact people in their personal lives, their family and their work environment. We are changing both

Copyright © 2005, Idea Group Inc. Copying or distributing in print or electronic forms without written permission of Idea Group Inc. is prohibited.

internally and externally due to technology and these changes are not in our best interests physically, socially or emotionally. (p. 1)

The environment must stimulate a network of interpersonal relationship that is part of an effective collaborative learning interaction, but it is very important to know how to manage time. Participants must feel comfortable to discuss any problem and solutions, to make choices about what they need. It is very important not only to manage one's own learning, but also one's own well-being.

It is essential to keep up face-to-face contact with persons - family, friends, and even virtual colleagues. The network of interpersonal relationships can go further than virtual learning environments. It can be noticed that a meaningful virtual interactions can enrich face-to-face relationships.

Both authors, Weil and Rosen (2001), emphasize that it is very important to learn how to maintain humanity in a technological world. Technology provides us with a range of options that can enrich and enhance our lives. However, to fight TechnoStress, we must learn to drive and not be driven by technology.

Pedagogical Mediation

How to engage co-learners into weaving a net of meaning collectively?

The most important aspect of networking theory is to understand how to deal with complexity and uncertainty in order to benefit from and elicit collective knowledge building. Concerning this aspect, pedagogical mediation is the key to guide the environment to deal with unpredictable challenges.

Figure 13: Pedagogical mediation and self-cognitive learning environments.

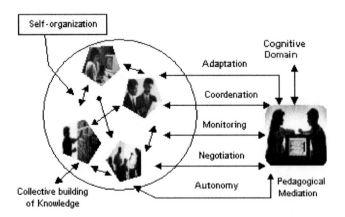

Copyright © 2005, Idea Group Inc. Copying or distributing in print or electronic forms without written permission of Idea Group Inc. is prohibited.

Concerning complex interaction among biological, social, and technological components, the importance of pedagogical mediation is extremely relevant in order to elicit collective building of knowledge.

Considering VLE as a living organism and the importance of pedagogical mediation, six important aspects suggested by Britain and Liber (1999) were analyzed:

- *Negotiation*: How do learners negotiate their learning process with their teacher in order to elicit the collective building of knowledge? Is this a one off or a continuous process?

  Frequent (virtual and face-to-face) discussion moments about the process were some great opportunities to identify problems and to promote reflections and agreements.

- *Coordination*: Can learners collaborate in creating their learning? How?

  Small working teams stimulated learning, particularly the challenge to go deeper into their preferred subject.

- *Monitoring*: How does a teacher monitor whether learning is happening, so that, if necessary, remedial action can be taken?

  The continuous support was very important, in particular the register of the trajectory of the groups, facilitating the accompaniment of difficulties and advances.

- *Autonomy*: How can each student find his or her own resources and advance his or her own learning independently of others? Can individual students contribute their discoveries to the group?

  All production shared among co-learners stimulated and inspired new productions. Autonomy occurs in such a way that changed the initial circumstances of the subject, and this occurred due to the involvement of the participants.

- *Self-organization*: What space or tools are available to let the learners organize themselves as a group, outside of the teacher's purview?

  Easily available and free-of-charge resources, as well as proximity among colleagues inside teams, have facilitated self-creation of collaborative learning environments.

- *Adaptation*: Is it possible for the teacher to adapt the course and its resources in light of experiences gained during its operations?

  In order to adapt to the needs of the participants and the proposals of the subject and the workshop, many changes were achieved in the structure of the environment concerning activities, rhythm, and period of accomplishment, contents, supporting materials; mainly the discussion of purposes and feedback were a great incentive.

Copyright © 2005, Idea Group Inc. Copying or distributing in print or electronic forms without written permission of Idea Group Inc. is prohibited.

Collaborative Learning Environments: Some Conclusions and Future Trends

It is essential to find new ways to organize what is relevant and meaningful within the collaborative learning environment and help participants manage their research time better.

This means thinking carefully not only about the best interfaces, software, and contents, but also about the best methodologies for enabling collaborative learning and knowledge co-construction. Knowledge is not produced just from the technology and informational resources, but from the attitudes of the people who are trying to establish what, how, and why.

Some critics believe that cyberspace has a more profound impact on social relations than it does on information processing. It affects both identity and community (Dodge & Kitchen, 2001). Using and reflecting on the interactions, interrelationships, and co-constructions in cyberspace, we can explore who we are and how we are changing.

The complex identity of cyberspace is defined by characteristics such as: fast updating, diverse information, multiple connections, open resources, and a hypertextual and fluid space for interactions. These characteristics are related to the six network theoretical aspects: metamorphosis (changes), heterogeneity (diversity), multiplicity (multiple levels), exteriority (outside), acentrism (no center), and proximity (close elements).

Table 3: The collective building of knowledge in virtual learning environments.

Network	Collaborative VLE	WebSite Structure	Pedagogical Mediation	Collective Building of Knowledge
Metamorphosis	Common & Clear purpose	Easy and simple Interfaces	Adaptation Promoting the VLE update continuously	Keeping curiosity
Heterogeneity	Self - Organization	Aesthetic design	Self-organization Creating circumstances for the participants to act.	Taking benefits from uncertainty
Multiplicity	Contextualization	Available resources	Coordination Guiding participants to go deeper in their projects	Connecting theory and practice
Exteriority	Co-Construction	Flexible architecture	Monitoring evaluating and self-evaluating by all participants	Reflecting from an opening view
Acentrism	Argumentative dialogue	Significant contents	Negotiation Managing process from many points of views	Reconstructing from questions and criticism
Proximity	Pleasure / well-being	Pleasant space to meet	Autonomy making the environment more pleasant and involvement	Discovering new alternatives

Copyright © 2005, Idea Group Inc. Copying or distributing in print or electronic forms without written permission of Idea Group Inc. is prohibited.

Considering these aspects, we can draw out some important principles about collaborative learning environments:

- *common and clear purpose articulated by the different participants*: when teachers and learners have a common and clear purpose, they become co-investigators.

- *self-organization*: learners and teachers need to be responsible for organizing the environment, making changes and updating when they want to.

- *contextualization*: it is important to know the contexts of all the participants to create interactivity and a group identity. Participants require situational and cultural contexts in order to understand the meanings negotiated in the environment.

- *co-construction*: when teachers and students are partners in collaborative learning, they can build knowledge together.

- *argumentative dialogue*: this is an essential condition for development of consensual and critical knowledge.

- *pleasure and well-being*: co-learners enjoy learning for pleasure. It contributes to self-esteem and self-knowledge.

Figure 14: The collective building of knowledge in virtual learning environments.

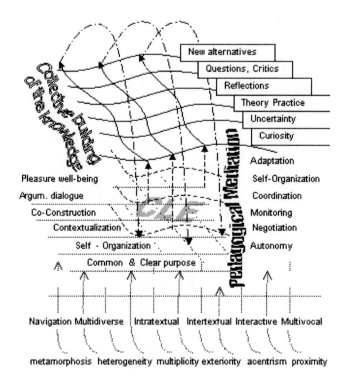

Copyright © 2005, Idea Group Inc. Copying or distributing in print or electronic forms without written permission of Idea Group Inc. is prohibited.

For that purpose, the CLE website should be:

- *Navigable*: learners need to explore the environment at their own pace, in a way that is comprehensible.

- *Multidiverse*: multidiversity improves communication and working. It enriches the co-construction with many points of view and diversity of information.

- *Intratextual*: structural links in our own website allow us to understand interrelated paths.

- *Intertextual*: links with other websites offer value-added information.

- *Interactive*: interactivity is the essence of communication. It is essential to create interactive spaces in which participants can engage (one verb is enough!), allowing a rich dialog between users and the environment.

- *Multivocal*: a variety of voices allow participants to make decisions, connections, and inferences.

Thus, CLE website designers should organize a structure with intuitive interfaces, aesthetic design, available resources, flexible architecture, significant contents, and a pleasant space to meet.

There are also requirements for teachers to facilitate the collective building of knowledge:

- *adaptation*: promoting the VLE update continuously;

- *self-organization*: creating circumstances for the participants to act collaboratively;

- *coordination*: guiding participants to go deeper into their projects;

- *monitoring*: evaluating and self-evaluating by all participants;

- *negotiation*: managing process from many points of views;

- *autonomy*: making the environment more pleasant, where learners can create collectively their own challenges.

These pedagogical mediations (actions) not only help students to interact more but also to reach their purposes in a meaningful way. They create a dynamic process to help students keep their curiosity, benefit from uncertainty, connect theory and practice, reflect on different points of view, and reconstruct ideas and thoughts from questions, reflections, and criticism.

So, it is probable that if, in the future, a professor suggests to his or her students using a virtual learning environment to promote discussion, collaborative learning, and collective building of knowledge, many of them will answer: "Yes!"

Copyright © 2005, Idea Group Inc. Copying or distributing in print or electronic forms without written permission of Idea Group Inc. is prohibited.

References

Britain, S. & Liber, O. (1999). A framework for pedagogical evaluation of virtual learning environments. University of Wales - Bangor. Retrieved online May 21, 2004 from: *http://www.leeds.ac.uk/educol/documents/00001237.htm.*

Colin, M. (1999). Delivering staff and professional development using virtual learning environments. Heriot-Watt University. Retrieved online May 21, 2004 from: *http://www.jiscinfonet.ac.uk/Resources/external-resources/jtap-044.doc.*

Deleuze, G. & Guattari, F. (1987). Mil platôs: Capitalismo e esquizofrenia, (Anti-Oedipus: capitalism and schizophrenia), v. 1, Rio de Janeiro, Editora 34. p. 15-25.

Demo, P. (2002). Complexidade e aprendizagem. São Paulo, Ed. Atlas. p 8-19.

Dodge, M. & Kitchen, R. (2001). *Mapping cyberspace.* London: Routledge.

Dodge, M. & Kitchen, R. (2002). The geography of cyberspace directory. Retrieved online May 21, 2004 from: *http://www.cybergeography.org/geography_of_cyberspace.html.*

Freire, P. (1987). *A pedagogy for liberation: Dialogues on transforming education.* South Hadley, Mass: Bergin & Garvey Publishers, p. 35-65.

Levy, P. (1994). *L'intelligence collective: Pour une anthropologie du cyberspace,* Paris: Éditions La Découverte, p. 22-30.

Machado, N. J. (1999). *Epistemologia e didática: As concepções de conhecimento e inteligência e a prática docente.* 3a.ed. São Paulo, Brazil: Cortez, p. 117-176.

Mason, R. (1998). Models of online courses. Sloan Consortium. A consortium of institutions and organizations committed to quality online education. Retrieved online May 21, 2004 from: *http://www.sloan-c.org/publications/magazine/v2n2/mason.asp.*

Maturana, H.R. & Varela, F. J. (1980). *Autopoiesis and cognition: The realization of living,* p. 65-85. Dordrecht, Holland: D. Reidel Publishing Co.

Moraes, M.C. (1997). *O paradigma educacional emergente.* Campinas, Brazil: Papirus Publishing Co., p. 20-40.

Morin, E. & Kern, A.B. (1999). *Homeland earth: A manifesto for the new millennium.* Cresskill, NJ: Hampton Press.

Nonaka, I.& Takeushi, H. (1995). *The knowledge-creating company: How Japanese companies create the dynamics of innovation.* New York: Oxford University Press, p. 61-85.

Okada, A. (2001). Nestor Workshop. Retrieved on May 21, 2004 from: *http://www.projeto.org.br/nestor/.*

Okada, A. (2002). The collective building of knowledge as a net of meanings in virtual learning environments. Master's Thesis. São Paulo PUC-SP University.

Okada, A. & Zeliger (2003). The building of knowledge through virtual maps in collaborative learning environments. In *Proceedings of EdMedia.* p. 1625-1628. Hawaii USA.

Copyright © 2005, Idea Group Inc. Copying or distributing in print or electronic forms without written permission of Idea Group Inc. is prohibited.

Weil, M. & Rosen, L.(2001). A conversation with TechnoStress authors. Retrieved on May 21, 2004 from: *http://www.technostress.com/tsconversation.htm.*

Whitaker, R. (1995). Self-organization, autopoiesis, and enterprise. Retrieved on May 21, 2004 from: *http://www.acm.org/sigois/auto/Main.html.*

Zeiliger, R., Belisle, C. & Cerrato, T. (1999). Implementing a constructivist approach to web navigation support. Retrieved on May 21, 2004 from: *http://www.aace.org/dl/files/EDMEDIA1999/edmedia1999p438.pdf.*

Appendix

Freeware interfaces available in cyberspace:

Table 1: Html editors.

Netscape Composer	http://cannels.netscape.com/ns/browser/download.jsp
FrontPage Express	http://microsoft.com/dowloads/search.asp
Nestor Web Cartographer	http://www.gate.cnrs.fr/~zeiliger/nestor/nestor.htm
Others	http://www.setarnet.aw/htmlfreeeditors.html

Table 2: Servers.

HPG	http://www.hpg.com.br
GEOCITIES	http://www.geocities.com
TRIPOD	http://www.tripod.com

Table 3: Chats.

CJB NET	http://www.cjb.net
ICQ	http://go.icq.com/

Table 4: Forum.

FÓRUMNOW	http://www.forumnow.com
INFORUM	http://inforum.insite.com.br/

Copyright © 2005, Idea Group Inc. Copying or distributing in print or electronic forms without written permission of Idea Group Inc. is prohibited.

Table 5: Group lists.

YAHOO	www.yahoo.grupos.com.br
GEOCITIES	www.geocities.com
ESCRIBE	www.escribe.com/

Table 6: Blogs.

BLOGSPOT	http://www.blogspot.com/
IG	http://blig.ig.com.Br
WEBLOGGER	www.weblogger.com.br

Copyright © 2005, Idea Group Inc. Copying or distributing in print or electronic forms without written permission of Idea Group Inc. is prohibited.

TOURO COLLEGE LIBRARY

Chapter V

Collaboration or Cooperation?
Analyzing Small Group Interactions in Educational Environments

Trena M. Paulus
University of Tennessee, USA

Abstract

This chapter illustrates how computer-mediated discourse analysis (CMDA) can be used systematically to investigate online communication. It argues that intended outcomes of learner interactions, such as meaningful dialogue and joint knowledge construction, must be identified and analyzed to better understand the effectiveness of online learning activities. The CMDA approach is illustrated through analysis of a synchronous chat held by a three-person graduate student group as it completed a course assignment at a distance. Findings from the analysis reveal that while all group members participated in the task and communicated with mutual respect, a cooperative rather than collaborative approach was taken, and group members did not challenge initial opinions. These findings can assist with the future instructional design of such online learning tasks. It is hoped that this chapter provides guidance to researchers in identifying intended outcomes of online collaboration and utilizing CMDA to determine whether the outcomes have been met.

Copyright © 2005, Idea Group Inc. Copying or distributing in print or electronic forms without written permission of Idea Group Inc. is prohibited.

Introduction

With an ever-increasing number of courses and programs being offered at a distance via the Internet, instructors and course designers are now faced with the challenge of determining what works best for teaching and learning in these environments. In residential educational environments, most interactions among learners and instructors occur in the classroom, during office hours, or even in the hallways. Obviously this type of contact is not possible at a distance, so teaching at a distance requires different instructional strategies for facilitating learner interactions.

A lack of interaction has frequently been cited as a major weakness of distance education. High attrition rates have always been a concern in distance courses (Simonson, 2000; Simonson, Smaldino, Albright, & Zvacek, 2000), and feelings of isolation and frustration have been attributed to the lack of interaction among learners who may be situated around the globe with minimal to no face-to-face contact (Hara & Kling, 2000; Vrasidas & McIsaac, 1999). Increasing the opportunity for interaction has thus been identified as a critical component for successful online learning (Hirumi & Bermudez, 1996; Moore, 1989; Roblyer & Ekhaml, 2000; Schrum & Berge, 1997; Vrasidas & McIsaac, 1999; Wagner, 1994).

The emphasis on interaction also arises from current theories of how people learn. Social constructivism emphasizes the negotiation of meaning and construction of shared understandings by learners through dialogue (Bonk & Cunningham, 1998; Bonk & Kim, 1998; Jonassen, Davidson, Collins, Campbell, & Haag, 1995). Interaction and dialogue are also the key components of social learning theory (Bandura, 1971). Vygotsky's (1978) view of learning as a social process occurring within the zone of proximal development also positions interactions as crucial to the development of patterns of thoughts and behaviors.

Kearsley (2000) argues that "the most important role of the instructor in online classes is to ensure that there is a high degree of interactivity and participation" (p. 13). Defining interactivity and interaction continues to be the focus of much discussion in the distance education field. Moore's (1989) distinction between learner-content interactions, learner-learner interactions, and learner-instructor interactions is quite useful in this regard. Learner-learner interactions have typically been the weakest in distance education environments. Today more substantial interactions among learners are possible through communication tools such as electronic mail, Web-based discussion forums, and synchronous chat. Of growing interest to researchers and practitioners is how students communicate and learn with these tools.

This chapter outlines types of learner interactions, cooperation, and collaboration that may be facilitated through the instructional design of online tasks. It then identifies some intended outcomes of these interactions and illustrates the use of a promising research approach, computer-mediated discourse analysis (CMDA) that can be used to systematically analyze interaction at a distance.

Copyright © 2005, Idea Group Inc. Copying or distributing in print or electronic forms without written permission of Idea Group Inc. is prohibited.

Background

Computer-mediated communication (CMC), "the communication produced when human beings interact with one another by transmitting messages via networked computers" (Herring, 2001, p. 612), results in faster and more frequent interactions among geographically separated learners, making activities such as small group projects increasingly popular in distance education courses. These activities are often called "collaborative learning activities." However, Roschelle and Pea (1999) fear the word collaboration is in danger of losing its meaning because "technology evangelists tend to label almost any web facilities for correspondence or coordination across distance as a 'collaboration tool'" (p. 23). This may reflect an element of technological determinism – a belief that because learners now <u>can</u> interact more frequently, they automatically will. In a similar vein, research studies that examine any type of interaction tend to be labeled as collaboration studies, with an assumption that participant interaction is the same as collaborative learning.

Henri and Rigault (1996) provide a needed distinction between the instructional design of group learning, cooperative, and collaborative activities. Group learning occurs when learners are together, often in larger groups, for "discussion, exchange, interaction and mutual assistance" (p. 46). Group learning is what occurs when distance students participate in an asynchronous discussion on the course readings. This is similar to what occurs in a residential graduate seminar. Collaborative and cooperative learning groups, on the other hand, are usually restricted to, at most, five or six participants working in heterogeneous groups to complete a particular task, according to Henri and Rigault. Hathorn and Ingram (2002) emphasize that for student groups to effectively collaborate they must have a common goal, have incentive to collaborate, and be independent from the instructor. These criteria can be met through the instructional design of the task.

There are several desired outcomes that educators may have for small group projects. One desired outcome of collaborative and cooperative learning activities is often the equal participation of all members of the group. An early hope for CMC environments was that the potentially anonymous, text-only nature of the communication would minimize the usual face-to- face power dynamics, creating more equal and democratic participation than is usually the norm (Harasim, 1993; Kiesler, Siegel, & McGuire, 1984; Sproull & Kiesler, 1991). This hope has not always been realized. Males, for example, tend to be less polite, express more opinions, and dominate online discussions (Herring, 1993). In multicultural environments, asymmetrical participation patterns have also been noted, with a small number of participants dominating the discussions (Stewart, Shields, & Sen, 1998). Differences in status can also affect participation in negative ways (Weber, forthcoming). In many ways, online interactions can mirror asymmetrical face-to-face interactions. Researchers and practitioners with an interest in the outcomes of a designed collaborative or cooperative learning activity may want to examine group member participation as one of these outcomes.

Roschelle and Teasley (1995) conclude that "collaboration doesn't just happen because individuals are co-present; individuals must make a conscious, continued effort to coordinate their language and activity with respect to shared knowledge" (p. 94). Likewise, in CMC educational environments, participant contributions alone do not result in effective collaboration. Educators would like discussions to be on task and

Copyright © 2005, Idea Group Inc. Copying or distributing in print or electronic forms without written permission of Idea Group Inc. is prohibited.

relevant to the learning goals. CMC discussions, however, may move quickly off-task, with participants discussing a wide variety of topics (Herring & Nix, 1997). In a review of the computer-supported collaborative learning research, Bonk and Wisher (2000) noted that even when computer-mediated environments are specifically designed to have learners connect course knowledge to experience through interactions, deep connections are not made as frequently as opinions are exchanged and social acknowledgements made. Examining the topics of online discussions may be a second focus of the analysis of small group activities.

A further distinction can be made between collaboration and cooperation. How participants approach a small group assignment can vary considerably. Henri and Rigault (1996) identify cooperative tasks as those that are divided up and completed individually. Division of labor, task specialization, and individual responsibility for part of the final product are characteristics of cooperative learning. Roschelle and Teasley (1995) define collaboration, in contrast, as "a coordinated synchronous activity that is the result of a continued attempt to construct and maintain a shared concept of a problem" (p. 70). Schrage (1990) describes collaboration as a "process of shared creation: two or more individuals with complementary skills interacting to create a shared understanding that none had previously possessed or could have come to on their own. Collaboration creates a shared meaning about a process, a product, or event" (p. 40). For this reason, no formal roles are assigned in collaborative groups. Collaboration, it is argued, improves learning because it creates awareness of one's own thinking processes as multiple perspectives are shared through discussion (Arvaja, Hakkinen, Etelapelta, & Rasku-Puttonen, In press). Meaningful, sustained dialogue related to the conceptual learning is key to a collaborative learning experience.

It may be the case, particularly in a computer-mediated distance education context where participants are geographically separated, that cooperative strategies are viewed as more efficient than collaborative ones. What is designed to be a collaborative project may be interpreted as a cooperative project by the students, who choose to divide up tasks, complete them individually, and then combine the independent efforts into a final product, as reported by Kitchen and McDougall (1998-9). While both cooperation and collaboration may be valuable, the underlying beliefs about and intended outcomes of the approaches differ. Methods that can help researchers determine how learners approach a task, whether collaboratively or cooperatively, can be very useful in this regard.

Hathorn and Ingram (2002) add that collaborative group members must be interacting with each other in a way that "mutually values" contributions of each member, drawing upon members' diverse skills and resources to meet the specific goals together. Part of the process of valuing the contributions of group members is through negotiation. Dillenbourg, Baker, Blye and O'Malley (1996) describe negotiation as "a process by which students attempt (more or less overtly or consciously) to attain agreement on aspects of the task domain . . . and on certain aspects of the interaction itself . . ." (p. 19). Groups get work done through effective negotiation, and researchers may be interested in more closely examining how this occurs online.

As important as it is for groups to be operating within a framework of mutual respect, it is also expected that members will constructively criticize and challenge initial opinions in order to achieve greater quality of learning (Johnson, Johnson, & Smith, 1991.) These

Copyright © 2005, Idea Group Inc. Copying or distributing in print or electronic forms without written permission of Idea Group Inc. is prohibited.

types of interactions have been referred to as "challenge and explain" cycles of interaction by Johnson and Johnson (1996) and Curtis and Lawson (2001.) Through challenge and explain cycles, group members create an effective synthesis of information – an intended outcome of collaborative efforts (Hathorn & Ingram, 2002). Arvaja et al. (In press) call such interactions critical joint knowledge building through exploratory talk. "In exploratory talk, statements and suggestions are offered for joint consideration. These are then challenged and counter-challenged with justifications and alternative hypotheses" (p. 2). Cumulative talk, on the other hand, leads to uncritical joint knowledge sharing, where "the participators build positively but uncritically on what the other has said, thus constructing common knowledge by accumulation" (p. 2). Cumulative talk would not be as conducive to learning (from a constructivist viewpoint) as would exploratory talk.

Curtis and Lawson (2001) found that while their online students did exhibit indicators of collaborative behavior, such as seeking input, contributing, and monitoring the group's effort, they did not engage in challenge and explain cycles. Kanuka and Anderson (1998) also found that existing information was shared rather than new knowledge constructed. In the few instances where knowledge construction did seem evident, it resulted from initial "social discord." If an instructional goal is "mutual support and learning through dialogue and argument" (Howell-Richardson & Mellar, 1996, p. 49) then there should be questioning, answering, challenging, and responding evident in the interactions with other members of the group. Identifying these challenge and explain cycles of dialogue can be another useful focus of research into online interaction.

To summarize, researchers interested in online collaboration may examine indicators such as individual participation and the topics of online discussions. They may also want to examine how groups approach the task, together or independently, and whether members engage in meaningful dialogue that includes challenge and explain cycles within a context of mutual respect.

Computer-Mediated Discourse Analysis

Part of the challenge in examining learning outcomes is that nearly all students tend to receive high grades (particularly at the graduate level); instructors at times may have difficulty articulating specific outcomes that are easy to measure. Researchers in computer-supported collaborative learning (CSCL) have suggested that we "focus more on the processes involved in successful peer interaction, rather than just on learning outcomes" (O'Malley, 1991, p. v.). Possible indicators of collaborative process have been identified in the previous section.

How can we determine whether these indicators of collaborative process exist? The advantage of examining group interactions in CMC environments is that the transcripts of discussions are readily available for analysis. Previous research designs and methods for examining online collaboration have ranged from qualitative case studies (e.g., Stacey, 1999) to experimental designs (e.g., Bernard & Lundgren-Cayrol, 2001.) Computer-mediated discourse analysis (CMDA) is "any analysis of online behavior that is grounded in empirical, textual observations . . . [I]t views online behavior through the lens

Copyright © 2005, Idea Group Inc. Copying or distributing in print or electronic forms without written permission of Idea Group Inc. is prohibited.

of language, and its interpretations are grounded in observations about language and language use" (Herring, In press, p. 3). This lens is particularly appropriate when examining interactions at a distance through computer-mediated, text-based communication because, in essence, interaction is occurring through language. Dillenbourg et al. (1996) point out that a focus on the task and communicative level of interactions is critical to an understanding of when collaboration is actually occurring, admitting that "deciding on the meaning of these expressions in a given dialogue context is thus quite complex, but necessary if we are to understand when students are really collaborating and co-constructing problem solutions" (p. 18). They point out that "a promising possibility for collaborative learning research is therefore to exploit selective branches of linguistics research on models of conversation, discourse, or dialogue to provide a more principled theoretical framework for analysis" (p. 19).

Herring (In press) emphasizes that CMDA is more of an approach than a theory or method, providing a "methodological toolkit" that draws upon "theoretical assumptions . . . of linguistic discourse analysis." There are several important assumptions underlying CMDA, according to Herring (In press). First, patterns are present in discourse and may be identified by the analyst, though speakers themselves may not be aware of these patterns. CMDA can help reveal these patterns (such as challenge and explain cycles). Second, "discourse involves speaker choices" that "reflect cognitive and social factors," enabling findings that may be non-linguistic (such as whether mutual respect exists among the members) as well as linguistic. Finally, "computer mediated discourse may be, but is not inevitably, shaped by the technological features of CMC systems" (p. 7). It is important, then, to consider the way that features of the technology (e.g., synchronicity) may "shape" the communication.

CMDA was used to analyze the synchronous chat interactions of one small group working together to complete an assignment in a completely online graduate course. Henri (1992) points out that "in a CMC learning situation, the educator can offer input at three levels: what is said on the subject or theme under discussion; how it is said; and the processes and strategies adopted in dealing with it" (p. 121). Chat transcripts were examined for the following desired outcomes of collaborative learning:

1. What topics do small group members discuss as they work on a task? Are they on-task?

2. How do group members interact to complete the task?

    a. Does the group take a cooperative or collaborative approach to the task?

    b. Do they demonstrate mutual respect?

    c. Are the group members participating equally in the conversation?

    d. Is there evidence of negotiation, challenge and explain, or other cycles of interactions?

Copyright © 2005, Idea Group Inc. Copying or distributing in print or electronic forms without written permission of Idea Group Inc. is prohibited.

Analysis of Synchronous Chat Interaction

Context and Participants

The context of this study was a 14-week graduate-level education course at a large Midwestern university taught entirely at a distance using the university's homegrown course management system. The 19 students had approximately ten days to complete each of the eight units of the course. Each unit included reading assignments, asynchronous discussions on the readings, and an individual or group project applying the readings. Students were assigned to work together in small groups of three to four students for at least two of the units. Students were then given the option of working together in a group for additional units. Groups were free to use any or all of the communication modes available: email, a Web discussion forum, and a synchronous chat tool. All groups operated independently from the instructor, though she was available to answer questions.

Group Blue was selected for this initial analysis as a group who was successful in its efforts. It was one of only three groups in the course to receive an A+ on its product. Group Blue communicated by holding two synchronous chats within the course management system and by exchanging electronic mail. Synchronous tools have typically been viewed as appropriate for the social aspects of distance courses, whereas asynchronous tools have been considered better for serious academic discussion (Motteram, 2001). However, researchers have begun to speculate on the important role that synchronous tools may play for small group collaboration due to the limitations of asynchronous tools (Armitt, Slack, Green, & Beer, 2002; Davidson-Shivers, Muilenburg, & Tanner, 2001; Fisher & Coleman, 2001-02). The graduate students in Kitchen and McDougall's (1998-99) study, for example, specifically requested synchronous tools to help them complete their collaborative learning tasks more quickly. Curtis and Lawson (2001) also discovered in their research that students had indeed been communicating synchronously, although these discussion transcripts were not analyzed because the researchers did not expect students to communicate in this medium. Because studies of synchronous educational chat are less prevalent in the literature, the first of Group Blue's two chat sessions was to demonstrate how CMDA can be used to answer the research questions.

Sally, John, and Pam were the members of Group Blue. Sally, a 41-year-old female, had taken previous distance education courses that required group projects. Sally was geographically located outside of the state of the university that offered the course. John, a 52-year-old male, had previous distance education experience with required group work. John was geographically located in the same state as the university, but not in the same town. Sally and John knew each other from previous distance education courses they had taken, as they were part of a cohort pursuing the same graduate degree. Pam, a 27- year-old female, was a residential graduate student at the university, and she had no previous distance education experience.

Group Blue's task was to write a three to four page analysis of how the Motivated Strategies for Learning Questionnaire (MSLQ) reflects elements of cognitive processing

Copyright © 2005, Idea Group Inc. Copying or distributing in print or electronic forms without written permission of Idea Group Inc. is prohibited.

theory and how it might be useful for instruction (see Appendix A for the complete description of the task.) At the end of the task, each student completed an individual reflection on his or her learning process, including a peer evaluation of the contributions of each group member.

The chat transcript was downloaded into a word processing document for analysis. The chat was held from 6:44:14 a.m. – 8:32:54 a.m. There were 232 messages (3954 words) exchanged during this time. Approximately 2.16 messages per minute were exchanged with an average message length of 16.83 words. The unit of analysis was the individual chat message. The chat messages were analyzed on several levels to answer the research questions. Each level of analysis is described here.

Topic Identification

To answer the first research question, all topical threads were identified and a comprehensive list of discussion topics addressed in the chat was created. Then, each message was coded according to which topic it addressed. Topics were then categorized by their purpose in the conversation. Henri (1992) suggested simply distinguishing between messages related to the formal content of the task and messages that are not. Howell-Richardson and Mellar (1996) in their study divided topics into group-focus (social), task-focus, and off-task focus. In this study, emergent categories were used to organize the topic threads by their purpose and focus.

This level of analysis provides an initial answer to whether and how the chat discussion was related to the course assignment. The analysis also provides an initial view of how the participants approached the task — whether by discussing the cognitive learning theories in detail, by dividing up the task, or by taking another approach.

Functional Moves

Functional move analysis can shed light on whether participants are mutually respecting each other's contributions to the discourse and whether they are challenging each other's thinking. As its name would suggest, a functional move is literally the function or purpose served by a particular segment of the conversational discourse. Herring (1996) points out that "electronic messages are internally organized texts" and that different types of text display "distinctive schematic organization, or conventional sequence of functional 'moves' into which the text can be chunked" (p. 83). Herring and Nix (1997) identified functional moves such as *inform, inquire, greet,* and *react*.

In the present study, functional moves were identified as they emerged from the data to create a complete coding scheme. Then, each message was coded according to the scheme. As found in Herring and Nix (1997), some messages contained more than one functional move. In this case, the messages were broken into smaller units of analysis - the functional move unit.

Copyright © 2005, Idea Group Inc. Copying or distributing in print or electronic forms without written permission of Idea Group Inc. is prohibited.

Participation

The total number of messages and average message length was calculated for each participant to determine participation levels for each of the three members. The number of functional moves was also calculated by participant, giving a view of not only the number but also the type of contributions to the discussion.

Sequences

Of particular interest to this study is whether and how participants negotiate with each other, particularly whether they participate in challenge and explain sequences. The final level of analysis was to identify any recurring patterns of functional moves related to negotiation or challenge and explanation. These recurring patterns are called sequences. Discourse management is defined by Condon and Èech (1996) as "the strategies that speakers employ to structure and sequence the routine (and non-routine) elements of their talk into successful discourses" (p. 2). Routine elements of talk are similar to what Francis and Hunston (1992) call exchange structures. Herring and Nix (1997) clarify: "Exchange structure refers to sequences of functional moves, or speech acts (question, answer, greeting, etc.) as they occur in every day conversation exchanges" (p. 3). The chat transcript was analyzed to identify any patterns of functional moves that resulted in these exchanges structures, or sequences. The entire corpus was then re-analyzed, coding for these sequences.

In addition to analysis of the chat transcript itself, the group's final paper, peer evaluations, and individual reflections were all used to triangulate findings. Documents were read for themes related to the research questions. A code-recode procedure was used to establish stability in the coding of functional moves in the present study, since this is a potentially subjective aspect of the analysis (Herring, In press). Stability (also known as "intra-observer reliability" or "consistency") "is the degree to which a process is invariant or unchanging over time" (Krippendorf, 1980, p. 130). The data was re-coded for functional moves with 85% agreement between the first and second coding.

Findings and Discussion

Topics

Twenty-one topical threads were identified in the chat transcript. The threads over-lapped and revealed that different topics were discussed simultaneously. The threads were classified according to their primary focus, or purpose, in the conversation. Four categories fit the data: social, logistical, technical, and conceptual. Social topics consisted of greetings and closings. Logistical topics addressed setting up discussion times, procedures for exchanging documents, and establishing deadlines for completing the task. Technical topics concerned the use of the chat tool, word processing, and email.

Copyright © 2005, Idea Group Inc. Copying or distributing in print or electronic forms without written permission of Idea Group Inc. is prohibited.

Table 1: Topic identification and focus.

Topic of Thread	Focus	Number of Messages
1. Exchanging greetings	social	4
2. Wondering whether the chat tool allows scrolling	technical	4
3. Clarifying the assignment	conceptual	6
4. Wondering if the chat thread is automatically archived	technical	2
5. John offering to send Sally a copy of the MSLQ instrument	logistical	14
6. Expressing frustration that their messages are disappearing from the screen	technical	4
7. Wondering what happened to Pam, who has temporarily disappeared from the screen	technical	2
8. Proposing idea to use a matrix organizational framework for the assignment	conceptual	13
9. Approaching the analysis using their different backgrounds as a starting point	conceptual	16
10. Dividing up sections of the matrix	conceptual	23
11. Arranging next chat meeting time	logistical	9
12. Clarifying document exchange procedures, specifically, which word-processing program to use and which email address to send to	technical	14
13. Deciding which concepts from the book to include	conceptual	3
14. Deciding how to add commentary to the matrix	conceptual	15
15. Setting timeline for completion	logistical	7
16. Working through one example together to clarify the matrix structure	conceptual	49
17. Wondering if their chat archive is open or private	technical	4
18. Confirming that they need to cite sources	conceptual	2
19. Identifying and dividing up the theme areas	conceptual	19
20. Continuing to set timeline for completion	logistical	14
21. Closing and farewells	social	8
Total		232

Conceptual topics were those that directly addressed how to complete the assignment. Table 1 lists the topics and foci in the order in which they were discussed.

The number of messages per topic ranged from two messages to 49 messages. Sixty three percent of the messages were conceptual, 19% logistical, 13% technical, and 5% social.

Findings from this level of the analysis shows that the group members were focused on the task at hand and were not using chat mainly for social interactions, as was found by Motteram (2001) and Davidson-Shivers, Muilenburg and Tanner (2001). The communication environment was a supportive one, with polite greetings and closings. There were a few technical difficulties that needed to be discussed, but primarily the participants concentrated on their approach to the assignment. It is interesting that the group members spent very little time discussing the underlying concepts of cognitive processing learning theory that the assignment was designed to teach. Rather, the primary concern was how to organize their approach to completing the assignment. During Thread 16, the participants came close to discussing the MSLQ document itself and what

Copyright © 2005, Idea Group Inc. Copying or distributing in print or electronic forms without written permission of Idea Group Inc. is prohibited.

theories it may represent. Here the group started to analyze one of the MSLQ items together:

John: "In a class like this, I prefer course material that really challenges me so I can learn new things." That's an issue of motivation, right? And attention. And semantic networks, relating new information to old information. A challenging class would have a preponderance of new information, but tied well to what the learner already knows.

Sally: It definitely fall under the motivation area.

Pam: It looks as several items overlap in many different categories.

John: I think you're right, Pam. So we could have questions showing up in many areas.

Pam: I think it is going to depend on what type of relationships we want toshow [sic]. Either by main theme or interconnecteness [sic].

However, the discussion quickly turned to how the final paper should be organized and how much they would be able to finish before the deadline:

John: How would we characterize "interconnectedness" in a table? Any ideas? Can we show both main theme and interconnectedness in the same table, perhaps by using some sort of cross-referencing?

Pam: Should we also look at what is feasible to complete for the deadline?

The participants were also highly concerned with how to divide up the task among themselves. Threads 10, 12, and 19 explicitly addressed these concerns, as this excerpt illustrates:

Pam: Do we want to reconveniene [sic] or send eachother [sic] the matrix and then pool our reflection of them into a final paper?

John: How would we individually work on the matrix, Pam? Any ideas?

Pam: Anyway, I thought of two plans of attack. Either we could divide the items in the MSLQ and do chart them, or one of us could chart the items. Can you think of any other ways?

Sally: I was thinking we would take each cluster that is mapped in the MSLQ and apply strategies for each of the clusters. What do you guys think? Any other ideas?

John: I like the idea of divide and conquer – the matrix is a big piece of work, with implications for the entire analysis – but how do we divide it up?

Elsewhere, too, it was assumed that "divide and conquer" was the way to go. For example, from Thread 8:

Copyright © 2005, Idea Group Inc. Copying or distributing in print or electronic forms without written permission of Idea Group Inc. is prohibited.

John: If we each take a consideration, one of us could begin the matrix (I'd be happy to volunteer for that), one could talk about how the MSLQ applies to instruction, and one could include that "short reflection" [the instructor] cites.

Sally: The questions could definitely mapped into clusters. They all fell into a couple different categories.

John: What's your take on how we should tackle this, Pam [sic]? Should we discuss specifics this a.m., or carve up the task, do some work, and then regroup to discuss?

Pam: I think that the matrix is a great idea. I think it will focus the items into areas that were specifically identified in the readings.

Sally: I will take either two or three it doesn't matter. If you want I can take two since I teaching is my profession.

These results illustrate how topical analysis of small group interactions online can provide an initial view of how participants are working together. The findings here provide some initial evidence, consistent with Kitchen and McDougall (1998-9), that rather than engaging in a dialogue to develop a shared understanding of the concepts being learned, the group took a more cooperative approach by dividing up the task for individual completion.

Functional Moves

Of primary interest to this research is how group members work together to reach an understanding about the content of the assignment itself, rather than how they discussed technical and logistical matters. Therefore, the next phase of analysis was to examine the nine conceptual topics (146 messages) for functional moves and sequences. A total of 15 types of functional moves were identified in the data, listed in Table 2 from highest to lowest frequency. There were 215 functional moves in 146 messages.

The types of functional moves used by the group members shed light on how the participants interacted as they completed the task. Functional moves such as *agreeing, suggesting* (rather than *dictating*), *eliciting opinions* from others, and *offering to act* (rather than *directing*) point to an environment of mutual respect created by the group. No functional move for explicit *disagreement* or *challenge* was found; however, there were functional moves used by the participants to very indirectly express another point of view. One such move was the *counter-suggestion*, by which a speaker would suggest something different rather than explicitly disagree with a previous suggestion. For example, here Pam makes a *counter-suggestion* to John's initial *suggestion*.

John: We could have a fourth column on "applications" that would make the table less confusing.

Pam: Maybe we should try one after we have the table compiled. There may be a lot of overlap that would just be redundant.

Copyright © 2005, Idea Group Inc. Copying or distributing in print or electronic forms without written permission of Idea Group Inc. is prohibited.

Table 2: Functional moves.

Functional move	Code	Example	Total	%
Agree/support: express agreement with or support of another person's statement	agr	We're on the same wavelength…	37	17
Explain: provide elaboration or explanation of a previous statement	exp	This would address the first question above "Some things."	28	12
Request clarification: ask for a previous statement to be clarified/explained further	rcl	Can you explain what you mean by individually work on the matrix?	22	10
Request confirmation: check with others to confirm that his/her understanding is correct	rco	Correct?	20	9
Suggest: suggest a possible course of action or interpretation	sug	Maybe briefly discuss each section and then we can proceed individually	17	7
Provide clarification: provide clarification of a previous statement, may follow request for clarification	pcl	We're each taking a third and building a matrix that will map each question (or cluster) against all of the key CIP concepts.	15	7
Provide information: contribute information to the discussion, usually follows request for information	pi	I assume we do need to cite sources.	15	7
Restate: paraphrases anyone's previous statement (including own) or provides a summary of the discussion	res	So just to review, we are each taking our sections, and mapping them into clusters.	12	5
Provide confirmation: follows request for confirmation	pco	Yes.	12	5
Request information: ask for information	ri	When we're doing our individual write-ups, do we need to cite sources from our readings?	10	4
Elicit opinion: ask what others think about a proposition	eo	What do you guys think?	9	4
Counter-suggest: make an alternate suggestion, one different than a previously stated suggestion and/or one currently being considered	csug	Maybe we should try one after we have the table compiled.	6	3
Offer to act: offer to do something	oa	I'll take the first 33, okay?	5	2

Copyright © 2005, Idea Group Inc. Copying or distributing in print or electronic forms without written permission of Idea Group Inc. is prohibited.

Table 2: Functional moves (continued).

Functional move	Code	Example	Total	%
Suggest alternatives: offer several alternate suggestions for consideration	sa	I thought of two plans of attack. Either we could divide the items in the MSLQ and chart them, or one of us could chart the items.	4	2
Request action: ask individual or group as a whole to do something	ra	John, could you do maybe two questions in your table and then send so we are all consistent.	3	1
Total			**215**	**100**

John in particular used a certain strategy three times to express, in a rather indirect way, his disagreement or possible "challenge" of another member's view. His strategy was to utilize question formation, start by first expressing his understanding by "yes" or "well," then followed by his own (somewhat different) view.

Example 1:

Sally: I was thinking #2 meant how we would apply this to general instruction and study strategies that would be used from both long-term and short-term memory. What do you guys think?

John: Well, it's broader than that, right?

Example 2:

Sally: Is everyone going to use the concepts that are outlined on pg. 74?

Pam: Sure. The concepts outlined on p. 74 would work well.

John: Yes, but we also need to draw on Bandura, Keller, Weiner, right?

Example 3:

Sally: For the clustered themes are you using the outline on pg. 74?

John: Yes, but also p. 301-302, right?

Through use of this strategy the participants are encouraged to continue the conversation. The result is a very indirect form of possible disagreement, perhaps chosen so as not to threaten the other participants.

Other common functional moves were *requesting* and *providing clarification, information,* and *confirmation.* These moves suggest a group that is negotiating with each other, an important indicator of a collaborative process. Yet, as revealed by the topical analysis, this negotiation usually focused on dividing up the task rather than negotiating conceptual understanding of the material. In fact, six of the *suggests* functional moves specifically suggested dividing the work. For example, as the group decided how to approach the analysis of the MSLQ instrument, Pam posted: "My only fear of that is that

Copyright © 2005, Idea Group Inc. Copying or distributing in print or electronic forms without written permission of Idea Group Inc. is prohibited.

some items may not be identified or may get left out. I thought we could divide by number. For example, one person could take 1-20, etc." and later, "Do we want to divide up the theme areas now or once the table is constructed and completed?"

Other parts of the conversation suggested that substantial collaboration would come later, as shown by Pam's *restatement* of what decisions have been made:

Pam: I need to code the 33 questions according to the chart that John designs and will send by Wendesday [sic] (morning?). After I code my items, then I need to send a copy to both of you. After we have exchanged all items, then we begin discussing how the themes relate to our individual fields of practice.

Here the expectation seemed to be that after the individual work was completed, there would be some discussion among the group as a whole about the conceptual content of its work.

Similarly, this *suggestion* by John implied that soon the group would become involved in exchanging feedback on the ideas presented by each group member: ". . . We could each compile a list of theme areas during the day today and send it around for comments before I get started on the table." It could be that the group's overall strategy was to start with individual contributions to the group, followed by feedback cycles and an eventual synthesis into a final product. Of interest is that the final draft of the completed assignment included a final section entitled "Personal History with Study Strategies." Each group member independently contributed one paragraph to this section, and each paragraph was clearly labeled with the group member's name. This underscores the individual approach utilized by the group members as they tackled the assignment.

Participation

Equal participation is often a desired outcome of collaboration. Of the 232 messages exchanged by Group Blue members, Sally contributed 67 (29%) and 1,182 words, an average of 17.64 words per message. Pam contributed 65 messages (28%) and 1102 words, resulting in a 16.95 word average. John, contributed 101 (44%) messages and 1670 words, a 16.53 word average. All members participated in the conversation, though John posted more messages. While Sally and Pam posted fewer messages to the discussion, their messages were longer.

The functional move analysis provides additional insight to the type of contributions made by each group member (see Table 3).

All three participants made many *agreeing/supporting* comments, again suggesting an environment of mutual support. John contributed 45% of the functional moves; the majority of his moves were to *explain*, *provide clarification*, and *provide information*. John made more *suggestions* than Pam and Sally combined. Pam contributed 28% of the functional moves, and she made more *requests for information* than either John or Sally. She also *agreed* frequently. Sally contributed 27% of the functional moves, and many of her moves were to *agree*. Sally also made *restatements* more than the other group members.

Copyright © 2005, Idea Group Inc. Copying or distributing in print or electronic forms without written permission of Idea Group Inc. is prohibited.

Table 3: Functional move by participant.

Functional move	John	Pam	Sally
Agree/support	11	12	14
Explain	16	8	4
Request clarification	5	12	5
Request confirmation	11	3	6
Suggest	10	3	4
Provide clarification	13	0	2
Provide information	12	2	1
Restate	3	3	6
Provide confirmation	7	4	1
Request information	1	4	5
Elicit opinion	3	4	2
Counter-suggest	1	3	2
Offer to act	2	1	2
Suggest alternatives	1	3	0
Request action	0	0	3
Total	**97 (45%)**	**61 (28%)**	**57 (27%)**

Roles often tend to emerge in groups, whether assigned or not. While roles were not formally assigned for this task, each participant played a role in the discussion. Roles such as *initiator* and *information giver* (John), *information seeker* (Pam), and *coordinator* (Sally) can be seen here (as described by Chandler, 2001). As John commented in his self-reflection, "I was probably the most active in terms of pushing things along and coordinating [sic] the whole effort, partly because much of Sally's time this weekend was already spoken for with family sports activities." Even though there was no instructor present to dominate the interactions, there seem to be other power dynamics at work. In terms of gender, men tend to be less polite, express more opinions, and dominate the discussion (Herring, 1993). John, while not less polite, may fit this pattern.

Sequences

The final level of analysis, the sequence analysis, revealed a decision-making sequence that incorporated the desired outcomes of *mutual respect* and *clarification*, but not the *challenge and explain* cycles thought necessary for joint-knowledge building. Table 4 outlines the phases of this decision-making sequence. S1 refers to the first speaker, and S2 or S3 refers to the second or third speakers.

Sequences were found to overlap, be interrupted and at times not completed. The presence of *counter-suggestions*, for example, at times resulted in the sequence starting over again. The decision sequence occurred twelve times in the data as follows: all five phases occurred twice completely; a Phase 1, 3, 4, 5 sequence occurred six times; Phase 1, 2, 3 occurred twice; and a Phase 1, 3, 4 and Phase 1, 4, 5 sequence both occurred once. Table 5 is an example of a complete sequence.

Previous studies have identified sequences including *initiation-response-follow-up* (Francis & Hunston, 1992) and *move-response-other* (Condon & Cech, 1996). Herring and

Copyright © 2005, Idea Group Inc. Copying or distributing in print or electronic forms without written permission of Idea Group Inc. is prohibited.

Table 4: Phases of decision sequence.

Phase 1. S1 makes suggestion or offer to act (with/without explanation)
Phase 2. S1 elicits input/opinion of others
Phase 3. S2 or 3 agrees/expresses support (with/without explanation)
Phase 4. S2 or 3 requests clarification, information or confirmation
Phase 5. S1 provides clarification, information or confirmation

Table 5: Example of complete decision sequence.

Phase 1(a). S1 offers to act	John	I'll take the first 33.
Phase 2(a). S1 elicits opinion	John	OK?
Re-start Phase 1(b). S1 offers to act	Pam	Should I take 34-66?
Phase 3(b). S2 agrees	Sally	That is fine Pam. I will take the last section.
Phase 3(b). S3 agrees	Pam	That would be fine.
Phase 4. S3 requests confirmation	Pam	Can we confirm what we are doing before we all leave? I just want to make sure I have it written down. The short-term memory is not yet awake.
Phase 5. S1 provides confirmation	John	We're each taking a third of the questions and building a matrix that will map each question (or cluster) against all of the key CIP concepts.

Nix (1997) identified an example of this sequence that they called *inquire-inform-receive*. The identified decision sequence is similar to exchange structures previously identified by Francis and Hunston (1992) and Condon and Cech (1996), as outlined in Table 6.

The embedded clarification subsequence is of interest because of the roles that the participants played in the discussion, as mentioned in the previous section. Toward the end of this chat, Pam or Sally was nearly always the one to *request clarification,* and John *provided clarification*, as evidenced by this exchange (Table 7):

The *counter-suggestion* would seem to be close to a *challenge*. It could be hypothesized that a sequence following a counter-suggestion would resemble a *challenge and explain* cycle. However, as seen in Table 8, even when a counter-suggestion was made, the same decision- making sequence occurred. In this example, John made a suggestion about how to divide up MSLQ items and Pam counter-suggested.

Though Pam was making a suggestion counter to that made by John, John did not explain his own position further; rather he readily agreed and the conversation moved on.

These findings reveal that Group Blue used negotiation and decision-making sequences as part of its group process. There was a dearth of challenge and explain cycles in the group discourse, consistent with findings of Curtis and Lawson (2001). This could be because most of the topics in this first chat conversation were related to figuring out how to approach the assignment and divide up the work. It is possible that challenge and explain cycles came into play later in the group process.

Copyright © 2005, Idea Group Inc. Copying or distributing in print or electronic forms without written permission of Idea Group Inc. is prohibited.

Table 6: Comparison with other sequences.

Decision sequence	Condon and Čech, 1996	Francis and Hunston, 1992
S1 makes suggestion or offer to act	Move: suggest action	Initiation
S1 elicits input/opinion of others	Move: requests validation	Initiation
S2/3 expresses support/agreement	Respond: agree	Response
S2/3 requests clarification, confirmation or information	Move: requests information	Initiation/response
S1 provides clarification, confirmation or information	Respond: complies with request	Follow-up

Table 7: Clarification subsequence.

Phase 4. S1 requests clarification	Pam	So, are we only charting the questions?
Phase 5. S2 provides clarification	John	Charting with commentary, right?
Phase 4. S1 requests clarification	Pam	Can you explain what you mean by commentary?
Phase 4. S3 requests clarification	Sally	Am I still taking # 2 question and pam taking # 3 or is # 2 going to be built into the matrix?
Phase 5. S2 provides clarification	John	By commentary, I meant something like citing the question, tying it to a concept, and then providing an explanation of why it exemplifies the concept.
Phase 4. S1 requests clarification	Pam	That's the part I'm confused on too. John, do you remember
Phase 5. S2 provides clarification	John	The commentary would lead to #2: How the MSLQ is useful for instruction, right?

Table 8: Counter-suggestion sequence.

Phase 1. S1 counter-suggests	Pam	My only fear of that is that some items may not be identified or may get left out. I thought we could divide by number.
+ explanation		For example, one person could take 1-20, etc. Although, that would require another chat.
Phase 2. S2 agrees	John	Your approach sounds the [sic] most sensible ... and will prevent duplication and confusion.
Phase 4. S2 requests confirmation	John	Divide them up by thirds, then?
Phase 5. S1 provides confirmation	Pam	Thirds are fine with me.

Copyright © 2005, Idea Group Inc. Copying or distributing in print or electronic forms without written permission of Idea Group Inc. is prohibited.

Group Member Reflections and Evaluations

Group Blue submitted a seven-page document as its final product, which received a grade of A+ from the instructor. Several themes emerged from analysis of the final product, the individual reflections, and peer evaluations.

First, these documents provide further evidence of the group member's feelings of mutual respect for each other. They felt the group experience was worthwhile. Pam stated, "Even though I had computer difficulties at the beginning . . . both members were supportive, and we solved the problems within the time constraints of the group." John's feelings were that:

John: We . . . relied on each other, which was apparent in the online chats and many email exchanges we had during the week . . . It was a pleasure to work with Pam and Sally. The work was less onerous because we were able to share insights and pool resources quickly. I am convinced that we ended up with a better paper than any one of us could have done solely.

All members rated each other with the highest possible rating on the peer evaluations, reflecting their opinions that members participated equally. Probably the best indicator that the group members truly enjoyed working with each other was that they requested to work together again in a later unit and did so with success.

Second, individual reflections provided further evidence for the cooperative division of labor. The division of labor is described here by Sally: "I was responsible for working on questions 67-100 and placing the information in the matrix . . . Pam's role was to complete the matrix covering questions 34-66 . . . John's role was to complete the matrix covering questions 1-33." Within this division of labor, the group's diverse experiences contributed to its satisfaction with its final product, and reflected a strength of its approach. The members' decision to draw upon their diverse work experiences was first proposed by Pam during the chat:

Pam: It may be interesting to address the use of these tools from the different professional backgrounds we have. We all have different backgrounds. It may be interesting to see if the items are all applied in a similar manner across disciplines.

John: I like that because my experience in the past five years has been corporate training.

Sally: Pam, I think that is an excellent ide3a [sic]

Sally commented on this decision in her reflection: "We all used our work experiences to list real-life examples of how we thought the themes could be applied in both the academic (high school and college) and corporate areas." John added, "My more recent corporate training experience was an excellent companion piece to the academic classroom experience shared by Pam (college) and Sally (high school)..." The instructor seemed please with this approach, mentioning in her feedback "good application to academic and corporate situations."

Copyright © 2005, Idea Group Inc. Copying or distributing in print or electronic forms without written permission of Idea Group Inc. is prohibited.

There is evidence, then, that despite a lack of direct challenges to other points of view through dialogue, multiple perspectives were shared solely by virtue of individual contributions to the final product.

Conclusions

This chapter illustrates how a computer-mediated discourse analysis (CMDA) approach can be used to analyze interaction for indicators of collaborative learning through the lens of language. Group Blue regarded itself as successful, as did the course instructor. During the chat, the group clearly focused on discussing the assignment, though there was little discussion of the cognitive processing learning theory to be learned in this unit. Rather, the focus was on completing the assignment as efficiently as possible. This reflects more of a cooperative learning strategy than a collaborative approach as defined by Henri and Rigault (1996).

All group members participated, but in qualitatively and quantitatively different ways. Even though roles were not assigned, each member played a different role in the discourse. Overall, the group members operated within a framework of mutual respect. The group interacted through a negotiation of meaning and clarification style of discourse; no challenge and explain cycles were found. The findings provide evidence for uncritical joint-knowledge sharing at this stage of the group process, rather than the critical joint-knowledge building hoped for in collaboration.

It is heartening that a three-person group, working only at a distance through Internet tools, succeeded in interacting within a framework of mutual respect to complete a learning task to its own, and to the instructor's satisfaction. Educators hope that through a dialogic process of sharing diverse perspectives and challenging other's ideas, new knowledge can be created. However, the group in this study chose to cooperate through individual contribution to the task rather than collaborate through sustained dialogue about the concepts to be learned.

While cooperation itself has value, it is often used for teaching group-process skills and has not been viewed as the most effective strategy for adult learners. "The cooperative learning process might, in striving to achieve very interdependent group functioning, worsen the distant learners' constraints . . . The collaborative approach for its part, seems to be more flexible and meets the requirements of distance education for adults" (Henri & Rigault, 1996, p. 50). Adults who already have the skills for coordinating their face-to-face activities in an efficient way may be spending so much time on coordination of tasks at a distance that they are not learning the content that the task was originally designed to teach.

Yet Group Blue did choose to cooperate rather than collaborate in this instance. Perhaps cooperation is simply more efficient at a distance and multiple perspectives can be shared through individual contributions to the product in a way that is different than what was anticipated. Dillenbourg et al. (1996) point out that "collaboration is in itself neither efficient or inefficient . . . it is the aim of research to determine the conditions under which collaborative learning is efficient" (p. 8). It is hoped that this chapter provides guidance

Copyright © 2005, Idea Group Inc. Copying or distributing in print or electronic forms without written permission of Idea Group Inc. is prohibited.

for researchers and educators in their design of online learning activities to foster collaboration, in their identification of intended outcomes of online collaborative activities, and in their use of the CMDA approach to determine whether the outcomes have been met.

References

Armitt, G., Slack, F., Green, S., & Beer, M. (2002). *The development of deep learning during a synchronous collaborative on-line course.* Paper presented at the CSCL 2002, Boulder, CO.

Arvaja, M., Hakkinen, P., Etelapelta, A., & Rasku-Puttonen, H. (In press). Social processes and knowledge building in project-based face-to-face and virtual interaction. In J. Bobry, A. Etelapelto, & P. Hakkinen (eds.), *Collaboration and learning in virtual environments.* University of Jyvaskyla: Institute for Educational Research.

Bandura, A. (1971). *Social learning theory.* New York: General Learning Press.

Bernard, R. M. & Lundgren-Cayrol, K. (2001). Computer conferencing: An environment for collaborative project-based learning in distance education. *Educational Research and Evaluation,* 7(2-3): 241-261.

Bonk, C. J. & Cunningham, D. (1998). Searching for learner-centered, constructivist, and sociocultural components of collaborative educational learning tools. In C. J. Bonk & K. S. King (eds.), *Electronic collaborators: Learner-centered technologies for literacy, apprenticeship, and discourse,* pp. 25-50. Mahway, NJ: Erlbaum.

Bonk, C. J. & Kim, K. A. (1998). Extending sociocultural theory to adult learning. In C. M. Smith & T. Pourchot (eds.), *Adult learning and development: perspectives from educational psychology,* pp. 67-88. Mahwah, NJ: Lawrence Erlbaum.

Bonk, C. J. & Wisher, R. A. (2000). *Applying collaborative and e-learning tools to military distance learning. A research framework.* (Technical Report 1107). Alexandria, VA: U.S. Army Research Institute for the Behavioural and Social Sciences.

Chandler, H. E. (2001). The complexity of online groups: A case study of asynchronous distributed collaboration. *ACM Journal of Computer Documentation,* 25(1); 17-24.

Condon, S. L. & Cech, C. G. (1996). Discourse management strategies in face-to-face and computer-mediated decision making interactions. *Electronic Journal of Communication,* 6(3), Available at: *http://www.cios.org/getfile/CONDON_V6N396.*

Curtis, D. & Lawson, M. (2001). Exploring collaborative online learning. *Journal of Asynchronous Learning Networks,* 5(1): 21-34.

Davidson-Shivers, G. V., Muilenburg, L. Y., & Tanner, E. J. (2001). How do students participate in synchronous and asynchronous online discussions? *Journal of Educational Computing Research,* 25(4): 351-366.

Dillenbourg, P., Baker, M., Blaye, A., & O'Malley, C. (1996). The evolution of research on collaborative learning. In E. Spada & P. Reiman (eds.), *Learning in humans and*

Copyright © 2005, Idea Group Inc. Copying or distributing in print or electronic forms without written permission of Idea Group Inc. is prohibited.

machine; Towards an interdisciplinary learning science, pp. 189-211. Oxford, UK: Elsevier.

Fisher, M. & Coleman, B. (2001-2002). Collaborative online learning in virtual discussions. *Journal of Educational Technology Systems,* 30(1), 3-17.

Francis, G. & Hunston, S. (1992). Analysing everyday conversation. In M. Coultard (ed.), *Advances in spoken discourse analysis,* pp. 123-161. London: Routledge.

Hara, N. & Kling, R. (2000). Students' distress with a web-based distance education course. *Information, communication & society,* 3(4).

Harasim, L. M. (1993). Networlds: Networks as social space. In L. M. Harasim (ed.), *Global networks: Computers and international communication,* pp. 15-34. Cambridge, MA: MIT Press.

Hathorn, L. G. & Ingram, A. L. (2002). Online collaboration: Making it work. *Educational Technology,* 42(1), 33-40.

Henri, F. (1992). Computer conferencing and content analysis. In A. R. Kaye (ed.), *Online education: Perspectives on a new environment,* pp. 115-136. New York: Praeger.

Henri, F. & Rigault, C. (1996). Collaborative distance education and computer conferencing. In T. T. Liao (rd.), *Advanced educational technology: Research issues and future potential,* pp. 45-76. Berlin: Springer-Verlag.

Herring, S. C. (1993). Gender and democracy in computer-mediated communication. *The Electronic Journal of Communication/La Revue Electronique de Communication,* 3(2): 1-29.

Herring, S. C. (1996). Two variants of an electronic message schema. In S. C. Herring (ed.), *Computer-mediated communication: Linguistic, social and cross-cultural perspectives,* pp. 81-106. Amsterdam: John Benjamins.

Herring, S. C. (2001). Computer-mediated discourse. In D. Schiffrin, D. Tannen & H. Hamilton (eds.), *Handbook of discourse analysis,* pp. 612-634. Oxford, UK: Blackwell.

Herring, S. C. (In press). Computer-mediated discourse analysis: An approach to researching online behavior. In S.A. Barab,. R. Kling & J. H. Gray (eds.), *Designing for virtual communities in the service of learning.* New York: Cambridge University Press.

Herring, S. C. & Nix, C. G. (1997,). *Is 'serious chat' an oxymoron? Pedagogical vs. social uses of Internet Relay Chat.* Paper presented at the American Association of Applied Linguistics, March 11, Orlando, Florida.

Hirumi, A. & Bermudez, A. (1996). Interactivity, distance education, and instructional systems design converge on the information superhighway. *Journal of research on computing in education,* 29(1): 1-16.

Howell-Richardson, C. & Mellar, H. (1996). A methodology for the analysis of patterns of participation within computer mediated communication courses. *Instructional Science,* 24: 47-69.

Johnson, D. W. & Johnson, R. T. (1996). Cooperation and the use of technology. In D. H. Jonassen (ed.), *Handbook of research for educational communications and technology,* pp. 1017-1044. New York: Simon and Schuster Macmillan.

Copyright © 2005, Idea Group Inc. Copying or distributing in print or electronic forms without written permission of Idea Group Inc. is prohibited.

Johnson, D. W., Johnson, R. T., & Smith, K. A. (1991). *Cooperative learning: Increasing college faculty instructional productivity* (ASHE-ERIC Higher Education Report #4). Washington, DC: The George Washington University, School of Education and Human Development.

Jonassen, D., Davidson, M., Collins, M., Campbell, J., & Haag, B. B. (1995). Constructivism and computer-mediated communication in distance education. *The American Journal of Distance Education,* 9(2): 7-26.

Kanuka, H. & Anderson, T. (1998). Online social interchange, discord and knowledge construction. *Journal of Distance Education,* 13(1): 57-74.

Kearsley, G. (2000). *Learning and teaching in cyberspace,* Retrieved online September 4, 2000, from: *http://home.sprynet.com/~gkearsley/chapts.htm.*

Kiesler, S., Siegel, J., & McGuire, T. W. (1984). Social psychological aspects of computer-mediated communication. *American Psychologist,* 39(10): 1123-1134.

Kitchen, D. & McDougall, D. (1998-9). Collaborative learning on the Internet. *Journal of Educational Technology Systems,* 27(3): 245-258.

Krippendorff, K. (1980). *Content analysis: An introduction to its methodology.* London: Sage Publications.

Moore, M. G. (1989). Three types of interaction. *The American Journal of Distance Education,* 3(2): 1-6.

Motteram, G. (2001). The role of synchronous communication in fully distance education. *Australian Journal of Educational Technology,* 17(2): 131-149.

O'Malley, C. (1991). Editor's preface. In C. O'Malley (ed.), *Computer supported collaborative learning,* pp. iv-vii. Berlin: Springer-Verlag.

Roblyer, M. D. & Ekhaml, L. (2000). How Interactive are YOUR distance courses? A rubric for assessing interaction in distance learning. *Online Journal of Distance Learning Administration,* 3(3).

Roschelle, J. & Pea, R. (1999). Trajectories from today's WWW to a powerful educational infrastructure. *Educational Researcher,* June-July, 22-25.

Roschelle, J. & Teasley, S. (1995). The construction of shared knowledge in collaborative problem-solving. In C. E. O'Malley (ed.), *Computer supported collaborative learning.* pp. 69-97. Heidelberg: Springer-Verlag.

Schrage, M. (1990). *Shared minds: The new technologies of collaboration.* New York: Random House.

Schrum, L. & Berge, Z. L. (1997). Creating student interaction within the educational experience: A challenge for online teachers. *Canadian Journal of Educational Communication,* 26(3): 133-144.

Simonson, M. (2000). Myths and distance education: What the research says (and does not say). *The Quarterly Review of Distance Education,* 1(4): 277-279.

Simonson, M., Smaldino, S., Albright, M., & Zvacek, S. (2000). Teaching and learning at a distance: Foundations of distance education. Upper Saddle River, NJ: Prentice-Hall, Inc.

Copyright © 2005, Idea Group Inc. Copying or distributing in print or electronic forms without written permission of Idea Group Inc. is prohibited.

Sproull, L. & Kiesler, S. (1991). *Connections: New Ways of Working in the Networked Organization*. Cambridge, MA: Massachusetts Institute of Technology.

Stacey, E. (1999). Collaborative learning in an online environment. *Journal of Distance Education*, 14(2).

Stewart, C. M., Shields, S. F., & Sen, N. (1998). Diversity in on-line discussions. *The Electronic Journal of Communication/La Revue Electronique de Communication*, 8(3/4).

Vrasidas, C. & McIsaac, M. S. (1999). Factors influence interaction in an online course. *The American Journal of Distance Education*, 13(3): 22-36.

Vygotsky, L. S. (1978). *Mind in society*. Cambridge, MA: Harvard University Press.

Wagner, E. D. (1994). In support of a functional definition of interaction. *The American Journal of Distance Education*, 8(2): 6-29.

Weber, H. L. (forthcoming). Missed cues: How disputes can socialize virtual newcomers. In S. C. Herring (ed.), *Computer-mediated conversation*. Greenskill, NJ: Hampton Press.

Appendix: Learning Task

The cognitive information-processing (CIP) model of learning seeks to describe how humans transform information into knowledge. Toward the end of Chapter 3, the authors outline some implications of CIP theory for instruction. One of these is "enhancing learners' self-control of information processing." The idea here is that if learners understand how they learn the best (how they process information), they can use this metacognitive awareness to create strategies for their learning.

One tool to help raise metacognitive awareness is the Motivated Strategies for Learning Questionnaire (MSLQ). Your task for this week's thought activity is to visit the URL below and take the MSLQ.

Then with your group, analyze 1) how the MSLQ addresses elements of the CIP theory of learning (in other words, how might the study strategies indicated tie to the basic processing model of CIP); and 2) how the MSLQ may be useful for instruction; and 3) include a short reflection on one strategy your group thinks might be useful for this course.

Some things to consider in the analysis:

1.   Does the MSLQ ask questions related to pattern recognition and perception? Rehearsal or chunking? Semantic networks?

2.   How do the study strategies relate to short-term memory, long-term memory, etc?

3.   Do you think this information could potentially be useful for learners? Any age restrictions? What do you think an instructor should do with this information? Is it useful to them? How?

Copyright © 2005, Idea Group Inc. Copying or distributing in print or electronic forms without written permission of Idea Group Inc. is prohibited.

4.   Were you ever taught study strategies? How? If not, did you find yourself inventing them? Are they similar or different to the ones on the MSLQ?

How this thought activity will be assessed:

1.   Please limit your analysis to 3-4 pages.
2.   Support your analysis with evidence from the readings.
3.   Make sure you consider the groups of items. You don't necessarily have to address each and every item on the MSLQ. But, how do the groups of items relate to the learning theory?
4.   Include your group or self-reflection on strategies suggested above. Have you tied this analysis to your readings?

Individual self-reflection:

When you complete the project, please answer the following questions as a reflection on your own learning process:

1.   What resources, people, websites, etc., did you find helpful in completing this project? How did you use the resources available to you?
2.   How did your understanding of the material in this unit change through the learning activities?

Peer evaluation:

Briefly describe your individual role on the thought activity you just completed. Also provide a concrete summary of each team member's contributions to the activity along with your numeric rating. Refer to the descriptions below as you make your ratings. Evaluate the contribution of EACH of your project team members, including yourself, on a scale from 1 to 5.

0 =   team member made no visible contributions to the project OR made significant and sustained negative contributions to the project
1 =   team member made minimal contributions to the overall project
2 =   team member made uneven contributions to the project - some positive, some negative
3 =   team member made reasonable contributions to the project
4 =   team member made significant and sustained positive contributions to the project
5 =   team member made significant and sustained positive contributions to the project AND supported every member of the group by actively bringing out the best in others.

Copyright © 2005, Idea Group Inc. Copying or distributing in print or electronic forms without written permission of Idea Group Inc. is prohibited.

Chapter VI

Mapping Perceived Socio-Emotive Quality of Small-Group Functioning

Herman Buelens
Katholieke Universiteit Leuven, Belgium

Jan Van Mierlo
Katholieke Universiteit Leuven, Belgium

Jan Van den Bulck
Katholieke Universiteit Leuven, Belgium

Jan Elen
Katholieke Universiteit Leuven, Belgium

Eddy Van Avermaet
Katholieke Universiteit Leuven, Belgium

Abstract

This chapter demonstrates the influence of the socio-emotional quality of small-group functioning in a collaborative learning setting. It reports a case study from a sophomore class at a Belgian university. The subjects were 142 undergraduates subdivided into 12 project groups of about 12 students each. Following a description of the collaborative learning setting, a longitudinal survey study focusing upon the evolution of the learners' perception of their own group's socio-emotional functioning is presented. The aims of the study were to map group members' perception of the socio-emotive quality of their own group functioning and to examine if and how problems in groups

Copyright © 2005, Idea Group Inc. Copying or distributing in print or electronic forms without written permission of Idea Group Inc. is prohibited.

of learners can be detected as soon as possible. Having demonstrated that
dysfunctionalities within groups can be detected rather early, the authors hope that
corrective interventions can be implemented when they can still have an effect.

Introduction

Students who collaborate in small groups on a common research project have abundant
opportunities to present and discuss ideas and to plan, organize, and carry out activities
related to the task at hand. Several authors attribute a long list of potential benefits to
the richness and the diversity of these learner activities. Because a collaborative learning
environment actively involves students in the learning process, educational theorists
believe that collaborative settings such as small project groups of co-learners are an
effective means of learning, and they therefore play an important role in knowledge
construction (Blumenfeld, Marx, Soloway, & Krajcik, 1996; Collins, Brown, & Newman,
1989). By expressing ideas into words, by formulating opinions, by externalizing tacit
knowledge, attitudes, approaches, values, and perspectives, learners are expected to
explore their own understanding in more detail (Johnson, 1971, 1974), to generate more
and better questions (Panitz, no date) and to develop higher level thinking skills
(Johnson, 1971; Vygotsky, 1978). It is hoped that vague mental conceptualizations of an
idea become internalized into more concrete representations (Resnick, Levine, & Teasley,
1991) resulting in a long-lasting, firmly rooted understanding (Kulik & Kulik, 1979).

Because cognitive activities of learners become visible during group work, these
activities also become subject to intervention and coaching. Hence, the externalized
ideas of the learner provide a means for other learners and their teachers to react to,
negotiate around, and build upon what they heard from the learner's side (Arias, Eden,
Fischer, Gorman, & Scharff, 1999). Consequently, the conceptualizations of co-learners
will gradually become fine-tuned and a common language and a common understanding
-or a "shared knowledge"- will be created (Scardamalia & Bereiter, 1994). Important as
they are, the cognitive benefits listed above are but a small portion of the advantages
attributed to collaborative learning. Panitz (no date), for example, presents a referenced
list of 67 theoretical advantages of collaborative learning, ranging from academic over
social to psychological and assessment benefits. Not unimportantly, some of the
cognitive benefits believed to be associated with collaborative learning have already
received direct empirical support. To illustrate, two recent reviews are positive with
regard to the effectiveness of various forms of small-group learning. Springer, Stanne,
and Donovan (1999) conclude that small-group learning is successful in promoting
greater academic achievement and more favorable attitudes toward learning. According
to the authors, these results are superior to most findings in comparable reviews of
research on other educational innovations. Comparing small-group and individual
learning in a context in which students learn to use computer technology, Lou, Abrami,
and d'Apollonia (2001) found significant positive effects of small-group learning on
student individual achievement, task performance, and several process and affective
outcomes. In view of the overwhelming number of theoretical arguments and of the
empirical support for the cognitive benefits associated with collaborative learning, it

Copyright © 2005, Idea Group Inc. Copying or distributing in print or electronic forms without written
permission of Idea Group Inc. is prohibited.

would therefore appear as if there is every reason to promote collaborative instructional formats.

However, this enthusiasm regarding collaborative work environments is not shared by everyone. Diehl and Stroebe (1987; 1991), for example, notice that several forms of "cognitive blocking" can hinder the cognitive processes of individuals, and mainly so during face-to-face synchronous communication sessions within a group. While brainstorming, some group members are talking too fast for others to react upon, theses are remembered imprecisely or they are quoted incorrectly, irrelevant or long meandering monologues enter group discussions, etc. These interactions interfere with and disrupt ongoing cognitive processes, thereby thwarting the learning outcomes intended by having students communicate with each other.

Aside from cognitive blocking effects, which are perhaps only detrimental with regard to individual learning outcomes in the short run, more serious and longer lasting negative effects of group work have also been described. Bales (1953) noticed that instrumental, task-related activities within a group of co-acting people cannot be considered apart from the socio-emotive context in which these activities take place. In the same vein, others have pointed out that collaborative work can have but little effect on students' learning outcomes, because teams (of collaborating learners) can fall prone to a long list of social inhibiting factors that impede participants from performing effectively (Brown, 2000; Hertz-Lazarowitz, Benveniste Kirkus & Miller., 1992;McGrath, 1984; Paulus, 2000; Paulus, Dugosh, Dzindolet, Coskun, & Putman, 2002; Salomon & Globerson, 1989). "Social inhibition" can result from group members' tendency to make self-favoring social comparisons by contrasting their own contributions with those of (somewhat) less performant group members (i.e., "downward comparison"). The resulting belief that one is doing quite well (an "illusion of productivity") may further inhibit the efforts exerted by group members (Paulus, 2000, p. 242). In the worst case, the vicious circle of downward social comparison might be consolidated in a group norm prescribing low achievement. Most attention however has been paid to the empirically sound observation that group members reduce or "inhibit" individual effort when their contributions to a common group task remain unidentifiable (i.e., "social loafing" and "free riding"; e.g., Williams, Harkins & Latané, 1981). Of course, individual group members who refrain from taking responsibility in fulfilling their part of the work slow down project work itself. More detrimental however is their long-term effect upon both socio-emotional group life and upon the development of trust between group members. This is particularly regrettable, because both intra-group socio-emotional stability and trust are important antecedent conditions for group members to learn from and with each other (Bruffee, 1994). It therefore seems as if the potential benefits associated with small-group projects will be a function of the group's capability not only to cope with task-related aspects, but also to develop and to maintain a constructive socio-emotive group life.

From a teacher's point of view, the question arises "How to coach a group of collaborative learners adequately?" Successive preventive, diagnostic, and curative actions might be considered. As an initial "preventive" step, teachers can try to design the collaborative environment such that the opportunities for a group to deal successfully with both task-relevant and socio-emotional aspects of group life are maximized. A deliberately designed collaborative environment, however, does not guarantee that all groups will do well. Therefore, teachers need a subsequent (second) evaluative phase in which groups that

Copyright © 2005, Idea Group Inc. Copying or distributing in print or electronic forms without written permission of Idea Group Inc. is prohibited.

go astray will be detected. Although both are necessary, the follow up of task-related group activities is a notably easier job than adequately scrutinizing socio-emotional and intra-group relational patterns.

The difficulty of monitoring socio-emotional aspects of group life is at the heart of a case study from a sophomore class at a Belgian university, the Media Studies Seminar (MSS), presented hereafter. First, the seminar itself will be introduced and some attention will be given to elements of the design that were explicitly incorporated in order to help groups deal successfully with both task-relevant and socio-emotional aspects of group life (i.e., the "preventive" step). Next, a longitudinal survey focusing upon the evolution of the learners' perception of their own group's socio-emotional functioning will be presented. The aims of the study were to map group members' perception of the socio-emotive quality of their own group functioning and to examine if and how problems emerging in a partly face-to-face, partly virtual group of learners can be detected as soon as possible (i.e., the "diagnostic" step).

In the present study, no attempt was made to proceed to the "curative" step based on the data gathered. Groups were thoroughly coached as usual, but the coaches were not informed about the survey data. The aim was to map the spontaneous evolution of the perceived quality of group functioning in a context where coaches cannot but count upon their experience and devotion to optimize in-group activity. It will be clear, however, that in the future, survey data will be put at the disposal of both the coaches and the groups, if it would turn out that this "diagnostic" information might constitute a useful instrument to guide "curative" interventions.

The Media Studies Seminar

The MSS is one of the ten courses students have to take in the second year of the undergraduate Communication Sciences program at the largest Belgian university (Katholieke Universiteit Leuven). The MSS takes the format of an ICT-supported business simulation covering the full academic year. It aims to make students familiar with empirical research in communication sciences. Students have to acquire the basic skills necessary to investigate a new problem within this science discipline independently, and they have to be able to deliver a final report of good quality. At the start of the seminar, students can indicate which of the approximately 15 available research topics they would like to work on (e.g., how do parents coach children in their media use; romance, relationships, and sexuality in popular TV shows; the meaning of media for the visually impaired, etc.). Taking into account their personal preferences, about twelve students with common interests are put together into the same project groups, and they will work together at the project during the full academic year. Since they have already spent one year together, most students will know each other. Typically, students within a project group met several times a week, each time for a period ranging from a few minutes up to several hours.

At the start of the project, each project group has to submit a research proposal in the form of a detailed business tender (including a time schedule, a budget, and staffing plan).

Copyright © 2005, Idea Group Inc. Copying or distributing in print or electronic forms without written permission of Idea Group Inc. is prohibited.

Following the approval of the business tender by the team of project coaches, each project group is subdivided into four smaller units. Every unit holds the main responsibility for group work during one of the four major stages of the MSS. In a first stage, students explore the available literature on the subject, and a central research question is derived. In a second phase, students construct a research instrument (e.g., a survey, a tool to analyze newspaper content). In a third stage, the actual research is carried out (e.g., interviewing people, analyzing content, conducting a telephone survey), after which the data collected are analyzed. In a final stage, a research report is written, and all the project groups present their own project to the other groups during a simulated academic conference. The latter activity concludes the business simulation.

Since the MSS was the first experience of these students with both empirical research and collaborative group work at the university, great efforts were made to help them to have a fruitful learning experience. First, attractive, professionally relevant, socially meaningful, and motivating research topics were presented (see the examples above). In addition, students were asked to apply for a specific topic. As a result, student motivation was enhanced, complaints about unfair allocation of topics were avoided and, perhaps most important, students knew that other group members would also be interested in the topic. Second, great care was taken to ensure that groups could start work as soon as possible. Therefore, all groups were provided with a written rationale covering all the stages of the group work. In a first collective meeting, this rationale was explained in detail. It was explicated why collaborative group work is required for this project and what learning outcomes it was hoped would be achieved. Deadlines and formal requirements were indicated. It was made clear how group work and individual contributions would be assessed (i.e., all subjects within a group will receive the same mark that can be slightly adjusted by means of a peer assessment procedure). Hints and helpful resources were added. It was explained what to do if the group had difficulties, what the potential risks of group work are and how to deal with them. An overview of when and where to meet with the teachers was included. Third, by partitioning group work into mutually connected sub-tasks, and by advising about role and turn taking within subgroups, the stage was set to create a fair division of labor, to install a relatively high degree of positive interdependence, and to keep the whole project manageable for the students. Fourth, to enhance individual accountability and responsibility, the task was subdivided in smaller units, and each student's contribution within the group was assessed by his or her peers at four points during the academic year. Fifth, group project work was sustained on a continuous basis by means of a Digital Learning Environment (DLE). Functionalities that enhance information delivery and information exchange between learners (such as digital drop boxes, group pages, and group calendars) were promoted when it came to writing reports and planning group activities. The use of asynchronous communication tools (such as group email and group discussion forums) was encouraged to prepare (and follow up) regular face-to-face meetings. Besides facilitating group work in a direct way, having a virtual group space at one's disposal was also intended to enhance a feeling of belonging to a group. Via the promotion of the use of asynchronous communication tools, teachers hoped to provide opportunities for students to collaborate in ways that lead to shared understanding (e.g., Brown, 1990; Harasim, 1990; Hiltz, 1990), and they hoped to prevent several forms of "cognitive blocking" (cf. supra). Finally, a great deal of monitoring and coaching moments was embedded in the design of the MSS. At

Copyright © 2005, Idea Group Inc. Copying or distributing in print or electronic forms without written permission of Idea Group Inc. is prohibited.

designated times, relatively informal meetings were organized with each group, and individual group members were free to contact their teachers at all times. On more formal occasions, small oral presentations including a report of group progress were scheduled. As indicated earlier, group members had to assess the contribution of each group member. These peer-evaluation data were also used by teachers as a monitoring tool.

However, despite all the preventive measures taken, year after year it turned out that about one-third of the groups suffered from an inferior socio-emotional atmosphere. Moreover, despite attempts to monitor groups closely, instructors found it hard to judge the socio-emotive aspect of group functioning correctly. After all, instructors always remain relative outsiders. In addition, groups of learners remained highly reluctant to report emerging problems in their group, partly perhaps because they worried about losing marks if teachers discovered that a group was confronted with difficulties that could not be solved by the group itself. Only at the end of the academic year did some individual students start complaining about how their group had been or was doing. At that late stage, teachers ran the risk of misjudging the complaint. Moreover, even when a correct diagnosis of the complaint would have been possible, no time was left for curative actions to be implemented.

Therefore, at the beginning of the academic year 2001-2002, we started investigating if and to what extent relational group (dys)functioning can be mapped at the very early stages of group work by means of an ad hoc constructed measurement tool. This diagnostic instrument (described below) is a rather broad-spectrum questionnaire reflecting socio-emotional aspects of group membership, as well as perceptions, evaluations, and feelings about the group as a whole, its members, and the student's own membership within the group. By administering the online questionnaire to group members at set times, it is our aim to obtain an evolutionary diagnostic group profile.

Mapping Perceived Socio-Emotive Quality of Group Functioning

In this section, first the measurement tool and the data collection process will be highlighted. Next, we will turn to a report and an analysis of the results obtained.

Measurement Tool and Data Collection

During the academic year 2001-2002, the MSS was attended by 142 second-year Communication Science undergraduates at the K.U.Leuven. Taking into account individual student preferences, 12 different project groups of 12 students each were formed. Each project group was subdivided into four sub-units of three students each. Every 1.5 months (November 2001, February 2002, March 2002, May 2002), following the completion of each major stage of the MSS, an online questionnaire (81 items) was administered to all 142 students. The questionnaire related to the project group as a

Copyright © 2005, Idea Group Inc. Copying or distributing in print or electronic forms without written permission of Idea Group Inc. is prohibited.

whole1 (12 students), and it comprised ten existing scales measuring different aspects of the quality of group functioning: "Interaction" (Watson et al., 1991; 8 items), "Equal Contribution" (Kramer, Kuo, & Dailey, 1997; 11 items), "Discussion Quality" (Kramer et al., 1997; 3 items), "Dominance" (Kramer et al., 1997; 2 items), "Solidarity" (Wheeless, Wheeless, & Dickson-Markman, 1982; 13 items), "Affect" (Freeman, 1996; 6 items), "Fairness of Equal Scores" (Freeman, 1996; 2 items), "Fairness of Contribution" (Freeman, 1996; 3 items), "Waste of Time" (Freeman, 1996; 3 items), "Surplus Value of Group Work" (Freeman, 1996; 6 items), together with some items that were constructed to indicate "Illusion of Productivity" (5 items), "Free Riding" (4 items), "Downward Comparison" (4 items), and "Within group communication" (11 items).

A few examples of questions are: "I am satisfied with how group members interact with each other"; "I feel we have good communication among group members"; and "Every member of our group deserves the same final grade." All 81 items were scored on a common six-point scale (1=*strongly disagree*; 6=*strongly agree*). Since the questionnaire was completed four times by each of the 142 subjects, a data matrix consisting of 142 subjects by 4 measurements by 81 items was obtained.

Analysis & Results

Socio-Emotive Quality of Group Functioning

The data matrix was restructured in a two-way table consisting of 568 rows (142 students x 4 measurements) and 81 columns (scores on 81 items). To detect like patterns of socio-emotive quality of group functioning (i.e., data within one row), a cluster analysis (Ward's method; squared Euclidian distances) was performed on the rows of the two-way table. The analysis clearly categorizes perceptions of students (at a set moment) in two distinct clusters. One "cluster" or "class" consists of those students who indicated their group was doing well during the preceding 1.5 months (the "functional" cluster). A second "cluster" contains those students who indicated that they were rather dissatisfied with their group and the way it was functioning during the preceding 1.5 months (the "dysfunctional" cluster).

Students in the "functional" cluster perceived their group as a coherent and harmonious entity and indicated that they performed more efficiently than if there were no groups (during the preceding 1.5 months). They believed that their interactions resulted in decisions of good quality. Group work was not perceived as a waste of time, and students were satisfied with both the final result of the group work and with the way group members interacted with each other. Students had the perception that all group members contributed evenly, that there were neither distinctly dominant group members nor free-riders. They judged it as fair that everyone in their group would receive the same score. Students in the "dysfunctional" cluster showed the reverse pattern.

Next, for each of the four periods preceding a measurement, the relative number of students in the "functional" cluster was used as an index of the perceived quality of socio-emotive quality of a group during that period. As it turned out, some groups consist exclusively of subjects from the "functional" cluster (see Figure 1: Group 1 before

Copyright © 2005, Idea Group Inc. Copying or distributing in print or electronic forms without written permission of Idea Group Inc. is prohibited.

Figure 1: The percentage of group members in the functional cluster at the four measuring moments.

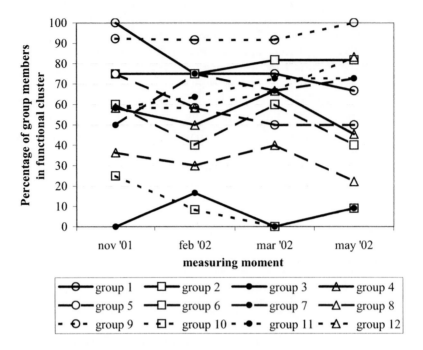

November 2001), while other groups consist only of students from the "dysfunctional" cluster (see Figure 1: Group 3 before November 2001). Clearly, the socio-emotive quality within such a group is very low. Furthermore, Figure 1 clearly shows that, by and large, the "perceived quality of group functioning" remained constant over the academic year. Groups that did not score well after the first stage of the project (November 2001) generally were classified as "dysfunctional" after completion of the other three stages as well. Similarly, groups that started well remained "functional" during the remainder of the project. That is not to say that no changes at all were observed. As can be seen in Figure 1, the most marked changes in socio-emotive quality of group functioning (Group 1 and Group 7) are noticed between the first (November 2001) and the second (February 2002) measurement.

Relation between Socio-Emotive Quality of Group Functioning and "Getting On With the Job"

At the end of the academic year, the final reports of the groups were graded by the faculty member responsible for the MSS, in consultation with the groups' instructors. It is interesting to observe that the two "dysfunctional" groups (Group 3 and Group 10) were the only groups failing to score higher than 10 on a 20 point scale. On the other hand, the most functional groups scored best on their final report. This important result

Copyright © 2005, Idea Group Inc. Copying or distributing in print or electronic forms without written permission of Idea Group Inc. is prohibited.

challenged us to look into the relationship between the socio-emotional and the task-related aspects of group functioning. As indicated above (Bales, 1953), getting on with the job and getting on with other people within the group seem essential for delivering a good final group result. The correlation between both was investigated.

The final grade on the seminar groups' reports was taken as an index of successfully coping with the job. For "getting on with people," it was assumed that the percentages of group members who belonged to the functional cluster were an adequate measurement unit. A Spearman Correlation between both scores showed a substantial relationship between "getting on with the job" and "getting on with your fellow team members." The correlation was $r = 0.7$, $p < 0.0001$. Project groups with a lot of students in the dysfunctional cluster (groups scoring low on "getting on with people"), consequently did not score as well for their final report as groups in which more students say that their group is functional.

Academic Achievement

One obvious factor that might moderate the observed relationship between socio-emotional and task-related aspects of group functioning is the student's level of academic achievement. Students' results, at the end of their first undergraduate year,

Figure 2: Estimated marginal means of the socio-emotional indices for the twelve groups at the four measurement moments.

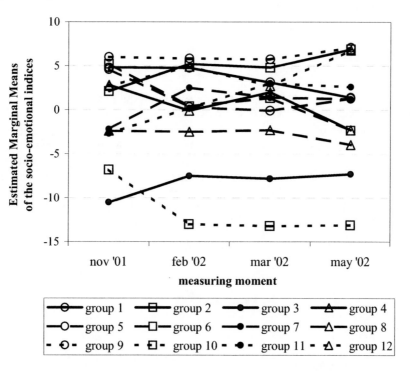

Copyright © 2005, Idea Group Inc. Copying or distributing in print or electronic forms without written permission of Idea Group Inc. is prohibited.

were taken as an index of academic achievement. In order to test the potentially moderating role of "academic achievement," a repeated measurement ANOVA2 was done using the subjects' academic achievement as a covariate, and with the four different moments of measurement as a within subjects or time factor (socio-emotional functioning), and group membership as a between subjects factor (having 12 levels). The within subjects or time factor was not at all significant ($F (2.44, 290.18) = 0.96$), indicating that —as expected— there were no differences between the indices at the four different moments. More important, the covariate academic achievement also did not reach any significance level ($F (1, 119) = 0.65$). This confirms our thesis that the differences in socio-emotional indices between the groups cannot be attributed to differences in overall academic achievements between these groups. The between subjects factor of group functioning, however, was highly significant ($F (11, 119) = 6.52, p < .001$), which is clearly illustrated in Figure 2. The estimated marginal means of the socio-emotional indices are clearly different between the 12 groups. Yet, most of them are situated between 0 and 5. Furthermore, there was a significant interaction between group functioning and the repeated measures time factor ($F (26.82, 290.18) = 4.10, p < .001$). When inspecting the figure, it can be seen that there is a great deal of variation in the fluctuations of the curves between the 12 groups. In our opinion, this interaction should not be over-interpreted. There are no clear patterns of differences between groups of either increased or decreased estimated marginal means over time. There are differences and crossings between curves, but for all groups, the estimated means remain at the same level in a rather horizontal manner. (The repeated measures or time factor was not at all significant.)

Summarizing, we feel that the differences in the groups' socio-emotional functioning are not influenced by the overall academic achievement of the members of these groups.

Discussion

Bringing about successful group work is not just a matter of putting students together. Students do not automatically become involved, thoughtful, tolerant, cooperative, or responsible when working with others. The ultimate learning effect of group work depends on what the tasks are, how the group is organized, who participates, and how the group is held accountable. Teachers must consider these factors in designing group work, and they must address potential problems of process if group work is to be successful. However, explaining the task and guiding the groups through their project, etc., do not, by themselves, seem to give any guarantee for a successful project. In this chapter, it was demonstrated that in collaborative learning, there is also a clear influence of the socio-emotional quality of group functioning.

Copyright © 2005, Idea Group Inc. Copying or distributing in print or electronic forms without written permission of Idea Group Inc. is prohibited.

Findings

"Stability"

One of the most astonishing results of our research is that, by and large, the "perceived quality of group functioning" remained constant over the academic year. Groups that did not score well after the first stage of the project generally were classified as "dysfunctional" after completion of the other three stages as well. Similarly, groups that started well remained "functional" during the rest of the project (see Figure 1). However, and against this general trend, some relatively small variations in socio-emotional quality were observed going from the first (November 2001) to the second (February 2002) measurement. In the language of group-development theories, perhaps groups are leaving (see Figure 1, Group 7) or entering (see Figure 1, Group 1) the "storming stage" in which conflict is the general theme and in which task resistance (such as missed meetings or reduced task focus) and relationship hostility between group members are common (Tuckman & Jensen, 1977).

"Applicable Instrument"

Having demonstrated that dysfunctional groups can be detected rather early using a questionnaire that measures different aspects of group functioning, an optimistic perspective is being offered here. Interventions can be planned at a time when they can still have an effect. Moreover, our rather elaborate questionnaire and the resulting group profile offer a first diagnosis of the (dys)functionality within a specific group.

Pending a more detailed analysis of the dysfunctions observed, simple feedback from the teacher, a group discussion, coaching, and pre-training for cooperation (listening and resolving conflicts; learning to appreciate skills and abilities of group members) constitute examples of potentially useful intervention strategies.

Future Research

Based on our research, which indicates that dysfunctional groups can be detected at a very early stage of group functioning, we suggested that interventions should also begin early in the process. Although it can be argued that early interventions stand a better chance of being successful than late interventions, in view of the stability findings, we don't know whether interventions will have an effect at all. Is there really a way to overcome these primacy effects and these stability effects? Moreover, there are different ways for a teacher to intervene in dysfunctional groups. Is simple feedback based on the questionnaire data collected enough, or will more radical coaching or collaborative skills training be necessary? Future research may give us some indication. Also, further research is needed at the level of the task analysis. There seems to be a serious shortage of models of task analysis in collaborative learning. Finally, our admittedly superficial

Copyright © 2005, Idea Group Inc. Copying or distributing in print or electronic forms without written permission of Idea Group Inc. is prohibited.

analysis of the relationship between the socio-emotional functioning of project groups, their "ability" and the groups' results should be supplemented by a more fine-grain analysis.

Limitations and Conclusions of the Study

This study obviously has a number of limitations. First, it is important to note that this chapter deals with rather large groups (12 people) of peers working together in a research seminar in media training. Although many of the factors involved may be exclusively in-group factors, it is obvious that factors external to the group can also be at work and have an impact on behavior inside the group. Furthermore, some of our participants may have had previously shared experiences, as they had studied together in their first under-graduate year. In addition, the students' motivation to participate in "collaboration" is partly extrinsic. Group work is not an option; it is a course requirement. Moreover, the setting of the students' collaboration is itself a potential intervening factor. Working together has become an important aspect of student life at our university. Students work together not only in study surroundings, but also in more informal surroundings such as students' residences. It can therefore be argued that the impact of the collaborative learning present in our research may differ from the impact of the setting for other forms of collaborative learning.

In addition, our research deals with very diverse forms of in-group communication. Part of the subjects' communication is asynchronous computer-mediated communication using the DLE; part is face-to-face communication. Project groups differ in their relative use of these two modes of communication. We noticed that some groups made almost no use of the DLE options. Other groups preferred to discuss their research using nothing but the DLE. The potential influence of the different communication modes was not studied in this chapter, but it provides an interesting topic for future research.

As a final limitation, although we monitored the subjects' socio-emotional functioning for nearly eight months, due to the length of this period it was almost impossible to deal with every possible factor that may have influenced the socio- emotional relations in the different project groups.

The goal of this chapter was not to argue for any particular view of the best prototypical scenario for group learning. If anything, our research made us aware of the vast differences that can be observed between collaborative settings. This is important, because what counts as collaboration within a group will differ widely. To our knowledge, there is still no agreed-upon framework to compare and to contrast studies on collabo-rative learning. It is our firm conviction, based on our findings, that task-related factors and socio-emotional factors should both occupy a place in this framework.

Copyright © 2005, Idea Group Inc. Copying or distributing in print or electronic forms without written permission of Idea Group Inc. is prohibited.

Acknowledgments

The authors express their sincere appreciation to Jos Feys and Martine Beullens for their statistical advice.

References

Arias, E.G., Eden, H., Fischer, G., Gorman, A., & Scharff, E. (1999). Beyond access: Informed participation and empowerment. In C. Hoadley & J. Roschelle (eds.), *Proceedings of the Computer Support for Collaborative Learning (CSCL) Conference*, Dec 12-15, Stanford University, Palo Alto, CA. Hillsdale, NJ: Lawrence Erlbaum Associates.

Bales, R.F. (1953). The equilibrium problem in small groups. In T. Parsons, R.F. Bales, & E.A. Shils (eds.), *Working papers in the theory of action*. New York: Free Press.

Blumenfeld, P.C., Marx, R.W., Soloway, E., & Krajcik, J. (1996). Learning with peers: From small group cooperation to collaborative communities. *Educational Researcher*, 24(11): 37-40.

Brown, J. (1990). Towards a new epistemology for learning. In C. Frasson & J. Gauthiar, (eds.), *Intelligent Tutoring Systems: At the crossroads of artificial intelligence and education*. Norwood, NJ: Ablex.

Brown, R. (2000). *Group Processes. Dynamics within and between groups,(2nd Ed.* Oxford, UK: Blackwell Publishers.

Bruffee, K. A. (1994). The art of collaborative learning: Making the most of knowledgeable peers. *Change*, 26(3): 39-44.

Collins A., Brown, S.J., & Newman, S.E. (1989). Cognitive apprenticeship: Teaching the craft of reading, writing and mathematics. In L.B. Resnick (ed.), *Knowing, learning, and instruction. Essays in honor of Robert Glaser*, pp. 453-494. Hillsdale, NJ: Lawrence Erlbaum.

Diehl, M. & Stroebe, W. (1987). Productivity loss in brainstorming groups: Toward the solution of a riddle. *Journal of Personality and Social Psychology*, 53: 497-509.

Diehl, M. & Stroebe, W. (1991). Productivity loss in idea-generating groups: Tracking down the blocking effect. *Journal of Personality and Social Psychology*, 61: 392-403.

Freeman, K.A. (1996). Attitudes toward work in project groups as predictors of academic performance. *Small Group Research*, 27 (2): 265-282.

Harasim, L. (ed.) (1990). *Online education: Perspectives on a new environment*. New York: Praeger Publishing.

Hatcher, L., Stepanski, E.J. (1994). A step-by-step approach to using the SAS(R) system for univariate and multivariate statistics. Cary, NC: SAS Institute Inc.

Copyright © 2005, Idea Group Inc. Copying or distributing in print or electronic forms without written permission of Idea Group Inc. is prohibited.

Hertz-Lazarowitz, R., Benveniste Kirkus, V., & Miller, N. (1992). Implications of current research on cooperative interaction for classroom application. In R. Hertz-Lazarowitz & N. Miller (eds.), *Interactions in cooperative groups. The theoretical anatomy of group learning,* pp. 253-280. Cambridge, UK: Cambridge University Press.

Hiltz, S. (1990). Evaluating the virtual classroom. In L. Harasim, (ed.) *Online education: Perspectives on a new environment,* pp.134-183. New York: Praeger Publishing.

Johnson, D. W. (1971). Effectiveness of role reversal: Actor or listener. *Psychological Reports,* 28: 275-282.

Johnson, D. W. (1974). Communication and the inducement of cooperative behavior in conflicts: A critical review. *Speech Monographs* 41: 64-78.

Kramer, M.W., Kuo, C.L., & Dailey, J.C. (1997). The impact of brainstorming techniques on subsequent group processes: Beyond generating ideas. *Small Group Research,* 28: 218-242.

Kulik, J.A. & Kulik, C.L. (1979). College teaching. In P.L. Peterson & H.J. Walberg (eds.), *Research in teaching: Concepts, findings and implications.* Berkeley, CA: McCutcheon Publishing.

Lou, Y., Abrami, P.C., & d'Apollonia, S. (2001). Small group and individual learning with technology: A meta-analysis. *Review of Educational Research,* 71 (3): 449-521.

McGrath, J.E. (1984). *Groups, interaction, and performance.* Englewood Cliffs, NJ: Prentice Hall.

Panitz, T. (no date). The case for student centered instruction via collaborative learning paradigms. Available online at: http://home.capecod.net/~tpanitz/tedsarticles/coopbenefits.htm.

Paulus, P.B. (2000). Groups, teams and creativity: The creative potential of idea-generating groups. *Applied Psychology: An International Review,* 49(2): 237-262.

Paulus, P.B., Dugosh, K.L., Dzindolet, M.T., Coskun, H., & Putman, V.L. (2002). Social and cognitive influences in group brainstorming: Predicting production gains and losses. In W. Stroebe & M. Hewstone (1991). *European Review of Social Psychology,* 12: 299-325.

Resnick, L.B., Levine, J.M., & Teasley, S.D. (eds.) (1991). *Perspectives on socially shared cognition.* Washington, DC: American Psychological Association.

Salomon, G. & Globerson, T. (1989). When teams do not function the way they ought to. *International Journal of Educational Research,* 57(2): 149-174.

Scardamalia, M. & Bereiter, C. (1994). Computer support for knowledge-building communities. *Journal of the Learning Sciences,* 3(3): 265-283.

Springer, L., Stanne, M.E., & Donovan, S.S. (1999). Effects of small-group learning on undergraduates in science, mathematics, engineering, and technology: A meta-analysis. *Review of Educational Research,* 69(1): 21-51.

Tuckman, B. W. & Jensen, M. A. (1977). Stages of small group development revisited. *Group and Organizational Studies,* 2 (4): 419-427.

Vygotsky, L.S. (1978) *Mind in society.* Cambridge, UK: Cambridge University Press.

Copyright © 2005, Idea Group Inc. Copying or distributing in print or electronic forms without written permission of Idea Group Inc. is prohibited.

Watson, W., Michaelsen, L.K., and Sharp, W. (1991) Member competence, group interaction, and group decision making: A longitudinal study. *Journal of Applied Psychology*, 76, 6, 803-809.

Wheeless, L.R., Wheeless, V.E., & Dickson-Markman, F. (1982). A research note: The relations among social and task perceptions in small groups. *Small Group Behavior*, 13 (3): 373-384.

Williams, K., Harkins, S., & Latané, B. (1981). Identifiability as a deterrent to social loafing: Two cheering experiments. *Journal of Experimental Social Psychology*, 40: 303-311.

Endnotes

[1]   The functioning of the smaller units of three is not the focus of this chapter. Even though a separate questionnaire for measuring the functioning of these units was used, we will not present the conclusions here.

[2]   Since the Greenhouse-Geisser epsilon was .81, which is greater than the .75 criterion proposed by Hatcher & Stepanski (1994, p. 237), there was no need to switch to a MANOVA. For the averaged tests of significance, this epsilon was used to adjust the degrees of freedom.

Copyright © 2005, Idea Group Inc. Copying or distributing in print or electronic forms without written permission of Idea Group Inc. is prohibited.

Chapter VII

A Constructivist Framework for Online Collaborative Learning:
Adult Learning and Collaborative Learning Theory

Elizabeth Stacey
Deakin University, Australia

Abstract

The purpose of this chapter is to review and discuss theoretical perspectives that help to frame collaborative learning online. The chapter investigates literature about the type of learning and behavior that are anticipated and researched among participants learning collaboratively and discusses how these attributes explain computer-supported collaborative learning. The literature about learning is influenced by perspectives from a number of fields, particularly philosophy, psychology, and sociology. This chapter describes some of these perspectives from the fields of cognitive psychology, adult learning, and collaborative group learning. Recent research into computer-supported collaborative learning that applies these theories will also be discussed.

Copyright © 2005, Idea Group Inc. Copying or distributing in print or electronic forms without written permission of Idea Group Inc. is prohibited.

Introduction

Computer-supported collaborative learning (CSCL) is an emerging paradigm gathering a research focus of discussion from a range of disciplines. Lipponen (2002), discussing the foundations of this new and emerging focus of research and its differing interpretations, traces its short history as an academic discourse and explains many of the differing concepts of collaboration that it represents. This chapter explores the theoretical background to collaborative learning, reviewed to frame and explain a research study into collaborative learning in a distributed computer-supported environment. The theoretical ideas explored here, also represented in the CSCL literature, are placed in a broader context of educational literature and discussed in detail. The chapter will focus on online collaborative learning from within an interpretive framework, from the perspective that knowledge is subjective and socially constructed. The constructivist and social constructivist viewpoints about learning and knowledge construction are presented here first through an investigation of relevant literature about constructivism.

The field of adult learning, particularly in higher education contexts, is also explored with examination of research into the nature of the facilitation of learning by group interaction and the theories that underpin this area of study. The significant contribution of the social nature of cognition as theorized by Vygotsky (1978) to the theory base underlying collaborative learning is reviewed, with the importance of dialogue within an online community of learners discussed.

Constructivist Perspectives about Learning

The literature about a constructivist approach to learning that is described here covers a diversity of ideas from cognitive developmental theory to research in adult learning, from collaborative and group learning to educational technology and instructional design. The theoretical perspectives of learning and knowledge through which these different disciplines and studies have been reviewed are the principles of constructivism and particularly social constructivism. Constructivism is perceived differently across the educational literature, ranging from being called a theory of epistemology or a theory of learning, to being described as a philosophy or approach underlying a range of theories of learning. Constructivism is considered here to be a set of beliefs about knowing that become a perspective for understanding learning.

Definitions of Constructivism

Within his discussion of autonomous and individualized adult learning, Candy (1991) described constructivism as "a broad and somewhat elusive concept" (p. 252) and wrote of the irony that the discussion about constructivism, with its multiple perspectives, has

Copyright © 2005, Idea Group Inc. Copying or distributing in print or electronic forms without written permission of Idea Group Inc. is prohibited.

emerged from the field of science, so long considered a positivist field of hard facts and laws. He quoted writers such as Feyerbrand (1975), Kuhn (1970) and Pope (1983), (cited in Candy, 1991) who wrote of science as people's multiple constructions of concepts based on a central principle, i.e., that knowledge cannot be taught but must be constructed by the learner. Many other disciplines such as anthropology, sociology, and psychology also reflected dramatic shifts in perspective about "how people invent, organize, and impose structures on their experiences, and have argued that essentially knowledge is a social artefact" (Candy, 1991, pp. 253-254).

Candy described constructivism as three interrelated domains: a constructivist view of people, a constructivist view of knowledge, and constructivism in teaching and learning. His constructivist view of people is that they are not shaped by circumstances beyond their control but continuously inquire and explore and are driven to interact with others to make sense of their experience and develop a schema for reality to guide their actions. The constructivist view of knowledge, unlike the positivist view of knowledge as "an accumulated body of empirically verifiable facts, derived directly from observation and experimentation" (Candy, 1991, p. 262), perceives its content as constructed by the learner who experienced it. This means that if knowledge is tentative and socially constructed, it cannot be taught but only learned (or constructed). Many of the constructivist ideas of learning originated from the work of cognitive psychologists such as Piaget, Bruner, and Vygotsky who shared a central notion of constructivism in which the learner had a representational model, a system of schema or personal constructs that provided an anticipatory scheme for the learner to make sense of any situation. Thus, constructivism in learning is concerned with "how learners construe (or interpret) events and ideas, and how they construct (build or assemble) structures of meaning. The constant dialectical interplay between construing and constructing is at the heart of a constructivist approach to education" (Candy, 1991, p. 272).

Another adult educator, from the field of teacher education, Fosnot (1988) developed a definition of constructivism from a Piagetian perspective, and she defined the term as having four main principles. Fosnot's first principle was that knowledge consists of past constructions; we can never know the world in a truly objective way, as if it is separate from ourselves and past experiences. Instead, we know it through our logical framework which transforms, organizes and interprets our experiences. This logic is constructed and evolves throughout our physical and cognitive development. Secondly, she described how constructions come through assimilation and accommodation, polar processes defined by Piaget (cited in Fosnot, 1988). Assimilation is our logical framework, and when it is insufficient we accommodate or develop a higher level theory or concept to encompass the new information. Thirdly, constructivism from this perspective assumes learning is an organic process of invention, not a mechanical process of accumulation. A learner-centred, active instructional model is one where the learner must construct knowledge. The teacher is a mediator, not a dispenser of knowledge. Finally, meaningful learning occurs through reflection and resolution of cognitive conflict and negates incomplete levels of understanding.

Hendry (1996) summarized a wide field of literature about constructivism, mainly from research studies in the field of math and science education, with the purpose of clarifying constructivism and identifying strategies for implementation in the classroom. Hendry drew on neo-Piagetian research to support the importance to the learners' construction

Copyright © 2005, Idea Group Inc. Copying or distributing in print or electronic forms without written permission of Idea Group Inc. is prohibited.

of meaning, of explaining their ideas and procedures to others in small groups, with the opportunity to agree and disagree. This social interaction led to children achieving higher levels of thinking than those not grouped (Kamii, 1990; Wheatley, 1991; as cited in Hendry, 1996, p. 29). Hendry quoted King's (1992) work in which he suggested that the process of explaining something to someone else led students to reconceptualize their views. This might be because they are able "to remember more acceptable knowledge because they generate and revitalize a greater variety of acceptable ideas which they have already constructed" (Hendry, 1996, p. 30). The discussion and feedback their explanations inspire may make them reconstruct their ideas as well as clarify them. Hendry described a range of teaching strategies not unlike those used in online adult learning context, based in real-life contexts, with students' questions and problems and a "problem-centred learning" process (Wheatley, 1991,) in which students were encouraged to collaborate in pairs and small groups to solve problems.

Constructivist Debate in Instructional Design

Teachers and course designers in the field of flexible and distance learning have adopted the constructivist approach to learning as an alternative to the more behavioristic model of learning that underpinned much of the earlier instructional design of distance learning materials. This paradigmatic change resulted in a debate that clearly defined the issues and understandings about the constructivist approach. The use of educational technologies, such as those used in CSCL, as a means of providing the interaction and feedback with teachers and fellow students that facilitate this way of learning means that the relationship between a constructivist approach, collaborative learning, and learning at a distance is a focus of this field.

Bednar, Cunningham, Duffy, and Perry (1992) described learning as an active process based on experience, with conceptual growth coming from sharing perspectives as well as from experience. They described the traditional objectivist view of teaching as that of transferring or communicating knowledge to the learner efficiently from a knowledge base. Such a perspective was incompatible with constructivism, which they defined as "a constructive process in which the learner builds an internal representation of knowledge, a personal interpretation of experience" (p. 21) that is constantly open to change as learners change their structures to add more structures of information and experience. A constructivist approach to instructional design means that content cannot be prespecified because the learner must construct his or her own understanding. Learning is not context-free but must be situated in a real-life context so the learner thinks as an expert in the field. Learners are not just efficiently processing information and remembering it to later retrieve it, but must learn to be reflexively aware of the process of their knowledge construction. They must be provided with authentic tasks and learn to think like the expert, not be given a version of information mediated by another viewpoint. The solution of Bednar et al. (1992) was to specify a core of central knowledge that *could* be defined, even though the boundaries of what may be relevant to the learner cannot be defined by the teacher.

This discussion of constructivism underlined the necessity for collaborative learning as a means of providing multiple perspectives to a concept. There was a need to see an issue

Copyright © 2005, Idea Group Inc. Copying or distributing in print or electronic forms without written permission of Idea Group Inc. is prohibited.

from different vantage points and to understand alternative views. Learners evaluated different viewpoints, identifying shortcomings and strengths through the creation of a collaborative learning environment. The goal of this process was not seen as coming to a consensus view but developing and sharing alternative perspectives on issues. The rigorous process of developing and evaluating the arguments in collaborative learning was seen as the goal. Such learning was not competitive but cooperative so students could understand multiple perspectives.

Strategies that the field of instructional design developed in response to the constructivist perspective include *situated cognition* (Brown, Collins & Duguid, 1989; Brown & Duguid, 2000). This strategy incorporates learning experiences that are situated in real-world experiences—not as isolated tasks but as part of a larger context— through projects and environments that are created to capture the larger context. Another strategy is the *cognitive apprenticeship* (Collins, Brown, & Newman, 1989), where the teacher models the process for students and coaches them to an expert performance. These are processes that can be achieved with CSCL, as it provides a tool for dialogue between teacher and learner. The teacher's responses are not scripted, so the students must have a dialogue in which the *process* of solving a problem can be seen as well as the solution.

The application of educational technology within a constructivist perspective has also been discussed by Jonassen (1995), who suggested the use of *situated learning*, which emphasizes conversation and context as an effective strategy. Jonassen argued that educators should observe students in informal learning situations and teach four areas: "domain knowledge, heuristic knowledge, metacognitive strategies and learning strategies" (p 60) in real-life useful contexts as cognitive apprenticeships. He assumed "the social constructivist perspective implied by communities of learners" (p 60) and described several attributes of meaningful learning. He wrote of meaningful learning as having the qualities of being active, with learners responsible for the result; constructive, with learners accommodating new ideas into prior knowledge to make sense; and collaborative, with learners working in learning and knowledge-building communities "exploiting" each others' skills while providing support and observing each others' contributions.

Jonassen (1995) believes learning should be intentional, with learners trying to achieve a cognitive object. It is conversational, because learning is inherently a social, dialogical process, contextualized in real-life meaningful tasks, and reflective, with learners articulating their learning and the process they undergo. His list of attributes, as described above, are a combination of many of the attributes that frame the rationale for online collaborative learning, and his discussion of the way technology should be used as cognitive tools that facilitate thinking and knowledge construction is supportive of the aims of CSCL. It can be suggested that CSCL meets his criteria for filling the proper role of technology in learning—first, as a tool for accessing information, representing ideas, and communicating with others or generating products; then, as an intellectual partner for supporting the internal negotiation of meaning making, constructing personal representations of meaning. Finally, it can be viewed as a context for representing beliefs, perspectives, arguments, and stories of others, defining a space for student thinking, and supporting discourse among a knowledge-building community of learners. Jonassen has written of technologies amplifying learning by "engaging learners in cognitive opera-

Copyright © 2005, Idea Group Inc. Copying or distributing in print or electronic forms without written permission of Idea Group Inc. is prohibited.

tions while constructing knowledge that they would not otherwise have been capable of" (p. 62) as they are used as knowledge-representation tools.

Another commentator in the field of instructional design and its use of technology to enable a constructivist perspective is Lebow (1993). In a comprehensive overview of the field of instructional systems design and its response to the principles and perspectives of constructivism, he argued that the philosophy of constructivism integrates the affective and cognitive domains of learning and offers another set of values to the field. He addressed the perceived incompatibility of the objectivist and constructivist aspects of instructional models, which he said was due to the perception that constructivism is a method, when it is a philosophy that supports the values of "collaboration, personal autonomy, generativity, reflectivity, active engagement, personal relevance, and pluralism" (Lebow, 1993, p. 5).

He maintained that instructional designers should attend more to the affective components of learning. His argument underlies an important assumption of online learning, "the process of acquiring new knowledge and understanding is firmly embedded in the social and emotional context in which learning takes place" (Lebow, 1993, p. 6). He incorporated these ideas into his principles of constructivism and wrote that "the feelings, intuitions, attitudes, values, interests, significant relationships and commitment of learners cannot be separated from the learning process" (p. 10).

His discussion of the principle that constructivism provides a context for learning that supports autonomy and relatedness is an important rationale for collaborative learning online. It encompasses the social constructivist perspective of valuing personal autonomy in learning as well as relatedness, through the use of methods of collaboration and interdependence that "emphasize personal responsibility and individual accountability" (Lebow, 1993, p.8). These values underlie the strategies of learning and assessment that can be achieved in small-group learning online. Lebow provided a rationale for why collaboration is integral to a social constructivist approach when he wrote: "Since constructivists believe that motivation cannot be separated from the social context in which it is embedded, they seek to structure student relations to promote collaboration" (p. 8). The social constructivist view of learning has developed an importance that requires examination and explanation.

Social Constructivism

The importance of the social perspective of constructivism is being increasingly considered in the field of group collaborative learning. Prawat and Floden (1994) wrote that, in the range of views about constructivism and how it can best facilitate the knowledge-construction process, the social constructivists' approaches were becoming more important than other approaches to constructivism. They defined social constructivists as "distinctive in their insistence that knowledge creation is a shared rather than an individual experience," with learners developing their knowledge by the interaction of their combined perspectives. The social constructivist approach is based on the assumptions that "knowledge evolves through a process of negotiation within discourse communities" (p. 48).

Copyright © 2005, Idea Group Inc. Copying or distributing in print or electronic forms without written permission of Idea Group Inc. is prohibited.

Jonassen, Mayes, and McAleese (1993) reiterated the idea that cognitive activity occurs in a social context before being integrated into the individual's construction of meaning. They concluded that the learner must participate in "cooperative learning in which the learner is exposed to alternative viewpoints that challenge initial understanding" (p. 234). Jonassen's (1999) model of Constructivist Learning Environments (CLEs), explains how technology can enable collaboration and social construction of knowledge. CLEs engage students in investigation of a problem, critique related cases, and review information resources. Learners develop needed skills and collaborate with others, using the social support of the group to learn effectively. Jonassen and Remidez (2002) describe an environment, such as that established to support CSCL through a web-based environment, that supports collaborative groups and facilitates a scaffolded discourse about problem solving.

From the constructivist perspective described so far, the need to provide adult learners with a social context for negotiation and construction of knowledge becomes more apparent. The literature of adult and group learning provides a context for this discussion.

Adult Learning: Major Perspectives

The conditions in which adults learn most effectively need to be understood before the process of adults learning collaboratively can be clearly defined. Viewing the field of adult learning historically must include the work of Knowles (1990) among the most influential early writers in the field. His theory of androgogy has had a wide influence on research and practice in training and higher education. His emphasis on contextualizing learning within the adult learners' experience and developing their motivated independence enables the development of the more constructivist approach described in the work of Candy (1991) and Foley (1995). Laurillard's (2002) more teacher-centered perspective provides another focus on adult learning.

Knowles (1990) long maintained that adult learners have different characteristics than young, developing, and school-age learners, and that the practice of adapting theories about children to adult learners was not satisfactory. From pioneering work in the area of adult learning by Lindeman (1926) and research by Houle (1984) and Tough (1979) that focused on adults, Knowles developed a data bank of characteristics of adult learners. He incorporated these into his principles of androgogy (adult learning), which he defined as different from pedagogy (children's learning), particularly in the motivation and independence of adult learners. He described adults as motivated less by their teachers and more by their own need to learn, being more independent and self-directed in their learning than children.

Knowles' (1990) key assumptions included ideas about motivation for learning: that adults are motivated to learn as they experience needs that learning will satisfy; that adults are oriented to life situations so this is the appropriate basis for an adult curriculum; that the core methodology for teaching adults should be an analysis of their experience; and that there be provision for differences in "style, time, place, and pace of

Copyright © 2005, Idea Group Inc. Copying or distributing in print or electronic forms without written permission of Idea Group Inc. is prohibited.

learning" (p.31). Knowles also used the findings of Tough's (1979) Canadian research that showed that adult learners preferred to have independent choice in details of their learning, including the content and style of teaching, and liked to learn collaboratively rather than independently.

Another commentator on adult learning, Foley (1995), also traced the sequence of learning theorists who influenced the concepts of adult learning, and his more recently written perspective described the change in educational research and practice that moved from focusing on effective teachers to studying what made effective learners. Foley's perspective also provided a framework for critique of the field as well as providing a critical theorist's perspective on adult learning. In describing the contribution to the interpretive understanding of learning and teaching that was made by cognitive psychology, he includes early Gestalt psychologists who described learners actively organizing knowledge into their own cognitive framework. These ideas he described as buried by behaviorist psychology—with its emphasis on scientifically observable responses and skills (including the work of Thorndike, Skinner, and Watson)—that dominated education until the 1950s. The exceptions to this were John Dewey's progressive education and Vygotsky's research and theory into child development in Russia (though this was not published in the West until the late 1960s), which were influential in representing a different approach.

Foley (1995) also stressed the importance to the field of adult learning of the work done with cognitive and learning styles, particularly the work of Kolb (1984). Kolb's theory of experiential learning underlying these styles integrated ideas from cognitive psychology, educational theory, social psychology, and psychoanalysis. His propositions incorporated ideas already informing this field, particularly Vygotsky's ideas about learning. Kolb emphasized that learning is social and that experiences influence the learning style a person prefers, while education and employment particularly affect the way a person learns. He described learning as an interactive activity between "individuals with their biological potentialities and the society with its symbols, tools, and other cultural artefacts" (Kolb cited in Foley, 1995, p. 39), and as a dialectical process involving people interacting with their environment. Foley, like Knowles, saw the understanding of such a variety of learning styles and epistemological positions as essential to helping adult educators understand the differences among their students.

Both Knowles and Foley described the significance of Rogers' (1969) influence on adult education in the late 1960s and early 1970s with his ideas of student-centered, self-initiated learning, which critiqued the didactic type of teaching prevalent at the time and encouraged the teacher into the mode of facilitator. This role is important in the type of adult learning possible and suited to the computer-mediated environment. Rogers maintained that we cannot teach a person but can only facilitate his or her learning, and that individuals will only learn things they perceive as being an enhancement of their structure of self. He supported an accepting and supportive climate for learning, with student responsibility for learning rather than predetermined outcomes devised by the teacher. The concept of facilitation "has been a dominant influence in adult education for the past 30 years" (Foley, 1995, p. 43) and has changed the didactic approach of many teachers. Foley described its importance in two main developments in recent adult education, self-directed learning and adult learning principles.

Copyright © 2005, Idea Group Inc. Copying or distributing in print or electronic forms without written permission of Idea Group Inc. is prohibited.

The work of Candy (1991), already discussed in defining the attributes of the constructivist approach to learning, has significantly contributed to the field of adult learning through his research on the adult self-directed learner. Candy critiqued Knowles' assumption that all adults are self-directing and found that the literature suggested that many adults do not feel self-directing. He, too, quoted Carl Rogers, who, as one of the strongest advocates for a student-centered approach, observed that "only a third or a quarter of learners are self-directing individuals, the majority being people who do what they are supposed to do" (quoted in Candy, 1991, p. 61). He suggested that students may lack the necessary knowledge of the subject to begin autonomous learning, and that a solution may be for the teacher to be specific and direct initially and then look to more student collaborative modes of learning as a way of helping the learner to more self-direction, a situation that can be addressed through the formation of collaborative groups.

Candy (1991) claimed that developing personal autonomy need not isolate the learner who is still part of a social learning environment, a fact often obscured in the discussion of self-directed learning. "Adult education is distinguished by its emphasis on socially relevant learning within contexts of mutual interdependence"(p. 123). He described how adult education literature emphasizes the social contexts and pressures of learning, and he argued that no matter how self-directed, most learning requires membership of social groups and takes place in group settings. The need for other people "against whom to measure their progress and with whom to share the experience" (p. 301) and to validate their ideas is basic to most effective adult learning.

Candy (1991) alluded to Brown, Collins and Duguid's (1989) work on cognitive apprenticeship, where the learner is introduced to this language and concepts by other practitioners and learners in his or her knowledge community. The teachers or experts in the field of study begin by providing a model and a scaffold and "as the learners gain more self-confidence and control, they move into a more autonomous phase of collaborative learning, where they begin to participate consciously in the culture" (Brown et al., p. 39). Brookfield (1986) has also defined the self-directed adult learner comprehensively.

Laurillard (2002), in her analysis of academic teaching and learning in higher education, acknowledged a lack of research and professional training at this level and an attitude that academic staff only required expertise and knowledge of their discipline. She described the early elitist view of university teaching: students should take responsibility for their own learning, and academic teachers were simply experts in their field of knowledge who imparted that knowledge, particularly at the undergraduate level. Academic teaching was imparting knowledge, and failure was seen as the student's responsibility. This perspective is gradually changing — "The aim of teaching is simple: it is to make student learning possible" (Ramsden, 1992, p. 5, cited in Laurillard, p. 13). Universities are becoming less elite and are catering to a wider range and larger number of students, and there is a greater responsibility on the teacher to mediate learning, particularly through the medium of the online environment.

Laurillard (2002) wrote that the tradition of pedagogy, from Dewey's rejection of the classical tradition of passing on knowledge in the form of unchangeable ideas, has always argued for active engagement of the learner in the formation of his or her ideas. Vygotsky, Piaget, and Bruner all describe active engagement, not passive reception of

Copyright © 2005, Idea Group Inc. Copying or distributing in print or electronic forms without written permission of Idea Group Inc. is prohibited.

knowledge. However, while these psychologists have influenced approaches to learning in schools, and primary schooling has now changed, many universities still relied on lectures and textbooks. Laurillard proposed that to have a rich understanding of a concept, knowledge must be used in authentic activity. She discussed the scope of what is authentic, the degree of embeddedness in the social or physical world. Students have to be taught to stand back and reflect on learning, but it cannot be assumed that students will transfer that knowledge and apply it to new situations. She argued that if formal education provided more naturally embedded activities, students could do their own sense-making, as knowledge is taken out of its context by teaching abstractions. Abstractions must be grounded in multiple contexts to transfer well, and academic learning should be an activity that develops abstractions from multiple contexts.

Laurillard (2002), in analyzing current theories and research findings, concluded that there are different ways of conceptualizing the topics we want to teach, and teachers and students must have a continuing dialogue that reveals all their conceptions and that the teacher continually analyzes to determine further teaching. She described the learning process as a dialogue between teacher and student and as discursive, adaptive, interactive, and reflective: discursive with teachers and students agreeing on learning goals and task goals, with an environment for acting on these goals and receiving appropriate feedback; adaptive with the teacher responding to the students' conceptions in determining the dialogue; interactive between students acting to achieve the task goal with feedback from the teacher; and reflective by students linking this feedback with each task goal. Laurillard described this as a conversational framework.

Though her conversational framework provided an important perspective on the learning researched in this study, Laurillard's approach demands a very active teacher-directive role that to some extent undermines the type of student group collaboration and interaction that this chapter describes. However, her framework provides a sound basis for computer-supported adult learning, with principles of a reflective and responsive curriculum negotiated through online discussion.

Adults Learning in Groups

As an overview to several decades of research and theorizing into group processes and their application for adult learners, Jacques (2000) comprehensively summarized and described group processes, particularly in higher education. He reviewed the findings of research and the development of theory about group interaction that contribute to the theory of learning groups found in CSCL. He defined a group very simply, as two or more people who interact for more than a few minutes, and described the classic group attributes developed from a range of research. These included the notion of *collective perception*, when members of a group are collectively conscious of their existence as a group, as well as group needs, when members join a group to satisfy a need or give them some rewards. To be a group, the members must have shared aims, which are common aims that bind them together with the goal of achieving these aims as their reward (in tertiary learning, these are often assessment requirements and learning support).

Copyright © 2005, Idea Group Inc. Copying or distributing in print or electronic forms without written permission of Idea Group Inc. is prohibited.

Groups become interdependent and are affected by, and respond to, events that affect the rest of the group. They devise social organization, with a group seen as a social unit with norms, roles, statuses, power, and emotional relationships. To be a group, members must interact, and this can be applied to the context of computer-mediated communication space as Jacques (2000) described — "the sense of group exists even when members are not collected in the same place" (p. 13). Their interaction requires some authentic purpose and will not take place without some need to "influence, share and be responded to" (p. 13), which gives them a reason to communicate. A group must be together long enough for a rudimentary pattern of interaction to occur, and cohesiveness develops when members want to remain in the group and contribute to its well-being.

Jacques (2000) wrote that the need to address the socio-affective side of learners is supported in group research and should be seen for its importance in educating students for the types of relationships they will deal with in the workforce. Such emotional needs that group work serve will also help learning, and these principles are also evident in the online environment, though mediated and without the influence of physical presence of the group members. However, even in this mediated form, social presence is an important factor in establishing effective grouping. The atmosphere or social climate of a group can affect the spontaneity of the behavior of individuals in a group and the group norms established within a group—their code of ethics about proper and acceptable behavior such as responsibility and courtesy determine the type of socio-affective group support the group will provide. The sociometric pattern of the group—who interacts with whom, who likes who, who annoys who— provides a picture of the nature of the group support system, and has been investigated through studies of social presence among electronically observed groups (Rourke, Anderson, Garrison, & Archer, 1999; Stacey, 2002).

Jacques' (2000) review of research showed that though groups are dynamic, there are predictable phases in their development, and he has summarized many classic pieces of research describing phases of dependence and interdependence (Bennis & Shepherd, 1956), flight, fight, and unite phases in group interaction (Bion, 1961), and the forming, storming, norming, performing, and adjourning phases of Tuckman and Jensen's (1977) work that have been widely integrated into studies of organizational behavior. Jacques' review of the body of research into group leadership concluded that it showed that "in normal situations, groups thrive best when the leadership functions are democratically shared among the members of the group" (p. 37).

Cooperative Learning

An influential strategy for group learning that has been implemented widely in the educational sector is *cooperative learning*. Researchers such as Slavin (1994) and Johnson and Johnson (1994) have developed strategies for teaching and learning in groups this way; e.g., Johnson and Johnson's social interdependence learning through which group members share common goals but rely on the actions of the other group members to achieve outcomes (Johnson, Johnson, & Holubec , 1998). Although the researchers mentioned above have been among the most active and influential in the field, Davidson and Worsham (1992) claimed that there is no one model of cooperative learning or one "guru" in the field. They found critical attributes that were required in

Copyright © 2005, Idea Group Inc. Copying or distributing in print or electronic forms without written permission of Idea Group Inc. is prohibited.

all methods, including the need for suitable tasks for group learning, for small groups structured for student-to-student interaction, and for individual responsibility and accountability. However, whereas cooperative learning encouraged cooperation through structured interdependence of group members having teacher- defined differing roles, the collaborative learning movement allowed a more autonomous attitude to group roles with less teacher direction or intervention.

Collaborative Learning Models

Collaborative learning has many similarities to cooperative learning principles and though in many cases the term is used interchangeably, it generally reflects a different philosophy to that of cooperative learning. Panitz (1996), in an Internet discussion about the difference in these terms, called collaboration a "philosophy of interaction and personal lifestyle" not just a classroom technique where cooperation is "a structure of interaction designed to facilitate the accomplishment of an end product or goal." Collaborative learning:

- respects and highlights individual group members' abilities and contributions

- shares authority and responsibility for group outcomes amongst the group

- has an underlying premise of consensus building through cooperation rather than competition.

Dillenbourg (1999), in analyzing the differences between cooperative and collaborative learning, focused on the difference in the division of labor, with cooperative learning often defined as splitting the work and then assembling it into its final output. In collaborative learning, partners do the work together and though some division of labor may well occur, the outcome is negotiated by the group. Collaborative learning is premised on a social constructivist approach with the understanding that knowledge is attained through the learner's construction of knowledge in the social context that the group process facilitates. Dillenbourg described computer-supported collaborative learning as a means of examining collaborative learning closely, and this has indeed become an intensive field of research (Koschmann, Hall & Miyake, 2001) which is explored in more detail in other chapters.

Bosworth and Hamilton (1994), though writing about face-to-face campus learning, claimed that "collaborative learning may well be the most significant pedagogical shift of the century for teaching and learning in higher education" (p. 2), because it can potentially change teachers' and learners' views of learning. Gerlach (1994) also de-scribed the college-based movement towards collaborative learning as being based on the idea that learning is a social activity in which participants talk together and, through that talk, learning occurs. He discussed Britton's (1970) ideas about conversation as the means of developing, exploring, and clarifying ideas and explored Vygotsky's (1978) ideas that "learners need to be active organizers who use language in continual interaction with the social world in order to change both the world and themselves" (p.

Copyright © 2005, Idea Group Inc. Copying or distributing in print or electronic forms without written permission of Idea Group Inc. is prohibited.

3). The social interaction meant students "talk to learn," and the affective and subjective aspects of learning are brought into play as students must articulate their viewpoints and listen to the views of other group members. This allowed them to work with other students to create knowledge and meaning and not rely on the one-way delivery of the teacher or their printed text. Gerlach saw well-managed grouping and a shift from a teacher-centered classroom to a learner-centered one as the main changes to traditional classrooms that would contribute to successful collaborative learning in higher education.

This movement towards a collaborative model of learning gathered momentum at a time when CSCL was being investigated as a means of distributed group learning. The models described adapt well to the online environment where teacher and students are able to use the flexibility of the medium to continuously negotiate the curriculum and online tasks towards the most relevant and authentic purpose for each group of learners. From Kaye's (1992) classic edited collection of studies into collaborative learning using computer conferencing to Salmon's (2000) guide to the teacher's role as e-moderator, the application of collaborative strategies into the online environment has been developed and discussed in the last decade (Harasim, Hiltz, Teles, & Turoff, 1995), though only slowly supported by a developing body of research. Paloff and Pratt (2001, 2003) have detailed such collaborative environments, and their discussion assumes an acceptance of the evolution of a learner-centered classroom when they write: "The virtual student needs to see the instructor as a guide who creates the structure and container for the course, allowing the students to co-create knowledge and meaning within that structure" (Paloff & Pratt, 2001, p.69).

Bruffee (1993), in his discussion of collaborative learning (though initially directed at changing the model of traditional face-to-face college learning), of on-campus teaching and learning that is particularly typical of undergraduate courses, has theorized and provided explanation for the possibilities of collaborative learning that can occur online, and his writing has become seminal to the CSCL discussion (Koschmann, 1999). He wrote of collaboration as a typical professional behavior where colleagues often ask colleagues to read a manuscript or draft a document together—reading and writing and discussing ideas together. He described this as reacculturation by collaboration, changing the models of teaching and learning education, particularly in higher education. He believed that if students are given experience in collaboration, they can develop an interest in interpreting tasks on their own, inventing and adapting language to negotiate consensus with other group members, and joining a community of peers in their construction of knowledge.

Bruffee's (1993) concept of collaborative groups is of groups that are "nonfoundational," i.e., not based on traditional positivist ideas of "giving" education from a knowledge base but on ideas of education as acculturation to a group process of learning. Teachers do not take over and tightly direct the group process but have a goal of productive collaboration among peers. This means that the teacher organizes students into groups, gives them their group tasks, and then backs off, not hovering over them or sitting in on their interaction, as this tends to encourage students to focus on the teacher's authority and interests. Finally, after analyzing and discussing the group consensus, the teacher compares it to the current consensus in the knowledge community that the teacher represents.

Copyright © 2005, Idea Group Inc. Copying or distributing in print or electronic forms without written permission of Idea Group Inc. is prohibited.

Bruffee's (1993) model included several criteria:

- An optimum number of five members for decision-making groups, as groups of nine, ten or more would "dilute the experience" (p. 32). Fewer than five would affect group dynamics in more obvious ways, as four will subdivide into two pairs, three would subdivide into one pair and one other, and groups of two (dyads) would sustain stress higher than other group sizes.

- Groups should not be too homogeneous (from the same place, or friends, or teammates), as there will not be the dissent necessary to provoke discussion, the conversation necessary to reach consensus that stimulates thought and learning if agreement comes too soon. Too much heterogeneity however may give no basis for consensus.

- Tasks have to be open-ended and require discussion and a seeking of consensus. The purpose is to generate discussion to reach consensus to help students organize collaboratively to work towards "membership in the discourse community that the teacher represents" (p. 38) without the teacher's help.

Bruffee conceptualized the effectiveness of collaborative learning as the fact that at the end of the sequence of consensus groups—first, the small group, then, the whole class, and finally, the knowledge community—the students have knowledge that is not "given" by the teacher but rather has been constructed by them in the course of doing the task set by the teacher. The authority of this knowledge increases with the size of the group consensus, from small group to the whole class group to comparing the consensus knowledge with the discipline-based community. Bruffee (1993) wrote that collaborative learning "models the conversation by which communities of knowledgeable peers construct knowledge" (p. 52) and that writing is fundamental to collaborative learning. As online learning requires a written conversation through the use of computer-mediated communication, Bruffee's points about social constructivism and writing are particularly relevant and important to the interpretation of this context.

Bruffee's philosophy about collaborative learning is premised on the assumption that knowledge is a consensus, something people construct interdependently by talking together. He also described education as initiating conversation which then initiates thought; therefore, people can think because they can talk with one another, and we all have membership of a knowledge community. The need for externalizing this conversation is not simple problem solving but people working within their "zones of proximal development" striving to "understand the world at the very frontier of their ability to understand it" (Bruffee, 1993, p. 123). They use a transitional language from whatever community they come, and eventually this leads to an agreed upon language of the knowledge community they are entering, the new community of knowledgeable peers. They internalize this conversation so they can continue it alone, but they need that step into conversation to make that conceptual change occur. Bruffee's ideas drew strongly from Vygotskian theory, which will be described in more detail later in the chapter. Bruffee's ideas were used to frame and theorize the model of online collaborative learning that emerged from a study of computer-supported collaborative learners described in detail below by Stacey (1999).

Copyright © 2005, Idea Group Inc. Copying or distributing in print or electronic forms without written permission of Idea Group Inc. is prohibited.

Cognitive Psychology, Constructivism and the Social Nature of Learning

The findings from cognitive psychology about the social nature of learning, particularly the work of Vygotsky, provide us with a theoretical understanding and a researched critique of the foundations of the learning through group processes that have been discussed so far.

Cognitive psychologists such as Piaget, Vygotsky, and Bruner emphasized the social nature of learning, particularly when learners are confronted with problems that they cannot solve on their own without the resources of a group. More important, the process of discussion—listening to other group members and receiving feedback on ideas— provides the cognitive scaffolding these constructivists see as essential to higher order thinking (Slavin, 1994).

Vygotsky studied children's development as a way of understanding complex human processes, and his research has been replicated and extended to include the study of learning that occurs in the social setting of a group of either children or adults. These ideas from cognitive psychology provide a basis for learning requiring social interaction because Vygotsky viewed learning as a particularly social process with language and dialogue essential to cognitive development.

Vygotsky's notion of a zone of proximal development has gained acceptance since his work was translated into English in the late 1960s. This is a zone in which a learner cannot achieve an understanding of a new concept alone and requires help from a teacher or a peer: "It is the distance between the actual developmental level as determined by independent problem solving and the level of potential development as determined through problem solving under adult guidance or in collaboration with more capable peers" (Vygotsky, 1978, p. 86).

Such a concept requires a learner to interact with other learners who will extend their understanding. Group interaction in the learning process is an important requirement for this condition and the exploration of Vygotsky's ideas can be used as rationale and explanation for the effectiveness of collaborative learning. Social interaction with its creation of a zone of proximal development enables learning that develops an internal process of cognitive thought that the learner can then construct independently. It also enables Vygotsky's notion of scaffolding, in which learners are given a great deal of support initially and then encouraged to become more independent and responsible for their learning as soon as possible. Vygotsky did not see learning as a developmental process but, properly organized, learning can result in mental development and can start other developmental processes that require learning. He refuted the traditional view that learning shows development but said that learning was the beginning of further development.

Vygotsky's concept of expert assistance has been influenced by the idea that this assistance has a vested interest in seeing that particular knowledge is acquired. The concept of the learner being active- a participant in the process-is emphasized in the post-Vygotskian research compared to the role of the adult in the learning process in the Vygotskian research. The motives of the learners are also to be considered as they are

Copyright © 2005, Idea Group Inc. Copying or distributing in print or electronic forms without written permission of Idea Group Inc. is prohibited.

not always enthusiastic receivers of expert assistance. A final challenge discussed by Goodnow (1993) was that of analyzing and describing interaction between peers and between the expert and the novice. The approach emerging from the literature is that of development being more than acquisition by one individual but acquisition of shared meanings. In the recent research on the social, affective, and cognitive benefits of cooperative and collaborative learning, Vygotsky is cited as one of the primary theoretical sources for the developmental approach to peer collaboration. However, according to Forman and McPhail (1993) who have reviewed collaborative problem-solving in comparison to other theoretical perspectives, researchers have interpreted Vygotsky's approach to peer collaboration as a peer-tutoring process that they considered incorrect. By describing Vygotsky's perspective as going beyond the process of transmission from expert to novice, they broadened the Vygotskian approach to peer collaboration.

Post-Vygotskian Research

Goodnow (1993), writing about the research inspired by Vygotsky, summarized the direction of post-Vygotskian research and reflected on the differing approaches and findings of theorists and researchers in this field from the 1960s to the 1990s. In the field of psychology in the 1960s, researchers found that the prevailing behaviorist views would not always fit their observations and that the effect of culture and context was important in cognitive development. Around 1970, as many developmental psychologists turned to other disciplines (social psychology, sociology, anthropology), other researchers focused more on social factors and looked to the works of Vygotsky and the Soviet psychologists. As context and culture were being researched and retheorized, so was cognition. There was recognition that when two people worked on a task, whether by talking to one another or solving the same problem, the critical point was not so much either individual's understanding as the presence of shared meanings or "intersubjectivity" (Goodnow, 1993, p. 374). The debate over whether cognition is general and transferable from one task to another or specific to the task reached a point of agreement through followers of Vygotsky's work: "Specificity now seems to be taken for granted by scholars working from a Vygotskian base" (p. 375). His work has meant that situations must be considered where learners work together as well as those that are individual.

Forman and McPhail (1993) critiqued psychologists who researched problem solving as an individual activity and who usually carried out this research in laboratory conditions, an approach they found less relevant than naturalistic settings. The research of those psychologists, educators, and anthropologists who have studied adults in naturalistic group problem-solving tasks showed a context in which "supports for, constraints on, and challenges to an individual's thinking occur." (Forman & McPhail, 1993, p. 213). They carried out a case study with adolescent girls that demonstrated that they could "establish, modify, reflect on, and refine their initial task goals and definitions so as to collaborate with their peers" (p. 224). They also provided a zone of proximal development for each other that facilitated higher mental functioning. They concluded that Vygotskian theory "supports and extends current debates on the benefits of collaborative problem solving" (, p. 225) and supported research that tried to establish the most effective social context and interactional processes for motivation for problem-solving collaboratively.

Copyright © 2005, Idea Group Inc. Copying or distributing in print or electronic forms without written permission of Idea Group Inc. is prohibited.

Vygotskian theory views cognitive growth as occurring when children are given an opportunity to set up their own goals and organize their own activities. This implies that teachers must give up some of their control of the learning situation for collaborative learning to be most beneficial. A shared means of communication is also essential so that learners are able to argue or share ideas and work collaboratively together and make collaborative learning a meaningful learning process. Learners should be interested in the task and share the goal of solving it, and they should receive immediate feedback. These last two factors are typical of the context of online collaborative learning. Students can have online access to question the teacher when needed and if motivated by interest in goal solving. If members of the group do not have this shared interest, their credibility could be questioned by the group who will check the accuracy of their statements. The electronic conferencing environment enables this questioning, and adult learners are usually confident in expressing their thoughts.

Collaborative Learning and Technology

This chapter has so far drawn together a theoretical basis for explaining the type of learning that is now possible through computer-supported collaborative learning. The last part of the chapter will briefly illustrate these principles through discussion of some recent research and will explore some of the current research discussions about online collaborative learning. Research into online groups has now become a meaningful field of inquiry intent on developing pedagogical models that take advantage of the possibilities of CSCL. Institutions worldwide are concerned with the value of this medium and the most effective ways of using its potential in teaching and learning.

Stacey's (1999) study investigated the experiences of 30 students over a year of their Master of Business Administration (MBA) course, focusing particularly on their use of group communication online as they studied Economics in small groups. Though initially meeting at a study school, their main communication was through the use of CSCL, which was researched as an ethnographic study with the context of the group formation and development and the process of their collaboration described through multiple research perspectives. The groups' ongoing processes of communication and interaction were researched by observation, recording, and analysis of the text of the electronic communication and analysis of the usage pattern of the participants. The learning processes the students experienced using this medium were described through their reflections during interview and through analysis of electronic observation of their communication. The students' process of learning was achieved through collaboration, and the attributes of the social construction of knowledge that emerged through collaborative learning via CSCL were through:

- the sharing of the diverse perspectives of the group members;

- their clarification of ideas via group communication;

- the feedback to a learner's ideas provided by other group members;.

Copyright © 2005, Idea Group Inc. Copying or distributing in print or electronic forms without written permission of Idea Group Inc. is prohibited.

- the process of seeking group solutions for problems;

- their practicing the new language of the knowledge community in discussion with other group members before using this language in the whole group or in the new knowledge community;

- the power of the process of group discussion either mediated by communications media or by through face-to-face contact; and

- the sharing of resources within the group.

The collaborative behaviors through the CSCL also provided socio-affective support that motivated learners. Learning online provided the students with a means of comparing their progress with other students, and the use of computer conferencing set up an environment that required collaboration in order for the group to function effectively. Group members helped each other become competent online users and supported the students who had no electronic access. Technical collaboration—working together to support each other while learning the skills of online access—provided a means of developing group cohesion, and the cohesive groups enabled a democratic system of group management, responsibility, and roles.

The groups in the study that used the group conferences to manage the work and administration of the group interaction had a central point of communication that could be read by all group members, and this meant that their interactive communication could flow smoothly and expectations of contributions could be clearly flagged, thus avoiding any difficulties. The group conferences were also used to ask for assignment and administrative help. The friendly social conversation appeared to provide a group cohesiveness in the face of shared concerns. Collaborating together motivated students to study effectively and to seek to continue the group collaboration over the continuing program. The study found that an effective online environment such as this provided the students with the benefits of reduced isolation and convenience through asynchronous communication, though it raised issues and challenges with the changes and technical hurdles of the electronic environment.

The notion of an online community has been identified by many writers and theorists in the field and has become a focus for recent research. Bernard, Rojo de Rubalcava, and St-Pierre (2000), in summarizing collaborative online learning developments, identified the need for the learner to feel part of a learning community where social interaction fostered community spirit. Garrison and Anderson's (2003) Community of Inquiry model, developed through their extensive research, identified factors of cognitive, social, and teaching presence as key attributes in analyzing online group interaction and learning. They challenge the rhetoric about online communities and see self-directed learning and critical thinking as essential attributes for participants to bring to a community of inquiry. The work of Wenger (Wenger et al, 2002) also provides a conceptual approach for understanding and investigating communities of practice, which he defines as those "groups of people who share a concern, a set of problems, or a passion about a topic, and who deepen their knowledge and expertise in this area by interacting on an ongoing basis" (p. 4). Wenger's conceptual explanations of communities of practice, though developed in studies of situated learning in workplaces, have translated easily into the

Copyright © 2005, Idea Group Inc. Copying or distributing in print or electronic forms without written permission of Idea Group Inc. is prohibited.

online learning environment as both workplaces and education and training have drawn people into communities whose participants are distributed geographically and dependent on communication technologies. In describing CSCL communities, Woodruff (2002) has identified four cohesion factors holding such communities together: the function or goal of the community, the identity or membership, the discursive participation or shared discourse online, and the shared values of the community.

Smith and Stacey (2003) mapped research into computer-supported collaborative learning and identified gaps and opportunities that have yet to be explored. Research into such CSCL communities can draw explanation for the learning that occurs from the theoretical discussion undertaken in this chapter. The chapter has reviewed literature about adult learning and collaborative group learning through a framework of a constructivist perspective to provide an understanding of computer-supported collaborative learning.

References

Bednar, A. K., Cunningham, D., Duffy, T. D., & Perry, J. D. (1992). Theory into practice: How do we link? In T. M. Duffy & D. H. Jonassen (Eds.), *Constructivism and the technology of instruction,* pp. 17-34. Hillsdale, NJ: Lawrence Erlbaum Associates.

Bennis, W. G. and H. A. Shepard (1956). "A theory of group development." *Human Relations,* 9: 415-37.

Bernard, R.M., Rojo de Rubalcava, B., & St-Pierre, D. (2000). Collaborative online distance learning: Issues for future practice and research. *Distance Education,* 21(2): 260-277.

Bion (1961). *Experience in groups.* London, Tavistock.

Bosworth, K. & Hamilton, S. (eds.). (1994*). Collaborative learning: Underlying processes and effective techniques.* San Francisco, CA: Jossey-Bass.

Britton, J. (1970*). Language and learning.* Portsmouth, NH, Boynton/Cook.

Brookfield, S. (1986). *Understanding and facilitating adult learning: A comprehensive analysis of principles and effective practices.* San Francisco, CA: Jossey-Bass.

Brown, J. S. & Duguid, P. (2000). *The social life of information.* Boston, MA: Harvard Business School Press.

Brown, J. S., Collins, A. & Duguid, P. (1989). Situated cognition and the culture of learning. *Educational Researcher,* 18: 32-42.

Bruffee, K. A. (1993). *Collaborative Learning: Higher education, interdependence, and the authority of knowledge.* Baltimore, MD: John Hopkins University Press.

Bruner, J. (1972). *Beyond the information given: Studies in the psychology of knowing.* London: Allen & Unwin.

Candy, P. C. (1991). *Self-direction for lifelong learning: A comprehensive guide to theory and practice.* San Francisco, CA: Jossey-Bass.

Collins, A., Brown, J. S., & Newman, S. E. (1989). Cognitive apprenticeship: Teaching the craft of reading, writing and mathematics. In L. B. Resnick (ed.), *Knowing, learning*

Copyright © 2005, Idea Group Inc. Copying or distributing in print or electronic forms without written permission of Idea Group Inc. is prohibited.

and instruction: Essays in honour of Robert Glaser. Hillsdale, NJ: Lawrence Erlbaum.

Davidson, N. & Worsham, T. (1992*). Enhancing thinking through cooperative learning.* New York: Teachers College Press.

Dillenbourg P. (1999) Introduction: What do you mean by "collaborative learning?" In P. Dillenbourg (ed.). *Collaborative learning: Cognitive and computational approaches,* pp 1-19. Oxford, UK: Elsevier Science Ltd.

Feyerabend, P. (1975) *Against method.* London: Verso.

Foley, G. (1995). Teaching adults. In G.Foley (ed.), *Understanding adult education and training,* pp. 31-53. Sydney: Allen & Unwin.

Forman, E. A. &. McPhail., J. (1993). Vygotskian perspective on children's collaborative problem-solving activities. In E. A. Forman, N. Minick, & C. A. Stone (eds.), *Contexts for learning: Sociocultural dynamics in children's development,* pp.213-229. New York: Oxford University Press.

Fosnot, C. T. (1988). *Enquiring teachers, enquiring learners: A constructivist approach for teaching.* New York: Teachers College Press.

Garrison, D.R. & Anderson, T. (2003). E-*learning in the 21st century: A framework for research and practice.* New York: RoutledgeFalmer.

Garrison, D.R., Anderson, T., & Archer, W. (2000). Critical inquiry in a text-based environment: Computer conferencing in higher education. *Internet and Higher Education,* 2(2-3): 87-105.

Gerlach, J. M. (1994). Is this collaboration? In K. Bosworth & S. Hamilton (eds.), *Collaborative learning: Underlying processes and effective techniques,* pp. 5-14. San Francisco, CA: Jossey-Bass.

Goodnow, J. J. (1993). Afterword: Direction of post-Vygotskian research. In E. A. Forman, N. Minick , & C. A. Stone (eds.), *Contexts for learning: Sociocultural dynamics in children's development,* pp. 395-407. New York: Oxford University Press.

Harasim, L. M., Hiltz, S. R., Teles, L., & Turoff, M. (1995). *Learning networks: A field guide to teaching and learning online.* Cambridge, MA: MIT Press.

Hendry, G. D. (1996). Constructivism and educational practice. *Australian Journal of Education,* 40(1): 19-45.

Houle, C. O. (1984). *Patterns of learning: New perspectives on life-span education.* San Francisco, Jossey-Bass.

Jacques, D. (2000). *Learning in groups: A handbook for improving group work* (3rd ed.). London: Kogan Page.

Johnson, D. W. & Johnson, R. T. (1994). Learning together. In D.W. Johnson & R.T. Johnson, (Eds), *Handbook of cooperative learning methods.* Westport, CT: Greenwood Press, 1: 51-65.

Johnson, D. W., Johnson, R. T., & Holubec, E. (1998). *Cooperation in the classroo*m. Boston, MA: Allyn & Bacon.

Copyright © 2005, Idea Group Inc. Copying or distributing in print or electronic forms without written permission of Idea Group Inc. is prohibited.

Jonassen, D. (1999). Designing constructivist learning environments. In C.M. Reigeluth (ed.), *Instructional design theories and models,* Hillsdale, NJ: Lawrence Erlbaum Associates.

Jonassen, D. H. (1995). Supporting communities of learners with technology: A vision for integrating technology with learning in schools. *Educational Technology,* 35(4):60-63.

Jonassen, D. & Remidez, H. (2002). Mapping alternative discourse structures onto computer conferences. In G. Stahl (Ed.), *Proceedings of CSCL 2002, Computer Support for Collaborative Learning: Foundations for a CSCL Community,* Boulder, Colorado. Hilldale, NJ: Lawrence Erlbaum Associates Inc., p. 237-242.

Jonassen, D., Mayes, T., & McAleese, R. (1993). A manifesto for a constructivist approach to uses of technology in higher education. In T. M. Duffy, J. Lowyck, & D. H. Jonassen (eds.), *Designing environments for constructive learning,* pp. 232–247. Berlin: Springer-Verlag.

Jonassen, D., Prevish, T., Christy, D., & Stavrulaki, E. (1999). Learning to solve problems on the Web: Aggregate planning in a business management course. *Distance Education,* 20(1): 49-63.

Kaye, A. R. (1992). Learning together apart. In A. R. Kaye (Ed.), *Collaborative Learning Through Computer Conferencing,* pp.1-24. London: Springer-Verlag .

Knowles, M. (1990). *The adult learner. A neglected species.* Houston, TX: Gulf Publishing.

Kolb, D. (1984). *Experiential learning: Experience as a source of learning and development.* Englewood Cliffs, NJ: Prentice-Hall.

Koschmann, T. (1999). Computer support for collaboration and learning. *Journal of the Learning Sciences,* 8(3/4): 495-497.

Koschmann, T., Hall, R. & Miyake, N. (eds.) (2001). *CSCL2: Carrying forward the conversation,* pp 157-168 Mahwah, NJ: Lawrence Erlbaum Associates. Kuhn, T. S. (1970). *The structure of scientific revolutions* (2nd ed.) Chicago, IL: University of Chicago Press.

Laurillard, D. (2002). *Rethinking university teaching. A framework for the effective use of educational technology.* 2nd edition. London: RoutledgeFalmer.

Lebow, D. (1993). Constructivist values for instructional systems design: Five principles towards a new mindset. *Educational Technology Research and Development,* 41(3):4-16.

Lindeman, E. C. (1926). *The meaning of adult education.* New York, New Republic.

Lipponen, L. (2002). Exploring foundations for computer-supported collaborative learning. In G. Stahl (ed.), *Computer support for collaborative learning: Foundations for a CSCL community.* Proceedings of CSCL 2002, pp 72-81. Boulder, CO., USA. Hillsdale, NJ: Lawrence Erlbaum Associates.

Paloff, R. & Pratt, R. (2001). *Lessons from the cyberspace classroom: The realities of online teaching.* San Francisco, CA: Jossey-Bass.

Paloff, R. & Pratt, R. (2003). *The virtual student: A profile and guide to working with online learners .* San Francisco, CA: Jossey-Bass.

Copyright © 2005, Idea Group Inc. Copying or distributing in print or electronic forms without written permission of Idea Group Inc. is prohibited.

Panitz, T. (1996). *Active and collaborative learning listserve. <L-ACLRNG@ psuvm.psu.edu>*.

Prawat, R. S. & Floden, R. E. (1994). Philosophical perspectives in constructivist views of learning. *Educational Psychology*, 29 (1): 37-48.

Ramsden, P. (1992). *Learning to teach in higher education*. London, Routledge.

Rogers, C. (1969). *Freedom to learn*. Westerville, OH: Merrill.

Rourke, L., Anderson, T., Garrison, R., & Archer, W. (1999). Assessing social presence in asynchronous text-based computer conferencing. *Journal of Distance Education* 14(2): 50-71.

Salmon, G. (2000). *E-moderating: The key to teaching and learning online*. London: Kogan Page.

Slavin, R. E. (1994). Student teams-achievement divisions. In S. Sharan (ed.), *Handbook of Cooperative Learning,* pp. 3–19. Westport, CT: Greenwood Press.

Smith, P.J. & Stacey, E. (2003). Quality practice in computer-supported collaborative learning: Identifying research gaps and opportunities. In G. Davies & E. Stacey (eds.), *Quality education @t a distance. Proceedings of the IFIP TC3/WG3.6 Working Conference*, February 3-6, 2003, pp. 119-128, Geelong, Australia. Dordrecht: Kluwer.

Stacey, E. (1999). Collaborative learning in an online environment. *Journal of Distance Education,* 14(2): 14-33.

Stacey, E. (2000). Quality online participation: Establishing social presence. In T. Evans (ed.), *Research in distance education 5,* pp.138-253. Deakin University, Geelong, Australia. Available at: *http://www.deakin.edu.au/education/RIPVET/RIDE_ Papers/RIDE_Book.htm.*

Stacey, E. (2002). Social presence online: Networking learners at a distance. *Education and Information Technologies* , 7(4): 287-294.Tough, A. (1979). *The adult's learning projects*. Toronto: Ontario Institute for Studies in Education.

Tuckman, B. and M. Jensen (1977). "Stages of small group development revisited." *Group and Organisational Studies,* 2: 419-427.

Vygotsky, L. S. (1978). *Mind in society: The development of higher psychological processes.* (M.M. Cole,. A. R. Lopez-Morillas, A.R. Luria, & J. Wertsch, Translators.). Cambridge, MA: Harvard University Press.

Wenger, E. (1998). *Communities of practice: Learning meaning and practice*. New York: Cambridge Press.

Wenger, E., McDermott, R., & Snyder, W.M.. (2002). *Cultivating communities of practice*. Boston, MA: Harvard University Press.

Woodruff, E. (2002). CSCL communities in post-secondary education and cross-cultural settings. In T. Koschmann, R. Hall,. & N. Miyake, (Eds.), *CSCL2: Carrying forward the conversation*, pp 157-168Mahwah, NJ: Lawrence Erlbaum Associates, pp.157-168.

Copyright © 2005, Idea Group Inc. Copying or distributing in print or electronic forms without written permission of Idea Group Inc. is prohibited.

Chapter VIII

The Real Challenge of Computer-Supported Collaborative Learning:
How Do We Motivate ALL Stakeholders?

Celia Romm Livermore
Wayne State University, USA

Abstract

This chapter starts from the premise that, to be effective, computer-supported collaborative learning (CSCL) has to be intrinsically motivating. However, in contrast to much of the literature in the field, which focuses almost exclusively on the needs of students, this chapter discusses three groups of stakeholders whose concerns and motivation have to be considered: students, instructors, and institutions. Following a critical review of the literature on online education in general, which highlights some of the major themes that have attracted research so far, the chapter proceeds to introduce a paradigm that integrates the needs of the above three groups of stakeholders. The model is followed by a description of the Radical Model, an innovative approach to computer-supported collaborative learning that is an example of applying the proposed paradigm in practice. The chapter concludes with a discussion of the research implications from the model.

Copyright © 2005, Idea Group Inc. Copying or distributing in print or electronic forms without written permission of Idea Group Inc. is prohibited.

Introduction

In a recent editorial (Emurian, 2001), the author hails online education as a revolution that would make the dream of "management of individual differences among learners" come true (pp. 3-5). In his editorial, Emurian lists a number of rhetorical questions that relate to the issues that he believes will be addressed by the advent of online education, including:

1. Where is it written that the pace of life must be controlled by an academic institution?

2. Where is it written that a course grade must be frozen in time forever?

3. Where is it written that a student must be limited to a single evaluation occasion, without the opportunity for additional learning to achieve an intellectual criterion of excellence?

4. Where is it written that the scale of an intellectual unit must be a traditional semester- long course?

The above questions reflect a series of issues that are of importance to students and that online education could address. Once these issues are addressed through the design of courses that can be started and finished at any time and assessment procedures that allow students to repeat tasks indefinitely, the end result could, indeed, be a highly individualized learning experience for students. But is this what online education is about, particularly in the context of universities?

The underlying premise of this chapter is that this is not the case. In order for online education to succeed, it has to cater to all of the stakeholders. Creating an environment that is motivating to students is one of the major objectives of any educational technology. However, for such a technology to be sustained over time, it has to be intrinsically motivating not only to the learners (students), but also to those who manage the teaching resources (instructors), and those who administer and resource them (institutions). This chapter introduces a paradigm that integrates the needs of students, instructors, and institutions. The paradigm is followed by a case study that details the Radical Model, an innovative approach to computer-supported collaborative learning that is an example of applying the paradigm in practice. The discussion section of this chapter concludes with an outline of the research implications from the model.

Background

The literature on online education to date seems to emphasize a number of themes. Following is a necessarily short review of those themes:

First, there seems to be a debate over the TYPES of approaches to online teaching. One of the central models in this area, the Typology of Dispersion (Johansen, 1992), differentiates between online teaching that occurs at the same place and at the same time

Copyright © 2005, Idea Group Inc. Copying or distributing in print or electronic forms without written permission of Idea Group Inc. is prohibited.

(Synchronous/Proximate), teaching that occurs at the same time but in different places (Anytime/Virtual), teaching that occurs at the same place but at different times (Synchronous/Dispersed), and teaching that occurs at different times and different places (Asynchronous/Dispersed). Other writings discuss specific technologies that can support the various teaching situations in the above model, such as presentation technologies (Leidner & Jarvenpaa, 1995) to support the same time/same place teaching, video conferencing to support same time/different place teaching (Alavi, Wheeler, and Valacich, 1995), Web page presentation, email, and other Internet-based technologies to support different time and different place teaching (Chizmar & Williams, 1996; Kuechler, 1999).

Second, there is a growing literature on underlying PHILOSOPHY of online teaching. One of the central models in this area, the Dimensions of Learning Theories approach, has been proposed by Leidner and Jarvenpaa (1995). The model differentiates between two broad philosophies of teaching—objectivism, which holds that learning occurs in response to an external stimulus, and constructivism, which holds that knowledge is created in the mind of the learner. As a result, while the objectivism approach would lead to learning situations where knowledge is "delivered" to passive learners by an active instructor, the constructivist philosophy would result in learning situations where active learners create knowledge through interaction with each other.

There is an emerging body of literature that looks at the implications of this model to online teaching (Passerini & Granger, 2000). The findings from this research seem to suggest that the objectivist approach does not result in significant benefits, namely, there are no significant differences between face-to-face and video conference lectures (Alavi, Yoo, & Vogel, 1997), and there are no significant differences between website- and audio-supported learning and face-to-face learning (LaRose, Gregg, & Eastin, 1998). However, the constructivist approach does seem to have relative benefits in that GSS-supported classes seem to do better than face-to-face ones (Alavi, 1994), particularly in areas relating to critical thinking (Alavi, Wheeler, & Valacich, 1995). Interestingly, while the quality of learning for the IT-supported students seems to be about the same as for the face-to-face ones, they appear to be less satisfied with the learning experience (Ocker & Yaverbaum, 1999).

Finally, a third prominent theme in the literature on online education is the discussion of its STRUCTURAL antecedents. Here we find, on one hand, the claim that online education is a necessary evil imposed on universities because of declining resources and the necessity to reduce costs and expand markets (Alavi, Yoo, & Vogel, 1997) and ,on the other, the fear that once universities embrace this innovation, it could result in a "second-rate" education for students and a transformation of university instructors from creators of new knowledge (researchers) into assembly-line laborers, delivering educational services to masses of virtual students (Klor de Alva, 2000).

Copyright © 2005, Idea Group Inc. Copying or distributing in print or electronic forms without written permission of Idea Group Inc. is prohibited.

Toward an Integrative Paradigm of Online Education

The above review suggests a need for integration of what appears to be several distinct bodies of research. While the first body of research, on the technologies that support online education, is important in terms of understanding the tools that can be applied in this area, and while the second body of research, on the underlying philosophies of computer-supported education, can help assess the effectiveness and efficiency of online education in terms of how it meets a given set of goals, both bodies of research are student-centered in that they focus primarily on the needs of students.

What is currently lacking in the literature is more emphasis on the needs of the two other stakeholders in the online education game, namely, instructors and institutions. Figure 1 presents a depiction of the three-dimensional integrative paradigm that we are proposing as a basis for a future research agenda in this area. The three dimensions of the model reflect the motivational needs of students, instructors, and institutions that are yet to be described.

The following are some issues that, based on current research, are likely to emerge as motivating factors for the three stakeholders and that could be the content of future versions of the proposed integrative paradigm.

Figure 1: Integrative model for online education.

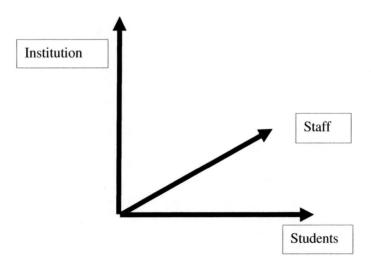

Copyright © 2005, Idea Group Inc. Copying or distributing in print or electronic forms without written permission of Idea Group Inc. is prohibited.

Students

Some of the issues that are likely to motivate students to engage in computer-supported collaborative learning are the perception that this experience has resulted in the acquisition of relevant knowledge and skills, satisfaction with the interaction with the instructor and other students during the learning process, receipt of ample amounts of feedback on progress, and the receipt of a fair grade at the end of the course.

Instructors

Some of the issues that are likely to motivate instructors to engage in computer-supported collaborative learning are the perception that this experience has resulted in an enhancement of the instructors' knowledge and skills, satisfaction with the interaction between the instructor and the students during the learning process, the perception of the effort invested in the teaching of course as reasonable, and the perception that the instructor receives appropriate recognition and rewards for his/her investment in online education.

Institutions

Some of the issues that are likely to motivate institutions to engage in computer-supported collaborative learning are the perception that the institution is likely to gain from investing in this innovation, satisfaction with the process of changing the organization to one that engages in CSCL, once it has been undertaken, the perception that computer-supported collaborative learning is cost-effective, and the perception that the organization is likely to benefit from the investment in online education in the future.

Applying the Integrative Paradigm in Practice

In the following sections, we discuss an approach to online education that is currently applied to a range of courses (Cooke & Veach, 1997; Roberts, Jones, & Romm, 2000; Romm & Taylor, 2000) at Central Queensland University, Australia, including small post-graduate courses (with up to 20 students) and large undergraduate courses (with up to 100 students). The students are a combination of on-campus and distant learners, with both groups treated as one homogenous class.

To date, this approach has been used to teach courses in Management of Information Systems and Electronic Commerce. Student responses to this approach have been very positive. One indication of this is that registration for the two elective courses that pioneered this approach has gone from zero to over 300 students in just over two years.

Copyright © 2005, Idea Group Inc. Copying or distributing in print or electronic forms without written permission of Idea Group Inc. is prohibited.

The teaching materials for this approach (irrespective of which area is being taught) include:

- A video that contains detailed explanations on how the course is run;

- A ten-page booklet "Course Outline" that describes all necessary information about the course (it is available online as part of the course's website and is provided to the students on a CD-ROM and on hard copy);

- A textbook; and

- A class email list.

Students are informed that they must initially:

- read the course outline,

- watch the video, and

- subscribe to the class email list.

They then introduce themselves to the class online so they can be divided into weekly presentation groups. The allocation to groups is completed by the second week of the semester. By this time, students are expected to establish contact with their virtual group members and start working on their assessment tasks. On Week 3 of the semester, the first group makes its presentation to the class online. The presentation consists of an article that the students have to enclose (as text), attach to the email (as a Word or rtf file) or provide a hyperlink to (as a webpage), and a detailed critique that links the article with the reading in the book for the week.

The presentation is made on Monday of each week. By Friday, each of the groups in the class is expected to comment on the presentation. On Sunday, the presentations for the week are read by the instructor along with the comments that were made by all the groups. All groups are graded every week for either their presentation or their comments about other students' presentations. This procedure is repeated for ten weeks until the end of the semester, with each week dedicated to an in-depth discussion of a different topic, with each of the topics being directly related to the reading for that week.

The students' grade for the course consists of 50% group mark for their performance in the group and 50% an individual exam. To make sure that students do not take advantage of their group membership, all groups are invited to submit a consensus opinion of their members. Students are told right at the beginning of the semester that if the members of a particular group are in agreement that one member did not pull his or her weight or in any other way contribute successfully to the submission, the mark of that student can be reduced by 10 points.

The Radical Model makes efficient use of the students' interactions with each other. Even though students have some private interaction with the lecturer ("one to one") and some interaction as a group, when the lecturer communicates with them on the class list ("one

Copyright © 2005, Idea Group Inc. Copying or distributing in print or electronic forms without written permission of Idea Group Inc. is prohibited.

to many"), the bulk of their interactions in this approach is in the "many to many" mode, with the other students in their presentation groups and with the rest of the students in the class through the class email list.

Throughout the semester, students are assessed on 11 assessment tasks (including their group presentation, comments on other students' presentations, and an end-of-term exam). In a class of 100, they get 18 comments that represent the views of their own group members (nine members), as well as nine group comments representing the other 90 students in the class. Since this procedure is repeated every week, the students can receive over 100 units of input from their group members, the other groups, and from the lecturer by the end of the semester. Note that most of the feedback on one's performance comes from the other students - not the lecturer.

It should be noted that even though class interaction is the means through which teaching takes place, the Radical Model does not result in the list being flooded by email messages. As indicated in the previous sections, students are instructed to refrain from using the class list for unlimited expression. The place for such interaction is supposed to be the small presentation groups that they establish to support their group work. The messages that end up being posted on the class list are messages from the list moderator (the lecturer), "formal" presentations of the students' work, and comments by the other groups about these presentations.

The Radical Model helps develop students' communication and other "soft skills". In addition to learning about the content area for the semester, students learn important on-line skills such as how to set up their e-mail lists, how to be citizens of an on-line community, and how to contribute to a virtual team, including dividing the work between the team members, resolving conflicts, developing ideas and projects, and providing positive feedback to others about their work.

Through the involvement of students from diverse backgrounds (many of whom are fully employed) students learn about how organizations use the abstract concepts that are mentioned in the readings. They also learn about relevant legislation and ethical issues.

The Radical Model is "flexible" for both the instructor and the students. This approach increases flexibility for students, because the students don't have to submit hard-copy assignments (hence, nothing can get lost through the system). They get to know if their submission was successful immediately when they see it posted on the class list. In addition, if something happens to preclude an individual student's contribution during the semester, time out and compensation work can be negotiated within groups. In fact, students don't need to ever negotiate with the lecturer on late submission, special consideration, etc. All negotiations on these issues are carried out within the group.

Students have further flexibility in not having to download large amounts of data from the class website (there is nothing on the website other than the Course Outline). They don't need to buy any books other than the course textbook, and even this book can be shared between them up until the end of the semester, as all assessment tasks are group based. Because all learning is facilitated by the class list, the students can engage in class activities from home, work, or while travelling. Further flexibility to the students is provided through the students' selection of supplementary readings for class discussion by themselves. As a result, students get to read quite a large number of articles on topical issues that are of interest to them rather than be forced to read articles selected by the instructor.

Copyright © 2005, Idea Group Inc. Copying or distributing in print or electronic forms without written permission of Idea Group Inc. is prohibited.

Lecturer flexibility is also an enormous advantage of the radical model. Since the package for this course does not include a Study Guide, there is no need to update one every semester. Since the course is in no way dependent on a textbook, there is no need to modify or change it in any way if and when there is a need to change a textbook. In fact, preparing study materials for a new semester should not take more than a few minutes, given that nothing substantial has to change.

As for ongoing teaching; reading the weekly presentation and the comments by the other groups (students are restricted to two pages or two screens maximum per critique or comment on other people's presentations), takes two to three hours per week. This can be done from anywhere, including from home or from a conference. Theoretically, even if the lecturer is totally incapacitated, another person can easily take over and do the ongoing weekly assessment, without inconveniencing the students.

Note that this design is also advantageous from a legal perspective. Since articles by other authors are not used as part of the course website, there is no infringement on other people's copyrights.

The most important aspect about this model is that no matter how many students are in the class, the amount of work for the lecturer is the same. No matter how many students are in the class, 10 or 100, the lecturer ends up checking 10 presentations of one page each per week for ten weeks. If the class consists of 10 students, these 10 pages of text represent the work of each of them. If the class consists of 100 students, the ten pages will represent the work of the ten groups into which the students have been divided. Thus, the amount of semester grading for the lecturer remains the same, irrespective of the number of students in the class.

Why is the Radical Approach an Application of the Integrative Model?

The Radical Model works because it represents an integration of the three components of the Integrative Online Education model. To demonstrate this point, let's go back to the issues that were mentioned previously as contributing to the motivation of the three stakeholders to engage in online education.

Students

The Radical approach is motivating to students because in addition to acquisition of relevant knowledge and skills, they also receive a large amount of feedback from the instructor and their fellow students. Because of its "constructivist" philosophy, the model is also associated with ample opportunities for interaction between the students and the instructor and among the students. Since 50% of the mark in this course is based on an individual exam, the students feel that their efforts, both as individuals and as a group, are acknowledged and fairly rewarded.

Copyright © 2005, Idea Group Inc. Copying or distributing in print or electronic forms without written permission of Idea Group Inc. is prohibited.

Instructors

Instructors are motivated to use this approach because by allowing the students to "create" the course (through selection of the weekly readings and leadership of the class discussion), there is an opportunity for instructors to expand their own knowledge and skills as a result of teaching the course. Since most of the administrative issues that are associated with the teaching of the course (handling late submissions, appeals, etc.) are resolved WITHIN the groups without any input from the instructor, the overall experience of interacting with the class is exceptionally positive for the instructor. Since students are basically teaching each other, the effort that is involved in teaching the class is minimal, hence contributing the perception of instructors that they are not investing more time and effort in the virtual class than they would in a face-to-face class.

Institutions

The above case did not elaborate on the organizational context of the Radical Approach. However, from the list of tools that are used to support this approach, it is clear that this approach involves minimal investment on the part of the institution (the only requirement is to establish an email list and have the students subscribe to the list). At least from this perspective, this approach can be seen as highly cost-effective for institutions, and, as such, highly motivating.

Future Trends and Conclusions

The underlying premise of this chapter - that the success of e-learning should be assessed in terms of its motivating potential to students, instructors, and institutions- could be researched in the following ways:

1. Outcomes - Future research could compare different online teaching styles in terms of their effect on outcome variables such as students', instructors', and institutional satisfaction, quality of the learning process, etc. Once undertaken, such research could determine empirically the dynamics between the three stakeholders that produce successful e-learning.

2. Process - An analysis of the interactions in the online class and in organizations that use online education on a large scale, particularly from a qualitative longitudinal perspective, can reveal patterns of communication and group dynamics that are typical of effective versus ineffective online education environments.

3. Antecedents - The effect of a range of moderating variables on both the outcome and the process of effective online education can be explored. Mediating variables could include: demographic variables (gender, age, socio-economic class, ethnicity), attitudinal variables (learning style, preference to work in the distant mode), institutional variables (course, program studied), and global variables (national

Copyright © 2005, Idea Group Inc. Copying or distributing in print or electronic forms without written permission of Idea Group Inc. is prohibited.

culture). All these variables should, of course, be explored in terms of their effect on the perceptions of members of all three stakeholder groups.

References

Alavi, M. (1994). Computer- mediated collaborative learning: An empirical evaluation. *MIS Quarterly*, June, 150-174.

Alavi, M., Wheeler, B., & Valacich, J. (1995). Using IT to reengineer business education: An exploratory investigation into collaborative tele-learning. *MIS Quarterly*, September, 294-312.

Alavi, M., Yoo, Y., & Vogel, D. (1997). Using information technology to add value to management education. *Academy of Management Journal*, 40(6):1310-1333.

Chizmar, J. & Williams, D. (1996). Altering time and space through network technologies to enhance learning. *Cause/Effect*, 19(3): 14-21.

Cooke, J. & Veach, I. (1997). Enhancing the learning outcome of university distance education: An Australian perspective. *International Journal of Educational Management*, 11(5): 203-208.

Emurian, H. H. (2001). The consequences of E-learning. *Information Ressources Management Journal*, 14(2): 3-5.

Johansen, R. (1992). An introduction to computer augmented teamwork. In R.P. Bostrum, R. T. Watson, & S. T. Kinney (eds.), *Computer Augmented Teamwork: A Guided Tour*. New York; Van Nostrand Reihold.

Klor de Alva, J. (2000). Remaking the academy. *EDUCAUSE Review*, 35(2): 32-40.

Kuechler, M. (1999). Using the Web in the classroom. *Social Science Computer Review*. Summer, 144-161.

LaRose, R., Gregg, J., & Eastin, M. (1998). Telecourses for the Web; An experiment. *Journal of Computer Mediated Communication (online)*, 4 (2). Available at: *http://www.ascusc.org/jcmc/vol14/issue2/larose.html*.

Leidner, D. E. & Jarvenpaa, S. (1995). The use of information technology to enhance management school education: A theoretical view. *MIS Quarterly*, September, 265-291.

Ocker, R. J & Yaverbaum, G. J. (1999). Asynchronous computer- mediated communication vs. face-to-face collaboration: Results on student learning, quality and satisfaction. *Group Decision and Negotiation*, 8 (September).

Passerini, K. & Granger, M. J. (2000). Information technology-based instructional strategies. *Journal of Informatics Education and Research*, Fall, 37-45.

Roberts, T., Jones, D., & Romm, C. T. (2000). Four models of online delivery. In *Proceedings of the Technological Education and National Development (TEND2000) Conference*, Abu Dhabi, UAE. (CD ROM).

Romm, C. T. & Taylor, W. (2000). Online education: A radical approach. *Americas Conference on Information Systems*, Long Beach, CA. (CD ROM).

Copyright © 2005, Idea Group Inc. Copying or distributing in print or electronic forms without written permission of Idea Group Inc. is prohibited.

Chapter IX

Use and Mis-Use of Technology for Online, Asynchronous, Collaborative Learning

W.R. Klemm
Texas A&M University, USA

Abstract

Online learners are typically considered to be isolated learners, except for occasional opportunities to post views on an electronic bulletin board. This is not the team orientation that is so central to collaborative learning (CL) theory. Why does formal CL receive so little attention in online instruction? First, the teachers who do value CL generally are traditional educators and not involved in online instruction. Second, online teachers often have little understanding or appreciation for the formalisms of CL. In this chapter, electronic bulletin boards, although universally used, are shown to provide poor support for Collaborative Learning. As a better alternative, shared-document conferencing environments that allow learning teams to create academic deliverables are discussed. Finally, examples are given of well-known CL techniques, illustrating how these are implemented with shared-document conferencing.

Copyright © 2005, Idea Group Inc. Copying or distributing in print or electronic forms without written permission of Idea Group Inc. is prohibited.

Introduction

Can you list the reasons why so many traditional classroom teachers refuse to use cooperative or collaborative learning (CL)? Now add the reasons why even fewer distance educators use the formalisms of CL. It is easy to understand why there is great need for more books on computer-supported collaborative learning.

In traditional classrooms, collaborative learning is appreciated only by a relatively few hard-core devotees (Cooper, 1995). But, these devotees generally fail to use Internet technologies to enrich their use of CL. Distance educators, who can be counted on to use the Internet, seem least likely to appreciate CL. Why do these paradoxes exist?

Without data, there can be only speculation. In the case of the traditional classroom, many teachers may be technophobes. In the case of distance educators, it is possible that they think they already practice collaborative learning via the discussion boards that are almost universally used in distance education courses. An explanation will be given later as to why true CL cannot be accomplished easily on a discussion board.

Individual achievement in the real world typically depends on how well a person can work with other people. Some students are more effective group learners than others, but experience has shown that all students need improvement in this area. This deficiency is most conspicuous with students in competitive educational tracks, such as pre-professional (law, medicine) or graduate school. Such students became competitive to gain admission to selective professional or graduate schools. This does not mean that they cooperate well. Upon commencement of their professions, however, they may need to work collaboratively. Most young lawyers work for large law firms with a large stable of diverse clients. Physicians depend on a variety of staff and often the other physicians in a group practice. The professional working alone in an ivory tower is a myth — professionals typically work in teams and must always network with peers in their field. They cooperate and collaborate with their peers to cultivate a reputation, to be published in the quality journals, to secure prestigious positions, to garner awards, and to obtain grant funding. Communication skills therefore are often more important for success in life than expertise or intelligence (Goleman, 1995).

Team learning in online computer conferences is not widely practiced, but it can be very effective, even more so than face-to-face collaborative learning (Klemm, 1995, 1996, 1998). Klemm (1995, 1998b) suggested that asynchronous computer conferencing could make CL more effective than team learning in face-to-face traditional classroom environments. The reasons include:

- All students can find the time to do their share of the work. No longer do they have the excuse of conflicting work or study schedules.

- Thinking is more focused and clear because everything is done in writing.

- Everybody is more accountable. Everyone sees what everyone else is doing (and not doing).

- All inputs are organized and archived for later review and update.

Copyright © 2005, Idea Group Inc. Copying or distributing in print or electronic forms without written permission of Idea Group Inc. is prohibited.

Alan Altany (2000) makes the case for written CL as follows:

Collaborative writing helps people work with others, develop an ability to both hear and listen, find out what one really thinks and how much one will defend that thinking or be willing to change it. It develops friendships that transcend class periods and proximity of chairs, benefit from other perspectives, worldviews, and interpretations. Participants work out ways to solve problems caused by disagreement or lack of responsibility, learn more about who is doing the learning (oneself), write with precision, and realize that both the mentor and those in the group take one's ideas seriously. (para. 36)

Distance educators too often fail to develop the most appropriate and effective teaching strategies, especially with regard to optimizing the opportunities for improved learning afforded by CL. Effective computer-supported collaborative learning (CL) requires the appropriate use of technology. For online learning where students must interact asynchronously at different times and places, the available enabling software ranges from simple email to threaded-topic electronic bulletin board systems (BBS) to shared-document computer conferencing systems (SDCCS).

There is abundant literature supporting the notion that CL is a good thing when done properly. The purpose of this chapter includes making the case that technology can help assure that CL is done properly. It will further emphasize CL activities and the technology needs for their implementation online. Specifically, it will examine the following issues, in order:

1.   Role of conversation theory in online Collaborative Learning. Conversation theory deals with how people talk (or write) to each other. The way conversation is conducted determines the effectiveness of conversation. This applies especially to asynchronous online conversation.

2.   Comparison of technologies available for online CL. Each approach has advantages and disadvantages for this method of learning.

3.   Common CL techniques and the choice of technology for online collaborative learning. The CL techniques that teachers have traditionally used will have to be undertaken in a different way in the online world. Some techniques may not work online, while other techniques work better online than in face-to-face instruction if the right technology is used.

4.   Future trends. What is holding back the spread of CL in online environments? Here it is hoped that the chapter will show how those who believe in the pedagogical power of CL may be able to promote its use.

Copyright © 2005, Idea Group Inc. Copying or distributing in print or electronic forms without written permission of Idea Group Inc. is prohibited.

The Role of Conversation Theory in Online Collaboration

First, it should prove useful to ponder how communication occurs online and what is expected to be achieved through "conversations" that occur online. All collaboration requires conversation, which poses special problems as well as opportunities in an asynchronous online environment.

Conversation is central to making a position known, persuading and motivating others, exchanging information, constructing intellectual products, and to learning. This is especially true if we consider conversation to include the written form. This is because writing engages the author with the content of the conversation more rigorously than does mere speaking. Writing can document who said what, when, and in what context. Written conversation can be archived and filed in searchable databases. Writing promotes richer conversation, because everyone has time to reflect on the conversation of others and to plan and edit what they will say. Writing can help us all learn better in many ways:

- Re-reading provides rehearsal of factual information that enhances memorization.

- Seeing a broad range of information helps us to consider alternative points of view.

- Seeing the thoughts of others increases the stimulus for our own creative thought.

- Information management and processing skills are developed and enhanced, as is required by large volumes of written materials.

Categories of Conversation

Patrick Jenlink and Alison Carr (1996) have summarized the essence of contemporary conversation theory in the context of traditional classroom education. These categories are listed below, along with the authors' view of how each applies to electronic network communication.

1. Monolog - exchange of opinion and supposition. Positions are taken, sometimes rigidly. The monolog style of conversation dominates email and postings on electronic bulletin boards.

2. Dialog - a community-building form of shared viewpoints. Individual advocacy tends to be minimized. Different views and alternatives are presented, and the group tries to achieve consensus on one or a few positions. In electronic environments, this style of conversation appears most frequently in so-called group-decision support system software, which expedites brainstorming and group-based decision making. Some software exists that formalizes group-decision making with automated voting features. Teachers can, but usually do not, structure discussion on electronic bulletin boards to promote consensus.

Copyright © 2005, Idea Group Inc. Copying or distributing in print or electronic forms without written permission of Idea Group Inc. is prohibited.

3.   Dialectic - conversation aimed at distilling truth or correctness from logical argument. The focus is on analytical thought and factual information. Online dialectic has not been widely used. The classic classroom form of dialectic is the Socratic method of asking questions and answering questions with yet other questions. Electronically, this can be achieved with real-time electronic chat. However, many students cannot always participate in real-time chats because of time conflicts. Moreover, there is no time for reflection and research, which was no doubt a problem for Socrates' students as well. Asynchronous Internet technology solves this problem. In an electronic bulletin board, for example, the teacher can post a question that students independently answer over the course of a few days or longer. Upon reading the answers, the teacher can then post a follow-up question, and the process repeats. To extend the conversational element, students might "talk to" each other asynchronously to debate possible answers and combine their commentary into a single or a few "best answers." With the right kind of software, which does not include bulletin boards, the students can electronically "write in the margins" of each student's answer with in-context sticky notes or links to Web resources to help develop better answers. The teacher can do likewise.

4.   Construction ("Design") – conversation that creates something new, usually some kind of deliverable, such as a literature review, an analysis, the defense of a position, a plan or a recommendation. The other three forms of conversation are often integrated into constructive conversation as tools to achieve a specified purpose.

The Construction type of conversation best fits the definition of collaborative learning. For a group to produce a deliverable, best results occur when the teacher employs the standard CL formalisms: 1) a team task, 2) defined roles for each team member, 3) interdependence among team members and shared ownership of a result, 4) a process for information-gathering, assessment and organization, and 5) an efficient way to construct the deliverable, as, for example, in a shared, community document. For online CL, these requirements need to be supported by the asynchronous electronic communication environment.

The prerequisite for online Construction conversation is a group task that directs all commentary toward producing a desired deliverable. Example tasks include problem solving, case studies, insight exercises, portfolios, and projects of various sorts (see later commentary on approaches that have been used). The well-known Delphi process also illustrates Construction conversation (see the description under Group Decision Making).

In their analysis of conversation, Sherry, Billig, and Tavalin (2000) guide the reader to the conclusion that dialectic and construction forms are the "higher" and most educationally valuable. Monolog is a relatively degenerate conversation. It is self-conversation whereby one person makes proclamations. Dialog is better but still tends to be unfocussed, limited to opinion sharing, and not linked to achievement and production of a deliverable. Dialectic imposes intellectual rigor, though it too is usually an academic exercise that does not lead to a deliverable. Construction conversation, which can incorporate and build on dialectic processes, produces tangible results.

Copyright © 2005, Idea Group Inc. Copying or distributing in print or electronic forms without written permission of Idea Group Inc. is prohibited.

So, the issue for teachers is "How do we stimulate students to have Construction conversation?"

Action Verbs to Make Construction Conversation Happen

Left to their own devices, students, in the author's experience, do not naturally gravitate toward Construction Conversation. Many students have been conditioned by formal education to be passive learners. The college lecture method has tended to train students to absorb rather than create.

When you get such students in an online collaborative group, it helps to specify certain action verbs that require the active construction of understanding, knowledge, and insight (Klemm, 2002a). Words that promote Construction Conversation include:

- Identify
- Compare and contrast
- Explain
- Argue
- Decide
- Design/construct

Identify

Students can develop their ability to observe and discern when they are required to identify relevant facts or issues that are not explicitly disclosed in the learning resources. Examples: 1) Identify the root causes of the U.S. Civil War, 2) Identify the criteria by which we decide whether or not a given brain chemical is a neurotransmitter.

Compare and Contrast

A classical teaching device is to ask students to compare and contrast, in essence requiring students to recognize similarities and dissimilarities. It extends the "identify" requirement to further analysis. Examples: 1) Compare and contrast the way computers work and the way brains work. 2) Compare and contrast Newton's view of gravity with Einstein's view.

Copyright © 2005, Idea Group Inc. Copying or distributing in print or electronic forms without written permission of Idea Group Inc. is prohibited.

Explain

We all know that one of the best ways to learn something is to explain it to someone else. Examples: 1) Explain what a mathematical derivative is. 2) Explain why the Soviet Union collapsed.

Debate

John Chaffee (1998) contends that the central reasoning tool required to analyze complex issues is argument construction and evaluation. He does not mean to argue in the sense of quarreling. Rather, the purpose of constructing arguments is to muster evidence and logic that can withstand scrutiny. Examples: 1) Why should we consider nitric oxide to be a neurotransmitter, even though it is a gas? 2) Why should the United States embrace free trade?

Decide

What could be more important than the ability to make wise decisions? Making decisions often is the culmination of earlier steps to identify, compare and contrast, explain, and argue. Examples in academic curricula might include: 1) Decide on the most cost-effective way to build a light rail system; 2) Decide which line of research in molecular genetics shows the greatest promise for immediate benefit. Do we have any systematic way to teach decision making to young people in most academic curricula? Group-based decision making is taught systematically in Business colleges. Why then is group-based decision making not an important skill to learn in other curricula?

Design

Both creativity and critical thinking are stimulated when people are asked to design something. In higher education, the design tactic is intrinsic in such curricula as Architecture and Engineering. However, the learning benefits could also be available in other disciplines. Examples: 1) Develop a plan to test the hypothesis that . . . ; 2) Design a Table of Contents for a book on . . .

Responding positively to such action verbs takes conversation to a new level far beyond the recitation of fact and the mere expression of opinion. This is especially true when the activities are conducted by learner groups operating under true team conditions.

Teachers regard the teaching of critical thinking skills as among their highest calling, yet seldom understand the role that conversational style plays in critical thinking. Nor do teachers usually structure online discussions in ways that stimulate critical thinking. Chafee (1998) points out that critical thinking in group settings occurs when each participant does all of the following:

- Expresses views clearly and provides supporting evidence and logic;

- Listens carefully to others, weighing their evidence and logic;

Copyright © 2005, Idea Group Inc. Copying or distributing in print or electronic forms without written permission of Idea Group Inc. is prohibited.

- Stays focused on the issues raised by others rather than on his or her own position;
- Asks relevant questions and then tries to answer the questions; and
- Strives for increased understanding.

Sadly, these conditions are seldom met where online instructors expect students to perform via email, even in a bulletin board system (BBS) environment. The typical requirement is for the learner to make a minimum number of postings in response to topic statements made by the instructor. Such discussions are often conducted without an explicitly meaningful mission and group deliverable. Without a group mission and group-graded deliverable, each member is tempted to tout personal views and biases.

Comparisons of Available Technologies

In this section, the strengths and weaknesses of the three technologies: email, BBS, and shared-document computer conferencing systems (SDCCS) are reviewed. Although a similar comparison was made long ago (Klemm & Snell, 1994), too many online educators still fail to appreciate the differences.

Email

Email is the simplest of online communication technologies. Almost everyone who uses the Internet has some kind of email system and knows how to use it.

Advantages

- Email can also be free, especially with such vendors as Juno, Hotmail, or Yahoo (however, you do have to put up with seeing a lot of unwanted advertisements).
- Email allows one person to send copies of the same message to multiple people. Moreover, when using so-called mail list servers, it is possible to have group engagement where any person in the group can send a message to all others in the group (mail list).

Disadvantages

- Email systems sometimes crash, and mail gets lost.
- The mail is not organized. It arrives chronologically and is not grouped by topic or context. Any organization has to be created idiosyncratically by each student's computer by saving messages into user-created folders.

Copyright © 2005, Idea Group Inc. Copying or distributing in print or electronic forms without written permission of Idea Group Inc. is prohibited.

- Email messages also contain a lot of header garbage that nobody wants to read.

- Email systems are typically flooded with spam. In a collaborative learning environment, students should not be forced to work in an environment where their serious work is contaminated by unwanted advertising mail.

- Email documents are not easily shared. Each member of a group cannot see the mail that others get unless the sender makes it a point to send copies to everyone. Annotation is cumbersome. To respond to a message with in-context annotation, one must open the message, instruct the mail system to "reply" to the sender with all or a portion of the original message included, and then insert comments in the reply message (see example in Fig. 1). Again, a special point must be made to send copies to the appropriate fellow learners. When they reply to the reply, the process multiplies, and everyone's mailbox is cluttered with numerous copies of the original message.

Bulletin Board Systems (BBS)

BBS, commonly called discussion boards, store email messages on a central fileserver computer. Thus, there is only one copy of each message. It is not mailed to users. Rather, users go to the virtual bulletin board, usually a website, and view the messages (Figure

Figure 1: Typical discussion board. Left frame shows a list of email messages (by author and date) in an outline form of who is responding to whom. Right frame shows the message selected in the left frame (#13). Note that the reply message begins by cutting and pasting text from message 12 in order to explain the context for the response.

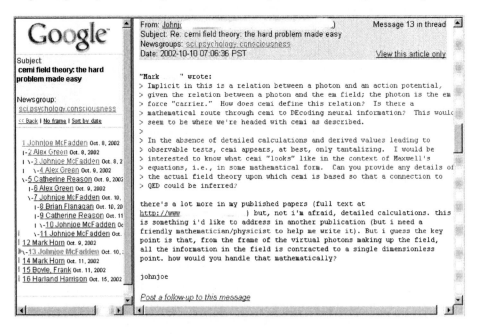

Copyright © 2005, Idea Group Inc. Copying or distributing in print or electronic forms without written permission of Idea Group Inc. is prohibited.

1). As the figure indicates, email messages are posted in one frame of an Internet browser, with an outline of all the messages and their relationships in another frame. This is a messaging environment, in which replies to a posted message have a fixed association with a single message.

Advantages

- Email is organized. Each message is posted according to message author (as in Fig. 1) or under a specific topic heading, creating so-called "threaded-topic" discussion. All messages and replies are grouped to make them relatively easy to find.

- Messages are all in one place, on a server computer - they do not have to be circulated.

Disadvantages

- Organization is rigid - in outline format. No way exists to link to items outside the proscribed position in the outline.

- Users may not know when to go to the BBS to view new messages. Some BBS systems have an email facility that notifies each group member that a new message has been posted. This, however, can become a nuisance if many new messages are being posted.

- Documents are not fully shared. Users may see each other's documents, but they cannot work inside each other's documents (notice the need to cut and paste in Figure 1).

- Boards encourage the expression of mere opinions because it is difficult to do much else without a shared document that can be built collectively.

- Working memory is limited. You have to view each item separately, and only one at a time can be opened in most systems. Thus, if there are ten notes for a given topic, you have to open each separately, and it becomes impossible to remember or easily check to see what is in the other nine notes.

- The better software products are commercial, can be expensive, and may require significant effort to learn.

Threaded-topic discussion boards support only a trivial form of CL, because it is difficult for a group to DO anything on bulletin boards. Few teachers have found a good way to use bulletin boards to help student learning teams make a decision, develop a plan, conduct a project, write a report, conduct a case study, construct a portfolio, or most of the other kinds of constructivist activities that rigorous Construction Conversation can enable.

There is a way for learner groups to share the same document, but it is not convenient. As shown in Figure 2, the author of a document can email it to everyone in the group for

Copyright © 2005, Idea Group Inc. Copying or distributing in print or electronic forms without written permission of Idea Group Inc. is prohibited.

Figure 2: Sharing of documents by email is do-able but cumbersome.

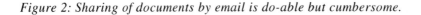

Document Sharing: Old Way

- One person initiates first document
- Document distributed by e-mail
- Four documents after editing/annotation
- No chance to respond to each person's input
- No archived master document that all can see

their revisions and edits. Now, instead of having one document, the group is faced with having to re-construct a single document from as many separate documents as there are members in the group. As a document goes through multiple edits by multiple group members, keeping track of the document versions may become impractical. But perhaps the greatest obstacle to effective CL is the inability for all students to see each group member's specific input and to respond to it.

Shared-Document Computer Conferencing Systems (SDCCS)

With SDCCS, message documents are stored on a central fileserver computer. But, unlike BBS messages, each "message" can be a full-featured, multimedia document that members of a group can check out for editing, insertion of new data and text, and annotation (Figure 3). This capability expands a teacher's options for group-learning activities.

Copyright © 2005, Idea Group Inc. Copying or distributing in print or electronic forms without written permission of Idea Group Inc. is prohibited.

Figure 3:   Shared document computer conferencing

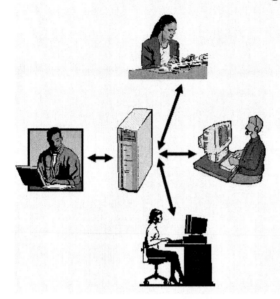

Document Sharing: New Way

- Anyone can initiate a document
- Document checked out from server for editing
- One master document
- Chance to see and respond to each other's input
- Archived master document
- Master document can evolve and be seen at all times by everyone

Advantages

- Documents and notes are organized.

- Messages are all in one place – they do not have to be circulated.

- More information from group members can be seen in one document.

- Messages can be expanded to full-featured, multimedia documents.

- Documents can be checked out by others, edited, and annotated.

- Messages are viewable with fewer mouse clicks to open and close messages.

- Working memory becomes more effective because more material can be seen in the same place.

- Context for inserts and annotation is self-evident.

Moreover, students can still participate in the equivalent of bulletin-board discussions and, at the same time, avoid the necessity of making so many separate mouse clicks to open and close messages. For example (Figure 4), students can put their messages in one document and indicate their authorship by name, initial, or color of text. However, the "replies" can be made as in-context links to pop-up notes. Thus, everything is in one place, with clear specific context.

Copyright © 2005, Idea Group Inc. Copying or distributing in print or electronic forms without written permission of Idea Group Inc. is prohibited.

Figure 4: Shared document format for threaded-topic discussions. Link anchors in one color go to Web pages, those in another color go to pop-up "replies" or comments. Initials indicate the different authors (they could also use different text colors).

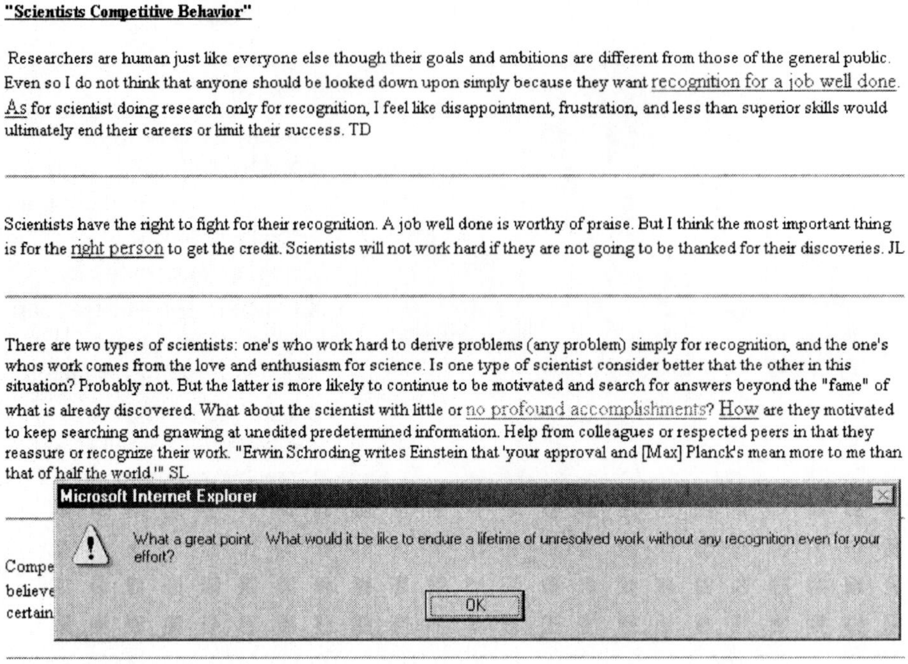

"Scientists Competitive Behavior"

Researchers are human just like everyone else though their goals and ambitions are different from those of the general public. Even so I do not think that anyone should be looked down upon simply because they want recognition for a job well done. As for scientist doing research only for recognition, I feel like disappointment, frustration, and less than superior skills would ultimately end their careers or limit their success. TD

Scientists have the right to fight for their recognition. A job well done is worthy of praise. But I think the most important thing is for the right person to get the credit. Scientists will not work hard if they are not going to be thanked for their discoveries. JL

There are two types of scientists: one's who work hard to derive problems (any problem) simply for recognition, and the one's whos work comes from the love and enthusiasm for science. Is one type of scientist consider better that the other in this situation? Probably not. But the latter is more likely to continue to be motivated and search for answers beyond the "fame" of what is already discovered. What about the scientist with little or no profound accomplishments? How are they motivated to keep searching and gnawing at unedited predetermined information. Help from colleagues or respected peers in that they reassure or recognize their work. "Erwin Schroding writes Einstein that 'your approval and [Max] Planck's mean more to me than that of half the world.'" SL

Microsoft Internet Explorer

⚠ What a great point. What would it be like to endure a lifetime of unresolved work without any recognition even for your effort?

Compe
believe
certain

[OK]

Disadvantages

·As with BBS, group members may not know when new material has been posted. On the other hand, SDCCS is most useful when a group is performing a learning task, and the group should have some kind of schedule that informs everyone when to expect new input.

- Commercial software is required - this can be costly.

- Many of the available products are complex, require significant support, and are hard for novice users to use

Copyright © 2005, Idea Group Inc. Copying or distributing in print or electronic forms without written permission of Idea Group Inc. is prohibited.

Collaborative Learning Techniques and Choice of Technology

Teachers should first choose the kind of collaborative learning techniques that they expect of online students. This decision will then usually dictate the kind of instructional technology needed. Many of the more engaging and active learning tactics cannot be implemented effectively in email or BBS. Some things can only be done effectively in an SDCCS.

This section will outline several kinds of commonly employed approaches to CL, such as explanation dyads/triads (Q&A, jigsaw, etc.), group quiz taking, insight exercises, brainstorming, group decision making, problem solving, case studies, projects. For each CL strategy, we then discuss needed information technology.

Explaining Protocol

A common CL strategy is to use two, sometimes three, students to explain a topic to another person. The listener responds with questions as needed until the explanation is satisfactorily achieved. In an asynchronous, online mode, email could accomplish this task. For example, the explainer could email a Word or WordPerfect document to the listener, who then places in-context pop-up comments in the forms of questions. The document is mailed back to the explainer, who then likewise adds comments in the form of answers and further explanation. However, if many dyads and triads are operating, and the teacher or other students want to "listen in," the mechanics could become too cumbersome. A BBS is much better, because, it organizes the explanation and responses. The explanation could be posted as a discussion topic, and the responses and responses to responses are shown as separate email notes, all of which are archived on a fileserver. An SDCCS, also accessible at all times from a file server, provides even more convenience, because each given document and the associated questions and responses can be contained in a single document.

Pairs Compare

This scenario, as described by Kagan and Kagan (1995), involves sets of two pairs of students. Each pair completes a task or assignment; then the two pairs compare results. The next step requires merging the two pairs into one team, in which they build a final deliverable, based on the independent work of the original two pairs. Finally, teams compare results.

These processes can be accomplished with email, by circulating copies of the documents. Management of team composition creates a problem in email, but much less of a problem in BBS or SDCCS, both of which have semi-automated ways of constructing and changing team membership. BBS would be a little less convenient than SDCCS, because there are no common pair documents that can be edited to remove duplications and insert ideas that only one pair or even no pair had thought of earlier.

Copyright © 2005, Idea Group Inc. Copying or distributing in print or electronic forms without written permission of Idea Group Inc. is prohibited.

Pair Note-taking

This technique was originated to improve the quality of note taking in a lecture hall (Johnson, Johnson, & Smith, 1995). Notes from each student in a pair are compared at the end of the lecture to determine what important items might have been left out or misunderstood.

Online lectures via streaming audio or VCR tapes could be approached in this way. Even assigned Web pages could be summarized by note taking in pairs. The technology issues are the same as those for a related pair activity, "pairs compare" (see above).

Think-Pair Share

This technique, as described by Millis and Cottell (1995), begins with the instructor asking a provocative question that demands significant thought. In the face-to-face situation, the instructor is supposed to allow significant time for each student to think of an answer; then the students form pairs, share their answers, perhaps consolidating each student's version. If learning teams of four or more have been established, the pairs may integrate answers across each pair in the team. The idea can be extended to require the team to come up with a one best answer. Finally, the instructor invites students or group spokesmen to share their responses with the whole class.

To implement this technique online, email would become unwieldy. BBS could organize the thoughts of each student as individual postings that the other pair member - and later - the learning team and whole class can see. However, many BBS do not have handy access permission-changing tools that would allow the instructor to block access to postings and then open them up sequentially first to the team members and then to the class. An SDCCS can achieve these requirements more readily, because the shared workspace makes it easier for pairs and teams of pairs to resolve discrepancies, consolidate thoughts, and generate a single group answer.

The author has used an online variant of this technique for some six years and calls it an "insight exercise" (Klemm, 1998)(see below).

Focused Listing

Students are asked to generate lists words or phrases to define or describe something (Millis, 1996). Then, acting as a group, they select a final list that all can agree on.

With email or BBS, students can independently generate suggested items for such a list, but the list must be generated by cutting and pasting list items from each mail message. In addition, to raise questions or suggest modifications of any list item, separate messages have to be posted. SDCCS systems can clearly expedite the list generation and modification.

Copyright © 2005, Idea Group Inc. Copying or distributing in print or electronic forms without written permission of Idea Group Inc. is prohibited.

Value Line

Millis (1996) also described the Value Line approach where the teacher presents an issue or topic and asks each student to rate on a numerical scale how he or she feels about the issue (for example, on a scale of 1-10, with 10 being strong agreement). Students would then be ordered according to their rankings for a given issue. A group for debating the issue could be created by pulling one person from each end of the value line and two from the middle (for example, with a class of 20 students, a group might consist of Students 1, 10, 11, and 20). Millis did not suggest this, but it seems that a similar thing can be done within a small group of five or six students. Each student ranks the options and then defends his or her first choice (or maybe even defends his or her last choice, which makes it more challenging). Grading could be based, in part, on how persuasive the arguments are in winning over others in a re-vote after the debate.

Email or BBS readily accommodate student postings of their ranks. However, the teacher and students cannot see all the rankings and defenses thereof in one document, complicating the process of selecting debate groups. The voting requirement cannot be accomplished conveniently with email, can be accomplished in some BBS software, but is not readily available in BBS or SDCCS software. Ideally, this requires an SDCCS system that can include a spreadsheet inside the main document.

Jigsaw

In this classic technique, the teacher requires each team member to master a segment of a lesson from a textbook or reference source. Then, after they become "experts" on their assigned topics, students from different groups who had been assigned the same topic are formed into a temporary group to compare findings. They then return to their home groups to present their findings. Finally, all students are quizzed on all topics.

Any of the technologies can handle the final dissemination of expert information. However, the first stage, where students assigned the same topic develop consensus, is most conveniently accomplished with an SDCCS system, because everything, including questions and critique commentary, is seen in one place.

Take A Stand

After presenting an issue, students are invited to choose a side that is either for or against the issue (Hellyer, 1994). They break into small groups with others who have chosen the same side. After consolidating the arguments of individual members, each group chooses a spokesperson who reports to the class. After both sides of the argument are presented, the whole class or spokespersons try to reach consensus.

Technology applications are similar to those mentioned above for Jigsaw exercises.

Copyright © 2005, Idea Group Inc. Copying or distributing in print or electronic forms without written permission of Idea Group Inc. is prohibited.

Memory Matrix

Memory matrix (Anderson & Specht, 1996) is a memory-enhancing technique in which students are given a blank matrix that organizes the informational content of a lecture or reference source. For a birth-control lesson, for example, columns might be labeled with the common methods (condom, diaphragm, IUD, etc.) and rows could be labeled with questions (How does it work? How effective is it? What are the advantages? What are the disadvantages?). Each student group is assigned to complete one column or one row. When all groups are finished, a spokesperson presents the results to the class, and after any questions and answers, each student completes his or her own copy of the matrix.

Most mail systems and BBS do not allow creation and display of tables. This collaborative learning tactic requires the students to share the same document, preferably with a text editor that supports table creation. SDCCS seems to be the obvious choice.

Guided Peer Questioning

This technique, as described by King (1995), requires students to question each other, and the process is guided to stimulate critical thinking. Students ask their own questions in this technique, but they have to be guided to promote thought-provoking rather than simple factual regurgitation questions. The first step is for students to generate questions, using a generic-question template that promotes critical thinking. As examples, such questions may take the form of "What does _____ mean?", "What would happen if _____?", " How does _____ relate to what we learned earlier?", or "Why is _____ important?" In the next step, each student poses such questions to the peer group, and the answers are discussed.

Implementing this approach online can be readily done in a BBS. The questions are posted as "topics," and the answers from various students in a group can be posted as a response. The process may bog down for very complex questions that require multiple iterations of responses, because it will be impossible to remember what is in every note and how all the notes relate. A similar problem occurs in SDCCS, but it is minimized because all responses can be contained within the same document, which is readily scanned and does not required multiple openings and closings of mail messages. This idea could also be extended as in the "insight exercise" (see below), so that each group is required to refine the answer to a particularly complex question and submit it as a group deliverable.

Peer Writing

The variants of this CL tactic are too numerous to explore here. But the essence involves students producing a written document that they submit as a group deliverable. Of necessity, one student in the group takes the lead as an editor, but ideally, everyone in the group contributes to the writing and to critiques of early versions.

Copyright © 2005, Idea Group Inc. Copying or distributing in print or electronic forms without written permission of Idea Group Inc. is prohibited.

Of the possible technologies for implementing collaborative or peer writing, only email or SDCCS are suitable. As shown in Figure 2, drafts can be created by one student and mailed to other students, who in turn send copies of annotated documents back to the originating author. The process repeats for various drafts of the document. Not only is this a cumbersome process, but the process is hidden from the teacher if only the final document is submitted for group grading. The teacher has no way to know if everyone was contributing their fair share to the enterprise.

No such limitations occur with SDCCS (Fig. 3), where the teacher can see all stages of the process because all materials are archived on a server. Additionally, the process is expedited for students, because they can insert text and graphics directly into an evolving document and make pop-up commentary to guide whoever is acting as editor in constructing the next version. Not only can students "write in the margins" electronically, but the teacher can likewise annotate the final document as part of the grading and feedback process.

Group Decision

Business schools in particular have a fundamental requirement for teaching student groups how to brainstorm and make group decisions. Some engineering curricula have a similar emphasis. While these processes are typically performed face-to-face, there are times when they must be conducted asynchronously. In the real world of business and industry, team members may often be located in different time zones, with conflicting schedules. Decision making commonly requires interruptions in the online interaction so that team members may gather data and study posted material before responding.

The iterative Delphi decision-making process, for example, requires asynchronicity and was in fact developed to accommodate situations where group members had to operate at different times and places (Turoff & Hiltz, 2002). Delphi processes focus on solving a problem via construction-level conversation. Team members exchange their expertise and judgment in iterative rounds that may begin with stating and clarifying a problem and the associated goals, followed by brainstorming, successive rounds of information input, survey questions, and a voting scheme that takes into account both an item's importance and its validity.

Clearly, email and BBS cannot support such activity efficiently. Even SDCCS software has limitations, because much of the Delphi process involves voting on survey items. However, survey questions and voting pages can be constructed to supplement SDCCS.

Project Development

Many types of projects can be employed online. Examples include science fair research projects, kiosks of various types, and presentations.

Not all of the activities involved can be accomplished online. The actual research of research projects, for example, would typically have to be performed by group members in a laboratory or field environment. However the planning, data collection, and

Copyright © 2005, Idea Group Inc. Copying or distributing in print or electronic forms without written permission of Idea Group Inc. is prohibited.

document preparation can be expedited by online collaboration. SDCCS is clearly the optimal environment, and if the SDCCS is Web-based, then presentations can be Web pages or slideshows that are hyperlinked from the Web pages.

Problem-Based Learning/Case Studies

These two collaborative learning tasks are related, but not necessarily identical. In traditional problem-based learning, the instructor requests student groups to solve a problem. Students typically respond with the following sequence of activities: identify the issues, review what they already know about the problem, identify what they need to find out, get the necessary information, integrate the old and new information in the context of the issues/problem, and resolve the issues/problem. Case studies, commonly used in engineering, business, and law education vary considerably, but they usually require student groups to demonstrate an understanding of the case, bring insight to it, and develop a method for resolution. The power of asynchronous online group activity is not usually exploited for this kind of learning, but Klemm (2002c) has recently published an SDCCS approach for using articles in scholarly journals as a basis for case study.

Accomplishing either kind of task clearly is not done conveniently with email or with BBS. Only SDCCS provides the completely shared document environment in which group members can share multimedia information, research findings, and insights to produce a cohesive way to solve a problem or understand a case.

SDCCS Examples of Construction-Level Conversation that the Author has used for CL

Shared-Document Discussion Board

The author has tested the idea of coalescing threaded discussions into common documents in the Biomedical Research course that is taught entirely over the Internet (http://classes.cvm.tamu.edu/bims470). In this course, students are asked to post an insight on assigned reading material, which they submit in a shared document. Then they create hyperlinked annotations. This way all of the commentary associated with a given document or topic is embedded in the document itself, and the context for each note is readily apparent. Participants in the conversation have the convenience of having everything in one scrollable place. Students in a learning team put their initials at the end of their text or use different font colors. After a stated deadline date, permission settings are changed so that each group can read but not edit the works of other groups.

A typical topic contains the postings from six students and six pop-up notes for four readings. That is 144 items. Imagine what that would look like on a bulletin board! It would take several screen displays just to list the topic titles for each of the 144 items (and each

Copyright © 2005, Idea Group Inc. Copying or distributing in print or electronic forms without written permission of Idea Group Inc. is prohibited.

would have to be independently opened and closed to see the contents). However, in this case, all of the actual commentary exists in an integrated single document of topic conversation that may be no longer than several word-processor pages. Can there be any doubt as to which approach is more convenient?

Biographies

In the Biomedical Research course, each student is required to write a short biography on the discovery process used by a famous scientist. These biographies have pictures, links to Web pages, and even some of the publications of the scientists. The best part of this exercise is that everybody can see all the biographies. If required, permissions could be set so that students could insert in-context questions and commentary on the biographies. Students not only learn more about the discovery process, but most of the time, they realize why some people received a better grade than others.

Web Quests

Students also conduct searches of Web pages covering certain topics. They put the hyperlink to the pages, along with a summary of what can be found at that website, all into one community document. Each topic can be covered in a separate document or related topics may be combined into the same document. Because everything is html-formatted, it is easy to build a hyperlinked Table of Contents.

Problem Solving

Some of the things that asynchronous student groups do online include solving statistics problems and reaching a consensus on bioethics problems. The work is made much easier because they are helping each other to understand the problems and the approaches to solutions. They use SDCCS, because their questions and answers can all be in the shared documents.

Insight Exercises

A common approach that is taken in the Neuroscience course (http:// classes.cvm.tamu.edu/vaph451) is to stimulate creative thinking (Klemm, 1998a). A great advantage of this approach is the requirement for both an individual and a group product. Every week each student is required to post into a shared document a creative, intellectually rigorous idea on that week's academic content. The insight is to take the form of a question, accompanied by a rationale and strategy for answering the question. Really good questions often do not have an answer, and in those cases, the task is to outline how to do experiments that could get to an answer. The document is secured so that students in other groups cannot see it.

Copyright © 2005, Idea Group Inc. Copying or distributing in print or electronic forms without written permission of Idea Group Inc. is prohibited.

Then, all the students in a given group evaluate each insight with pop-up notes that raise questions or provide additional information, and then they supply a ranking. Based on everyone's comments and rankings, the group reaches consensus on which insight provides the best opportunity for becoming their "Best Insight," which they synthesize from the commentary and submit for a group grade. Sometimes, students get especially creative and create their Best Q&A by combining two or more questions and answers. After a specified deadline, the permissions are changed so that each group can see the work of the other groups. This is a very demanding – and for some unpopular — exercise for many of the pre-medical students, presumably because they have spent their college years being programmed to memorize everything thrown at them. Creative thinking is not a prominent part of their curriculum.

Each group has a Group Leader (who assures that things get done on time and that everybody is pulling his or her share of the load), a Best Q&A Editor (who coordinates the debate and writes the revisions), and two or more Librarians, who do the library work to provide information. They often develop a team spirit and actually want to compete with other groups for the best grade.

Case Studies

In the neuroscience course, students have to become comfortable and reasonably competent in reading primary research literature. Toward this end, the assignment of papers for the group to read and analyze is part of the course (Klemm, 2002c).

Because the students are undergraduates and not familiar with research literature, guidance is provided in the form of questions that they are expected to answer as a group. The overall process takes them sequentially through the steps of understanding what they read and critiquing the rigor of experimental design/methods/results/interpretation, to the final stage of assessing the impact and generating new hypotheses. Instructions are supplied in a colored font, and students insert their information and analysis under each question. These questions have not yet been bolstered with the action verbs mentioned above, but that is on the drawing board.

Students usually approach this task by assigning each team member to write certain responses, and then they interact to correct any misunderstandings or add multiple insights. The students first generate a draft and then make inserts and annotations in the community document as needed to get a complete picture. Then, one member of the group acts as an editor to revise the original to generate a polished copy for grading.

Case Study with Interaction with Digital Library

Under development is a case study approach that will allow students to interact with a digital library. This approach illustrates how to integrate computer-based libraries or "expert systems" with SDCCS: in this case (Figure 5), the generation of a digital library on exotic and zoonotic animal diseases for veterinary medical students, and government agricultural and public health workers. The library has an interface that allows students to enter symptoms and field observations about a case.

Copyright © 2005, Idea Group Inc. Copying or distributing in print or electronic forms without written permission of Idea Group Inc. is prohibited.

Figure 5: Interfacing SDCCS with a digital library.

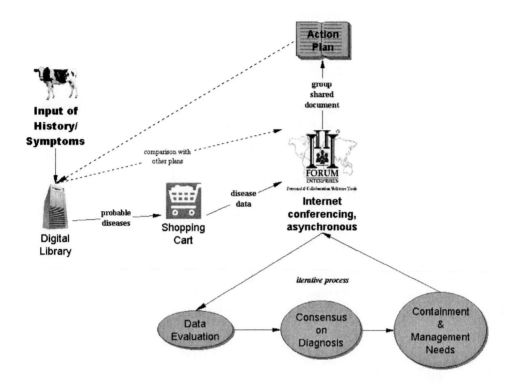

A search algorithm generates a list of diseases that best fit the input descriptors. Then the group can be guided toward a final diagnosis with a series of questions that they answer and debate in our SDCCS.

Note that the case study strategy can employ several of the standard CL techniques that are discussed in the Collaborative Learning Techniques section.

Sample questions include:

1.   Expert Summary - Each member of the group picks one or more of the tentative diagnoses. The student then posts a draft that explains which information about the circumstances surrounding the sickness, symptoms, and gross pathology provides a justification for considering this particular diagnosis. Other students make in-context comments and questions.

2.   Information Needed - All students in the group post and debate suggested calls for information that are not in the database that would clarify the diagnosis. Examples: What lab tests are needed? What tissues should be cultured or examined histopathologically? Students debate the postings with in-context comments and questions. The need for such information and its integration mandates that the deliberations be performed asynchronously in an SDCCS.

Copyright © 2005, Idea Group Inc. Copying or distributing in print or electronic forms without written permission of Idea Group Inc. is prohibited.

3.   Ranking and Debate - Each student ranks each tentative diagnosis on a scale of 0 to 10 (10 being most likely). Each student presents an argument for his or her top choice, which others critique with in-context comments and questions.

4.   Final Choice - As a group, a final differential diagnosis is made, along with the rationale and defense for that choice.

5.   Management and Containment plan - The group develops a comprehensive plan for containing and managing the disease. Exemplary plans can be put in the library for future reference by other groups.

6.   Comparison with Previous Work - The group can compare its plan with previous plans developed by others. This is especially valuable when previous plans have been developed by groups of experts, such as the USDA/APHIS plans for dealing with specific disease outbreaks.

In all of these teaching strategies, teacher feedback is easy and effective, because the educator can "write in the margins" just as in the good old days of paper and pen. Short notes are made in-context as pop-ups and extensive feedback is supplied in-context as an insert (using a different font or color for emphasis). By responding to a group rather than to each individual student, the teacher has less work and is more likely to be fully engaged in what the students are doing. When the same thing needs to be said to all groups, the teacher only inserts it once, referring other groups to that document.

On Sabotage

Many of these exercises were not formal team learning and therefore lacked its camaraderie and pedagogical power, but there have never been problems with sabotage. These are serious college students, and they seem to want to benefit from the ideas and input of fellow students. Where team-learning formalisms are involved, the built-in interdependence, bonding and group grading make sabotage even less likely (see below). In addition, in the small groups of five or six that are used, it should not be too hard to catch and punish any antisocial culprits who try to undermine the process.

The author has not been the only one to notice that given the proper online environment, students can develop a camaraderie that enhances group productivity (Barab, Thomas, & Merrill, 2001).

Software for Getting Beyond Messaging

Email or use of bulletin boards limit opportunities for applying online collaborative learning. What is needed is software that gets beyond little notes to group-created multimedia documents. The documents should be in highly transportable format, such as the html used by Web browsers. Moreover, students need software that allows them

Copyright © 2005, Idea Group Inc. Copying or distributing in print or electronic forms without written permission of Idea Group Inc. is prohibited.

to work on the same documents. They need software that allows them to insert text, data tables, spreadsheets, graphics, and sound or video clips into appropriate places in the documents. Software should allow the creation of multiple, in-context links to websites and for creating hyperlinked pop-up notes.

Some online teachers refer to the kind of software that is needed for shared-document conferencing as computer-supported intentional learning environments (CSILE)(Gay, 1996). The "intentional" part of the name signifies that students intend to reach a goal, as opposed to generally discuss an issue.

Several commercial SDCCS are available. The "mother of all SDCCS" is Lotus Notes, which also contains a BBS and is marketed by IBM. However, few schools use Notes because it is expensive, complicated, and requires significant support staff.

Three shared-document systems (Lotus Notes, Xerox DocuShare, and The Seven Mountains Integrate) have been reviewed by Eseryel, Ganesan, and Edmonds (2002). Examples of other systems that are potentially applicable to teaching include *E-room, Hummingbird, NextPage, Moveable Type, Blosxom, Manilla,* and *WebEx.* These systems were originally developed for corporations and government. They are called by different names: enterprise solutions, Web conferencing, meetingware, projectware, or peer-to-peer netware. A recent review (Long, 2002) asserts that the standard terminology is "blogware," derived from the idea of individuals creating logs of thoughts and links on websites (Weblog). However, Weblogs are not group created.

All of these systems are expensive and can cost in the tens of thousands of dollars. *WebEx,* for example, costs $6,000 to set up and $100 per user per month. In addition, some of these systems require extensive support infrastructure and some cannot be put on your own server. You must "rent" access from the vendor's servers.

Not all systems provide easy-to-use essential features. The key features needed for CL include:

- A database of users, user information, and login and password identifications

- Capability for grouping students into learning teams, each with separate workspace

- Permission system, with independent controls for individuals and for groups, that can be set independently on any document. Settings options should include No Access, Read Only, and Check Out for Editing.

- Server-side software made available for running on licensee's server. License can be purchased - not rented.

- Capability for all authorized users to create new documents.

- Mechanism for easy check out and check in of documents that can be jointly worked on by all students with access permission.

- Method for duplicating documents (for example, one copy as an original backup, another as a marked-up copy)

- Time-out setting that forces document check-in if user forgets to do so

- A graphic navigation tree that displays all documents in a collapsible hierarchy

Copyright © 2005, Idea Group Inc. Copying or distributing in print or electronic forms without written permission of Idea Group Inc. is prohibited.

- Provision of a Web editor that saves documents as Web pages

- A Web editor that lets authors insert multimedia content, in-context hyperlinks to websites, and links to in-context pop-up notes.

- Mechanism for showing metadata about each document (who created it, when, etc.)

- Direct access to user's email system and email addresses of participants

- Affordable (including individual "seat" pricing)

Using these criteria, the author and his colleagues at Texas A&M University developed a simple way to create an SDCCS for teaching, priced so that teachers could afford it. Their original software (*FORUM*®) allowed students to create community documents, provided all the in-context linking capability of Web pages, and did several things that Web pages cannot easily do: 1) accommodate independent teams of learners, 2) create workspaces for private individuals or groups, 3) provide variable levels of shared-access permissions to any given document, and 4) support pop-up, in-context sticky notes (writing in the margins). *FORUM* was limited in that it required client software installation that was cumbersome, and the documents were formatted in a non-standard word processor and not coded in html.

However, these key group-support features have now been incorporated into a new system called *Forum MATRIX* (www.foruminc.com). This Internet environment is designed to run on any server that can support a MYSQL database. In the latest version, students use their own Web browsers and word processors (MS Word). Students not only can view the scrollable documents in their Web browser, but most importantly, they can check out a document for inserting text and graphics, editing, or for making links (to websites, MATRIX documents, or to pop-up notes). Documents are downloaded into the user's own computer and after editing are saved to the Web server in html format for display on the Web and as a Word file for subsequent additions or edits. The "save as" feature can be used to save a local copy. Students can have their own login IDs and passwords. Documents can have a range of access permissions (no access, read only, full edit). Permissions can be set for individuals or groups, and permissions can be changed "on the fly," as for example when the teacher is ready for each group to see the work of other groups.

Students can create new Web pages (all *Forum MATRIX* documents are Web pages) in MS Word, which most of them have and with which they are familiar. We have installed macros that Word uses to control check of the document into the local Word executable and to save the edited document back to the server in both *.doc and *.html format.

Multiple items from different students can be put into the same document. All documents are archived on the Web server. Using their Web browser, students and teachers can scroll quickly through documents, recognizing quickly which inserts and pop-ups have special importance because of the context in which they occur. Unlike email messages on discussion boards, the inserts can be seen in context - without any opening and closing of files. Pop-up notes, also in-context, open and close quicker than email because they are stored as an integral part of the document that has already been opened. Mouse clicking is not needed to see the notes because notes pop up when the cursor rolls over the link anchor.

Copyright © 2005, Idea Group Inc. Copying or distributing in print or electronic forms without written permission of Idea Group Inc. is prohibited.

Future Trends

Today, many teachers of traditional face-to-face classes even use email and discussion boards as a key part of their courses. Use of these technologies is even more widespread among distance educators. However, these are only tentative, incomplete steps to true collaborative learning (CL).

What is the future of online asynchronous CL? The big problem is to get online teachers to use formal CL. Teachers tend to resist major change. Alan Altany (2000) describes the situation this way:

Many teachers continue to teach as if technologies were a passing fad and simply the latest technological idolatry ... some educators may even feel that their techno- or cyberphobia is even a sign of preserving the western intellectual heritage in the face of an electronic glitzy blitz ... (para. 1)

The pedagogically richer forms of CL that can be accomplished online require more creativity and effort from teachers than most have thus far been willing to expend. There are a few "early adopters," but it is by no means clear that CL will ever become as popular online as it is in the face-to-face classroom. Indeed, even face-to-face CL is not widely accepted. The lecture form of instruction still dominates at the college level and many high school classrooms. Unfortunately, few teachers appreciate the value of CL enough to change their behavior. Fewer still realize that CL can be more effective online than face-to-face.

Where some semblance of CL does exist online, as in BBS discussions, the popularity of this online activity is certain to grow to the point that all students will expect it. Students like to interact socially, and it is easy for them to have an opinion, which typically does not require much creative or critical thought. BBS are relatively easy on the brain. Teachers find it easy to think of questions or to post issues. Effective software for these boards is often free. Even though the teacher's learning curve can be about the same for administering a BBS as for an SDCCS, the use of BBS is so widely institution-alized that teachers often have support staff for the BBS.

Those who would like to use online CL know that email and BBS are not very suitable. Getting beyond discussion boards (see Klemm, 2002b) to the fully shared-document CL may not occur anytime soon. For teachers to embrace the more robust forms of CL, the following needs to happen:

- Teachers must be convinced of the value of more formal CL techniques. Too often, they are satisfied that requiring posts on a BBS is sufficient CL.

- Teachers, many of whom are still technophobes, have to invest a significant amount of time learning how to administer bulletin boards (which are often administered by third parties) or to administer an SDCCS.

Copyright © 2005, Idea Group Inc. Copying or distributing in print or electronic forms without written permission of Idea Group Inc. is prohibited.

- The technology, particularly that for SDCCS, needs to be made simpler to administer and use. In addition, it is easier to justify paying a third-party to administer the system, just as many schools now do for Web Masters and bulletin board administrators, if several teachers in an institution used the same SDCCS software.

- Prices will have to come down. We live in an age when many teachers want free software and are not willing to pay much for even powerful software.

- The author and his colleagues believe that *Forum MATRIX* provides simple, affordable SDCCS. In any case, the educational marketplace will not embrace SDCCS until teachers see software products that they perceive to be simple and affordable.

Conclusion

Collaborative learning theory has special educational applicability in an online environment. The online environment avoids some of the pitfalls that plague CL in face-to-face classroom environments. Many of the collaboration techniques used in the classroom cannot only be duplicated online, but may even be accomplished with better student learning and achievement. Some way must be found to convince teachers of the value of online CL and to train the technophobes so that they will feel comfortable in using the necessary technology.

Internet-based CL is most effective when the following conditions are met:

1. Group tasks are clearly defined in the form of an expected deliverable;
2. Group tasks require generation of some kind of intellectual product, which is group graded;
3. Group members use the formalisms of CL to produce the group's deliverable; and
4. Teachers and students have access to a shared-document SDCCS environment that supports CL better than email or a BBS.

In short, collaboration requires more than talking *at* each other via email and bulletin boards. Shared-document Internet environments create the opportunity for students to *work together* in a creative enterprise that produces a tangible and measurable outcome.

References

Altany, A. (2000). Shibboleths and the techniques of technological idolatries. *Invention*. 2 (2), 1-6. Available at: *www.doiiit.gmu.edu/Archives/fall00/aaltany.htm*.

Copyright © 2005, Idea Group Inc. Copying or distributing in print or electronic forms without written permission of Idea Group Inc. is prohibited.

Anderson, C. & Specht, P. (1996). Memory matrix. In S. Kadel & J.A. Keehner (eds.), *Collaborative learning. A sourcebook for higher education. Vol. II*, p. 127-128. University Park, PA.: National Center on Postsecondary Teaching, Learning, and Assessment.

Barab, S.A., Thomas, M.K., & Merrill, H. (2001) Online learning: From information dissemination to fostering collaboration." *J. Interactive Learning Research*, 12 (1), 105-143.

Chafee, J. (1998). *The thinker's way*. New York: Little, Brown & Co.

Cooper, J. (1995). You say cooperative, I say collaborative, let's call the whole thing off. *Cooperative Learning and College Teaching*, 5 (2), 1-2.

Eseryel, D., Ganesan, R., & Edmonds, G.S. (2002). Review of computer-supported collaborative work systems. *Educational Technology and Society*, 5(2), 1-8 Available at: *http://ifets.ieee.org/periodical/vol_2_2002/eseryel_ganesan.html*.

Gay, G. (1996). CSILE (Computer-Supported Intentional Learning Environments). Available at: *http://www.ldrc.ca/greg/greg/csile.htm*.

Goleman, D. (1995) *Emotional intelligence: Why it can matter more than IQ*, New York: Bantam Books.

Hellyer, S. (1994). Take a stand. In S. Kadel & J.A. Keehner (eds.), *Collaborative learning. A sourcebook for higher education. Vol. II.*, p. 116. University Park, PA.: National Center on Postsecondary Teaching, Learning, and Assessment.

Jenlink, P. & Carr, A.A. (1996). Conversation as a medium for change in education. *Educational Technology*, 36 (1), 31-41.

Johnson, D.W., Johnson, R.T., & Smith, K.A. (1995). Cooperative note-taking pairs. *Cooperative Learning & College Teaching*, 5 (3), 10-11.

Kagan, S. & Kagan, M. (1995). Pairs compare. A co-op structure for all classes. *Cooperative Learning and College Teaching*. 5 (3), 4-7.

King, A. (1995). Guided peer questioning: A cooperative learning approach to critical thinking. *Cooperative Learning and College Teaching*, 5 (2), 15-19.

Klemm, W. R. (1995). Computer conferencing as a cooperative learning environment. *Cooperative Learning & College Teaching*, 5 (3), 11-13.

Klemm, W. R. (1996). Enriching computer-mediated group learning by coupling constructivism with collaborative learning. *J. Instructional Science and Technology, 1*(2), March. Accessible at: *http://www.usq.edu.au/electpub/e-jist/docs/old/vol1no2/article1.htm*.

Klemm, W. R. (1998). Eight ways to get students more engaged in online conferences. *The Higher Education Journal*, 26 (1), 62-64.

Klemm, W. R. (1998a). New ways to teach neuroscience: Integrating two teaching styles with two instructional technologies. *Medical Teacher*, 20, 364-370.

Klemm, W. R. (1998b). Using computer conferencing in teaching. *Community College J. Res. & Practice*, 22, 507-518.

Klemm, W.R. (2002a). Software issues for applying conversation theory for effective collaboration via the Internet. *Proceedings of the 2002 International Conference*

Copyright © 2005, Idea Group Inc. Copying or distributing in print or electronic forms without written permission of Idea Group Inc. is prohibited.

on Advances in Infrastructure for e-Business, e-Education, e-Science, and e-Medicine on the Internet. July 29-Aug. 4, Rome, Italy.

Klemm, W.R. (2002b). Extending the pedagogy of threaded-topic discussions. *The Technology Source*, Sept/Oct.

Klemm, W. R. (2002c). FORUM for case study learning. *J. College Science Teaching*, 31 (5), 298-302.

Klemm, W.R. & Snell, J.R. (1994). Teaching via networked PCs: What's the best medium? *Technological Horizons in Education Journal*, 22, 95-98.

Long, P. D. (2002). Blogs: A disruptive technology coming of age? *Syllabus*, 8 -10, October 1.

Millis, B. & Cottell, P. (1998). Cooperative learning for higher education faculty. American Council on Education, Oryx Press [Now available through Greenwood Press].

Millis, B.J. & Cottell, P.G. Jr. (1995). A cooperative learning structure for large classes: Think-pair-share. *Cooperative Learning and College Teaching*. 5 (2), 13-15.

Sherry, L., Billig, S.H., & Tavalin, F. (2000). Good online conversation building on research to inform practice. *J. Interactive Learning Research*, 11 (1), 85-127.

Turoff, M. & Hiltz, S.R. (2002). Computer-based Delphi processes. In M. Adler & E. Ziglio (eds.), *Gazing into the oracle: The Delphi method and its application to social policy and public health*. London: Kingsley Publishers. Available at: *http://eies. njit.edu/~turoff/Papers/delphi3.html*.

Copyright © 2005, Idea Group Inc. Copying or distributing in print or electronic forms without written permission of Idea Group Inc. is prohibited.

Chapter X

The Personal and Professional Learning Portfolio:

An Online Environment for Mentoring, Collaboration, and Publication

Lorraine Sherry
RMC Research Corporation, USA

Bruce Havelock
RMC Research Corporation, USA

David Gibson
National Institute for Community Innovations (NICI), USA

Abstract

This chapter describes the Personal and Professional Learning Portfolio (PLP), a software application designed to provide a flexible learning environment connecting learners and advisors in discussions of posted works-in-progress and the relation of those works to mutually valued goals, standards, and rubrics. We describe the PLP's origins, structure, and pilot implementations across a range of educational settings including K-12 education, higher education, and professional agencies. We describe

Copyright © 2005, Idea Group Inc. Copying or distributing in print or electronic forms without written permission of Idea Group Inc. is prohibited.

in greater detail two higher education sites to illustrate the key issues surrounding PLP adoption. Through our discussion, we hope to bring awareness of the PLP to new audiences and expand consideration of its potential applications, while also shedding insight on the factors that influence adoption of collaborative technologies in institutional settings.

Introduction

The Personal and Professional Learning Portfolio (PLP) is an online environment for mentoring, collaboration, and publication built by the National Institute for Community Innovations (NICI), supported by the Preparing Tomorrow's Teachers to Use Technology (PT3) program, the National Science Foundation, and the Technology Innovation Challenge Grant program. The tools included in the PLP scaffold the process of creating local standards (or adapting existing standards for local use), developing and using rubrics, forming learners and advisors into various communities, and posting and collaboratively evaluating the work of participating learners. The PLP supports a range of "personal and professional learning" through three primary functions: a supported action planning framework, a "portal" to web resources, and a "portfolio space" for both working and demonstration collections of work. This chapter presents the history and a description of the PLP, implementation challenges, results of some pilot tests, and possibilities for future use.

History and Rationale

The lineage of the PLP comes from two sources. One source was an initiative by Montpelier High School, Vermont, which in 1993 placed "individualized educational plans for every student" into its long-term strategic plan. In 1995, this led to the creation and implementation of a school-wide program to place personal learning at the center of a continuous conversation involving all students, their parents or guardians, and caring adults in a school. The University of Vermont provided support and energy to this school-based development through the writings of students, researchers, and theorists such as Bentley (1999), Moffat (1998), Friedrichs (2000), and Gibson (1999, 2000). The PLP is based on a theory of dialogue recently articulated by Gibson and Friedrichs (Friedrichs, 2000; Friedrichs & Gibson, 2001). Friedrichs (2000) discusses four distinct dialogue states for which supports were explicitly built into the PLP:

- Sharing experience - listening to one's own and others' inner speech and natural attitude about a skill or concept;
- Expressing and examining diverse concepts - recognizing conflicts; analyzing old and new concepts, models, and beliefs; working in one's zone of proximal development;

Copyright © 2005, Idea Group Inc. Copying or distributing in print or electronic forms without written permission of Idea Group Inc. is prohibited.

- Articulating applications and understandings - practicing new skills; combining old and new concepts; using others' ideas; using scaffolds to renegotiate understandings; and

- Communicating new powers and creations - celebrating effects of critical analysis.

The PLP's second thread of lineage came from the pioneering work of The WEB Project, a Technology Innovation Challenge grant that used Web-based tools and networked communities to share and critique original student work online. The WEB Project provided a rich research base with which to explore online dialogue and design conversations within a virtual community of learners (Sherry, 2000; Sherry, Tavalin, & Billig, 2000). The WEB Project established a system that linked ten participating schools and districts (including Montpelier High School) and multiple cooperating initiatives in online discussions of student work. Art and music students posted works-in-progress and received constructive feedback from community practitioners and learners, based on their articulated intentions for their works-in-progress. Middle-school students from three schools across Vermont conducted book discussions, facilitated by staff from the Vermont Center for the Book and their teachers. Teachers discussed challenges, conducted action research, shared results, and co-developed rubrics to assess instructional processes, progress, and outcomes. Through these efforts, The WEB Project contributed substantially to knowledge of effective practice for conducting online dialogue and design conversations.

The premise of collaborative interaction as a basis for learning is consistent with research focused on authenticity, use of technology to create problem-centered learning teams, representation of complex dynamics in educational settings, and e-learning (Carroll, 2000; Gibson, 1999; Gibson, 2000; NSDC, 2001; Newmann & Wehlage, 1995; Sherry & Myers, 1998; Stiggins, 1997; Wiggins, 1989). The learner's productivity and self-efficacy are the ultimate goal of the PLP. Work samples are the critical source for evidence of learning, the documentation of progress, and the verification that high standards have been achieved. By placing learner-created work at the center of the PLP, the learner is pushed to a higher standard of personal accountability for the publicly visible quality of that work.

Structure and Implementation of the PLP

The PLP is a combination of a collaboration tool and e-portfolio builder, with an emphasis on the online dialogue and design conversations between learners and the people advising them. The underlying software program is written in Domino, an IBM data structure. The basic architecture was developed for intranets within corporations, but according to NICI's software engineer, "It fits the community-based systems where the PLP is being used." The tool is designed to be used either alone or inside a NICI Campus environment using *Campus*, an intranet software program developed by and licensed

Copyright © 2005, Idea Group Inc. Copying or distributing in print or electronic forms without written permission of Idea Group Inc. is prohibited.

from the National Institute for Community Innovations. The Campus software program supports asynchronous communication, synchronous chat and live chat-based user support, collaboration, mentoring, and professional development. The PLP can also function as a stand-alone toolset independent of the Campus; this is currently the predominant mode of implementation.

The PLP supports a process by which mentors validate learner-produced artifacts that demonstrate mastery of program, state, and national standards, and personal learning goals. The software includes tools for online survey building and administration, developing local standards and rubrics, organizing uploaded work in relation to those

Table 1: Programs using the PLP

Program Name	Program Description	Standards Employed
International Graduate Center	Low-residency master's and doctorate programs in education	National Staff Development Council (NSDC) Standards for Staff Development Interstate New Teacher Assessment and Support Consortium (INTASC) National Educational Technology Standards (NETS) Interstate School Leaders Licensure Consortium (ISSLC) IGC Program Requirements
University of Tennessee Urban Impact Program	In-school master's program for experienced urban teachers	INTASC National Board for Professional Teaching Standards – Core Propositions (NBPTS) Tennessee General Education Standards (TGE)
University of Tennessee Preservice Program	Initial licensing preservice teacher preparation program	INTASC NBPTS Core Propositions Teacher Work Samples TGE
Northfield Middle and High Schools	Public school program for students in grades 7 through 12	Vermont Framework of Standards
National Institute for Urban School Improvement (NIUSI)	National project providing technical assistance to large urban school districts	The Systemic Change Framework

Copyright © 2005, Idea Group Inc. Copying or distributing in print or electronic forms without written permission of Idea Group Inc. is prohibited.

Table 1: Programs using the PLP (continued)

Program Name	Program Description	Standards Employed
National Center for Culturally Responsive Educational Systems (NCCRESt)	National project providing technical assistance to state teams of special education leaders.	The Systemic Change Framework
Arrowhead Area Educational Agency (AEA5)	Regional Service Center serving hundreds of schools and thousands of teachers	AEA5 Comprehensive Goals ISSLC Iowa Professional Development Requirements Iowa Teaching Standards Key Concepts for Contemporary School Leadership 2001 NSDC Standards for Staff Development
New Hampshire Gates Project	Statewide school leadership development program	ISSLC Technology Standards for School Administrators (TSSA)
Ohio State University Special Education	Initial licensing teacher preparation program	Council for Exceptional Children (CEC) Standards

standards and rubrics, forming learners and advisors into various communities, and creating a completed e-portfolio. The learner is situated in an institutionally specific context of explicit standards and goals built into the PLP by each implementing educational program. The PLP provides a standards-based tracking, notation, and discussion forum centered on the learner. It also supports online mentoring, advising, and an improvement process for artifacts intended to be incorporated into e-portfolios. Interested colleges and universities, Professional Development Schools (PDSs), and other educational institutions with programs in teacher preparation and credentialing may customize the PLP in whatever manner best fits their own program or division requirements. Programs can use several sets of standards – national, state, or their own program requirements – to meet the various needs of individuals or program subgroups. A list of programs (Table 1) illustrates the range of organizations, projects, and programs using the PLP. In practice, learners in the PLP system relate their learning goals to standards for work or knowledge introduced into the PLP by program administrators, and upload computer files to the PLP server that exhibit their progress toward meeting these goals

Copyright © 2005, Idea Group Inc. Copying or distributing in print or electronic forms without written permission of Idea Group Inc. is prohibited.

and standards. Initial surveys provide mentors and administrators with information that is relevant to the learner's needs, goals, and priorities. The learner-whether a preservice teacher, inservice teacher, K-12 student, or school administrator-identifies the standards for which he/she wants to demonstrate mastery, together with other personal learning priorities. He/She then develops an e-portfolio of work demonstrating growing mastery of these learning goals and engages with online mentors who critique and assess the work and, to the extent the work demonstrates mastery of standards, validate it as performance-based evidence. These online mentors may comprise advisors, peer coaches, faculty, or content experts, depending on the structure and requirements of the program. Through a process of collaborative reflection, assessment, and several iterations of multiple work products, learners develop an e-portfolio showing their growth and abilities. This portfolio is then available to them as an exhibit of their growth and an aid to their future career progression.

Collaboration in the PLP occurs in threaded conversations centered on works-in-progress posted by the learner. Early in the process, learners and their program advisors select individuals to constitute a microcommunity of mentors centered on development of the individual learner and his or her collected works. As learners upload works-in-progress to their PLP sites, this microcommunity provides guidance, feedback, and validation through online messages discussing each new iteration of a posted artifact. Learners guide this discussion through their initial posts requesting feedback, by participating in the online conversation, and through judicious selection and incorporation of elements of mentor input into the reshaping of their evolving works. The learner decides when the work is ready to be published and what audience has permission to view it.

In prior research (Sherry, 2000; Sherry, Tavalin, & Billig, 2000; Tavalin, 1998; Tavalin & Boke, 1998) for The WEB Project, we found that, at their most effective, design conversations in microcommunities promoted continuous revision leading to products that met or exceeded the posted standards. To best promote this outcome, the dialogue around student work needed to respond to the original intent of the user and to address the specific areas for improvement on which the learner requested feedback. Thus, both the learner and his or her community of advisors need to take an active part in promoting reflection through their online dialogue and design conversations. The PLP explicitly provides structures intended to support these kinds of interactions.

The learner is in charge of his/her own evolving collection of work-in-draft stages, work-in-progress, including work receiving formal evaluation, and work that has been completed. The learner can create various collections of his/her completed works, which may then be exported to DVD, CD-ROM, a server, or paper format. The learner decides which pieces of work are in the various stages, which advisors are being asked for feedback, what criteria or sets of criteria are to be applied during feedback, and when the work is complete. The basic structure of the PLP uses a *plan-do-study-act* action research model. As originally envisioned, this tool enables preservice teachers to identify the skills and knowledge they want to strengthen in order to meet program graduation and certification requirements, professional standards, and personal learning aspirations; to share multimedia work samples that manifest the extent of their mastery of their learning goals; and to interact electronically with course faculty, academic advisors, content mentors, peer advisors, and others to (a) assess their learning goals, (b) critique their work samples

Copyright © 2005, Idea Group Inc. Copying or distributing in print or electronic forms without written permission of Idea Group Inc. is prohibited.

relative to graduation requirements, certification, and professional standards; and (c) strengthen their content knowledge and pedagogical skills.

Intended audiences include preservice programs and staff; inservice and teacher certification and recertification programs for students and staff, including state departments of education; Grades 6 to 12 students; professional organizations for ongoing staff development; training cadres, such as leadership cadres for national education efforts; and National Board for Professional Teaching Standards (NBPTS) teacher portfolio programs.

Pilot Testing

During 2002 and 2003, PLP was piloted or implemented at 50 sites, with learner populations ranging from middle-school students through four-year teacher preparation programs and learning teams of practicing teachers. Twenty-seven sites were described by the project leaders as deep users, meaning that they were extensively working with the PLP within their various programs and impacted approximately 50 teacher candidates per site per year. Several national organizations, including the National Staff Development Council (NSDC), introduced the PLP to their members. One of the project directors reported,

We have a lot of users, and the PLP is disseminated nationwide. About a dozen or so institutions of higher education are running programs that use our tools, with each program impacting around 50 students each. That's about 600 students per term. Preservice teachers are using tools such as the PLP for e-portfolios or some of the portal sites for research. [A colleague], the "guru of e-portfolios," has been making people aware of the PLP and promoting it as a good tool for developing e-portfolios. The surveys and portals both lend themselves to program assessment. The PLP also lends itself to program assessment and student assessment. [A teacher educator] in the New York City Public Schools wants to implement the PLP in his teacher induction program, and he also participated in the NSDC pre-conference on e-portfolios.

As of 2003, most of the use of the PLP took place within institutions of higher education that have teacher preparation programs at the bachelor's or master's level. For example, a teacher educator in Iowa used the PLP to scaffold the professional learning of over 300 teachers. The PLP was viewed as a useful tool for developing e-portfolios that could be used by students to demonstrate mastery of competencies required for credentialing, as a collection of products and artifacts that could be presented when applying for a teaching job, and as a means of providing data for National Council for Accreditation of Teacher Education (NCATE) re-accreditation. Initial feedback from pilot sites offered insight into the range of potential applications for this tool, as well as illumination of some of the issues associated with implementing the PLP in an institutional context. Examples are given below.

Copyright © 2005, Idea Group Inc. Copying or distributing in print or electronic forms without written permission of Idea Group Inc. is prohibited.

At Western Oregon University, faculty members used the PLP for team-based action planning activities that were intended to promote school-based change.

At the University of Nevada, the PLP was used as a mentoring vehicle for preservice teachers.

At the University of Colorado, at Denver, the PLP was piloted in three degree programs: for undergraduates in the Communications Department, students in the Initial Professional Teacher Education (IPTE) program, and doctoral candidates in the Educational Leadership and Innovation (EDLI) program.

In Iowa, one teacher educator used the PLP with over 300 teachers to support a class on e-portfolios and the action research cycle of 146 learning teams. She planned to expand the audience for the learning team experience to as many as 300 learning teams the following year. Through her own action research, she identified use of the PLP as one of four critical support variables needed to insure the success of a learning team. For this reason, though the PLP was an optional part of the learning experience for the teams she supervised during the previous year, in the future she planned to make PLP use mandatory. Although this administrator felt that she had initially lacked the resources to provide the level of support to learners that she would have liked, she nonetheless considered her PLP work to be successful and planned to take more deliberate steps to provide support for learners in future implementation. Providing a low enough ratio of students to instructors was another objective that this administrator felt would lead to more successful learning with the PLP. In her own words,

The PLP has surpassed my goals...the partnership [with the project directors] has been phenomenal. I needed something to help me manage the work of the learning teams and to support the needs of the learners...I wasn't able to get them all to use it deeply, but with the complex needs of adult learning, I am pleased with the learning of the teams and I do attribute [parts of their success] to technology... We've been gradually moving forward with our program for years, but as we listened to the needs of our educators, based on those needs I knew we needed something more. So based on that, I turned to technology. When you apply technology to comprehensive school reform, it accelerates.

In New Hampshire, a faculty member at New England College used the PLP in the context of a course on technology and education. Creation of a PLP was offered as an optional honors assignment, which three students chose to complete.

At the statewide level, an initiative in Vermont entitled *High Schools on the Move* proposed to use the PLP at the high-school level to support its vision for improving the state's schools. In its first phase, the project leaders planned to customize the PLP application for use in high schools. Over time, leaders planned to seek additional support through private foundations.

At its annual conference in November 2002, the NSDC experimented with the PLP as a way to extend the learning of the conference both before and after the face-to-face meeting, and to show the utility of personal documentation of conference goals and objectives for sponsoring school officials. At the request of the session facilitator, the project directors created a custom survey to support the goals of the conference session,

Copyright © 2005, Idea Group Inc. Copying or distributing in print or electronic forms without written permission of Idea Group Inc. is prohibited.

as well as creating a PLP site for uploading and discussing work related to the conference. In the end-of-session satisfaction survey, one respondent stated, "I believe the PLP will help me implement with feedback and support - I am looking forward to this experience."

The youngest students to work with the PLP to date were eighth graders in a Vermont public school system. A primary goal in this program was to "boost student engagement...By working through the goal setting and reflection process, [students] were able to articulate what they want to get out of school." Students in this group were undaunted by the technical challenges and posted more work than any other single group in this sample, often proceeding through several drafts. Comments from teachers, while initially somewhat superficial, began to show more substance as the process of providing online comments on student work became more familiar to them.

User Perceptions at Two Pilot Sites

During the summer of 2002, five of the pilot sites were selected to explore initial reception of and reactions to the PLP. To explore the boundaries of the PLP's flexibility, the sites were selected to illustrate the issues involved with customizing the PLP to meet the distinct needs of five very different learning communities. Interview data supported findings from survey data and an examination of artifacts, and lent insights into some of the implementation issues that surfaced during the pilot phase. Excerpts from interviews with a program administrator and a graduate assistant at two sites are presented below.

University of Tennessee

At the University of Tennessee, the PLP was piloted with two groups whose programs were funded by a grant from the Urban Network to Improve Teacher Education (UNITE), a preservice teacher preparation program, and an urban specialist program for practicing teachers. The program administrator described how her own philosophy of adult learning was compatible with the concept of a portfolio of works-in-progress. She also noted that in supporting awareness of and interest in the PLP among her university-based colleagues, their existing conceptions of a portfolio primarily as a showcase for completed work hindered her efforts to demonstrate the PLP's utility.

Several faculty members thought [the PLP] was too complicated. They didn't even understand the difference between a working portfolio and a showcase. They don't understand the power that this has...We were questioning our portfolio use. I saw an electronic portfolio as a way to pull it all together around standards and improve our portfolio program. I was more overtly committed to standards than some other faculty members...For one class, students were just making a scrapbook, which I didn't like. I didn't want my students to put in all that extra fluff. I wanted performance assessments and things that show their ability to impact student learning...When I saw the PLP, I

Copyright © 2005, Idea Group Inc. Copying or distributing in print or electronic forms without written permission of Idea Group Inc. is prohibited.

saw this as a way to do it efficiently, to [use] rubrics and standards so [students] would know that everything they were to do needed to be aligned with standards and linked and documented.

The program administrator saw the PLP as providing the ability to more closely track and advise students over the course of a multi-year program involving various classes and instructors, an approach contrasting with the focus on shorter-term outcomes or projects that characterizes many computer-supported collaborative activities.

In some assignments like inquiry into practice where we build their capacity as researchers over two years, they [will be able to] have a complete record of that so all the faculty members can access that work…The newest cohort of preservice teachers completed an assessment coming in, so we know immediately what they know, don't know, and need. All the faculty members can see it. We had been doing it on paper, and now the analysis is being done by the PLP, so we can see where they are at the beginning, midpoint, and end of year. We can see what they're thinking as they start the year, we can see what they wrote for other classes…It will help them see their own goal-setting better than they can see it now.

This user planned to expand her work with the PLP, both in the development of a new Urban Administrator's Academy and in helping other colleagues at her university adopt the PLP in their own programs.

The day that [one of the project directors] visited our site [to describe the PLP], a colleague approached me and said, "This is exactly what I want to do with my administrator preparation program." This is an example of how you use technology to improve learning as opposed to just doing stuff. There is a learning curve, but once you learn to ride that bike you really get to enjoy the views. Having gone through that process myself, I find it incredibly valuable to my work. I can see where others are and help them to understand it.

As a productive approach to resolving the issue of technical support, the program administrator employed a graduate student who had been a PLP user in the initial pilot period to support future implementations at that site. The presence of a technically knowledgeable staff member with personal experience using the PLP was perceived by the administrator as a powerful resource to facilitate future work with the PLP.

University of Colorado at Denver

At the University of Colorado at Denver, the PLP was adopted by the Educational Leadership and Innovation (EDLI) doctoral program with some reservations by some faculty members, but had limited success in the Communications Department and was

Copyright © 2005, Idea Group Inc. Copying or distributing in print or electronic forms without written permission of Idea Group Inc. is prohibited.

dropped in the Initial Professional Teacher Education (IPTE) program because the program no longer requires portfolios for exit. Instead, master teachers conduct performance-based assessments in the teacher candidates' classrooms. While no state standards existed for doctoral students in educational leadership and innovation, students developed their own goals and standards for graduation (four mandatory areas and three chosen by students) with input from their advisory committees, and made early steps toward organizing their works around those goals through the PLP. Students in the EDLI program requested to be able to give each other feedback, so all were given system rights as both learners and advisors. Some students started the process by "nonsense" feedback to test the system, but comments tended to develop more substance over time. Learners also started to personalize their PLP portal pages and requested more flexibility in doing so. Participants in the EDLI program felt that the program as a whole encouraged the process of getting feedback for revision, which increased receptivity to the PLP.

One of the graduate assistants who supported the work with the PLP at the University of Colorado during the 2002-2003 academic year felt that the PLP was very appropriate for teacher candidates because of its powerful connection to teaching standards and to assessment. She also provided training and technical support for other colleges and departments that wished to use the PLP within non-teaching programs. However, since other colleges at the university, such as Communications, Architecture, Pharmacy, and Nursing, did not have well-established standards like the programs in the College of Education, the PLP did not always suit the purposes of the students in those other colleges. Whereas the PLP supports communication and collaboration between the student and his or her program committee and emphasizes standards and assessments, some of the students in other colleges felt that they wanted more of a "showcase" portfolio as a capstone project at the end of their degree program, or to present to potential employers, rather than creating a developmental portfolio. Those who had sophisticated Web development skills felt that they wanted to customize the PLP in ways that were not originally intended for preservice teachers. The graduate assistant reported:

We were beta testing the PLP, but some of the students wanted to customize it. They were not using the strengths of the PLP. For example, the PLP supports reflection, but many of the students had not been taught how to reflect on their cognitive and metacognitive skills. We went through a lot of talk on assessment and standards, but some of those programs didn't have their standards in place. Moreover, the university wanted to use the PLP to address NCATE re-accreditation, to show that they were assessing and improving their programs, but they hadn't thought about how they would know whether a course had been successful or not. So we stretched the limits of the PLP when we began to use it for program assessment, and we found we could tweak the software in interesting ways.

The graduate assistant felt that the PLP was clear and appropriate for its intended audience of teacher candidates; it suited the purposes of students who wanted a developmental portfolio and were able to build it right from the beginning of their

Copyright © 2005, Idea Group Inc. Copying or distributing in print or electronic forms without written permission of Idea Group Inc. is prohibited.

coursework; the software was user-friendly; and the PLP was a very powerful tool for assessing students regarding a host of standards. However, experimenting with the PLP outside of the field of education presented cultural issues.

As the departments do get organized and begin to deal with standards and assessment, then the PLP is ready for them. The communication and organization features are very good, but presently, they are underutilized in the other departments. One-on-one training helped to overcome these challenges, not only with mastering the technology, but also understanding what people mean by assessment and what its benefits are. It also helped when I created a mock student portfolio, put content into it, and test drove the system and pushed it to its limits.

An administrator agreed that there were cultural compatibility issues that mediated adoption of the PLP within the doctoral program:

Will we adopt the tool? Does it match our way of doing things? As we move toward e-portfolio systems, it allows more accessibility and dialogue regarding student work. But that will only work with the faculty who are willing to do this. That is a cultural issue, and there are differences among faculty members. Some faculty members encourage students to use e-portfolios, while others still prefer their students to hand in paper copies.

However, there were deeper academic and epistemological issues that presented challenges to the implementation of the PLP:

[The PLP] relies on the program setting outcomes. It's a place where students can post their work. For student self-assessment, they need rubrics and program-based outcomes. There are [major] differences among the faculty members here. How should we, or should we, codify student performance? That limits their possibilities. So those who see a reason to work with rubrics do meet and work on them, but it's only those who agree with the idea of using rubrics...Others think that by specifying outcomes, you limit the ability of students to explore their universe of possibilities. It has to do with the nature of knowing, the nature of a doctoral program, and the whole belief system about setting standards.

This statement highlights the fact that, while the PLP can flexibly accommodate a variety of frameworks, the basic structures that comprise all PLP implementations do, in themselves, constitute an epistemological stance: that growth and mastery can be measured by evaluating the quality of processes and products against a determined set of standards, goals, or rubrics. While most undergraduate and Master's level programs are compatible with a view of knowledge embracing standards and documentation of their attainment, the focus of many doctoral programs on independent thinking and knowl-

Copyright © 2005, Idea Group Inc. Copying or distributing in print or electronic forms without written permission of Idea Group Inc. is prohibited.

edge construction along novel conceptual frameworks may not be compatible with a tool like the PLP.

Mediators of Progress

The experience of the pilot users offered substantial insight into some of the general factors that support a successful experience with this tool. Implementation in a number of varied sites and contexts confirmed that the structures of the PLP were flexible enough to meet the needs of every permutation of educational learning community encountered. Participants reported high levels of satisfaction with several aspects of the PLP's flexible design, including the easy customization of roles and groups; the ability to develop different sets of standards, goals, rubrics, and surveys for those groups; and the natural affordance of technology to transcend traditional boundaries of distance or time. Though four participants offered suggestions for technical improvements to the PLP, all contacted users expressed satisfaction with the intensive and personalized technical support and high degree of responsiveness provided by the project leaders and staff.

In each of the pilot sites, effective program ownership and advocacy to build interest, enthusiasm, and commitment around use of the PLP was perceived as a critical factor supporting implementation. Some sites experienced difficulty stemming from different personnel being responsible for the technical and the conceptual ownership of the PLP at those sites. Where these functions were undertaken by one person, that program tended to be successful. In several settings, both users and advisors were less active in their engagement with the PLP until it either became part of their programmatic requirements or its utility in helping meet their larger personal and professional goals was demonstrated. The presence, commitment, and informed outreach efforts are likely to play a key role in the PLP's future sustainability among existing and new learner communities.

Feedback from pilot sites spurred modifications increasing the PLP's flexibility to meet the needs of various audiences. Interviews with PLP users indicated high rates of satisfaction with the responsive feedback and support that they received from the project leaders. Program leadership provided for the needs of PLP users in tailoring the PLP to particular environments and developing the PLP in directions that meet user needs, e.g., to support group as well as individual learning, and by developing tools and processes to aid sites in PLP adoption.

Cultural Compatibility

One of the most powerful and challenging processes observed at each of the PLP sites was the intersection of existing cultural norms with some of the changes in thinking and practices implied by the approach to learning, assessment, standards, and mentorship embedded in the PLP (Rogers, 1995; Wilson, Ryder, McCahan, & Sherry, 1996). On one

Copyright © 2005, Idea Group Inc. Copying or distributing in print or electronic forms without written permission of Idea Group Inc. is prohibited.

level, the importance of considering participants' experience with portfolio systems prior to using the PLP was very evident among the pilot users (cf. Barrett, 1998). In a deeper sense, the cultural or institutional practices and prior experiences of different pilot groups strongly influenced their initial engagements with the PLP. Beyond their experience with paper portfolios, different groups had varying types of norms in place surrounding many aspects of their work with the PLP. This included at various times participant conceptions of mentorship, reflection, the purpose of a portfolio-like collection of work, the relevance of such a portfolio to their jobs and careers, the idea of assessment as entailing a fixed-point evaluation of a finished product, and familiarity with and ideas about content standards. As users became more familiar with the PLP, some tentative reconsideration of norms of teaching and learning, visible through more reflective comments, active engagement with PLP work, and descriptions of such changes by program administrators, were evident as risk-taking and experimentation with the PLP was supported and encouraged.

Participants' most effective initial engagements with the PLP, that is, those that led to increased buy-in and participation, centered on aspects of its design that were analogous to structures and practices with which participants were already familiar. Expectations and rewards for participation also tended to be closely tied to existing program structures.

All potential users of an innovation need to be persuaded of the viability and relevance of a new way of doing things (cf. Rogers, 1995). In these cases of PLP implementation, those selling points were exploited from various angles by those program advocates who recognized a match between these points of leverage and existing institutional values and conditions. In some cases, this constituted the four-step work cycle or the survey component; for a group that had more extensive experience linking work to standards, creating standards-linked individual goals was a logical first step.

Conclusions and Implications

The scaling up, dissemination, and continuous improvement of the PLP has been ongoing and successful. When used with its intended audience and for its intended purpose, the PLP is eminently successful. Early lessons from the field show that innovative education systems – networks of learners, mentors, and evaluators within and across institutions – are in fact able to create and productively develop and use standards-based performance reviews via this new online tool. Participant descriptions of future pilot and continuing implementations indicate that awareness and use of the PLP will continue to grow in the future. Results from pilot PLP implementations also suggest that future experiences with the PLP at existing pilot sites will bring further successes as program administrators learn more to effectively exploit the PLP's capabilities in support of learner needs. In several settings, different components of the PLP were adopted as integral parts of participating institutions' learning programs. Continuing lessons from the field will provide insight into the complex issues associated with adopting the assessment approach embodied in the PLP. User experiences continue to provide feedback that will contribute to successful future implementation.

Copyright © 2005, Idea Group Inc. Copying or distributing in print or electronic forms without written permission of Idea Group Inc. is prohibited.

A new implementation is being planned for New York City's new teacher induction and support program that will involve teacher experts using the PLP for their own growth in leadership skills as well as to mentor and support new teachers. The PLP will thus be used as a mentoring tool and an e-learning environment for teacher mentor trainers, teacher mentors, and new teachers in the field – a new context for testing its applicability as an action planning tool. Another new application of the PLP will soon be occurring in the International Graduate Center (www.nationalinstitutes.org), where it will be used to guide the development of doctoral dissertations in education.

The PLP embodies a relatively novel configuration for CSCL microcommunities: it is an ongoing rather than short-term collaboration; it integrates task-oriented and knowledge-construction activities; and it focuses on individual learners and their growth as the purpose of collaboration. Though promising experiments in learning communities demonstrate many of the first two features (cf. Riel & Polin, 2004; Schlager, Fusco, & Shank, 2002), we have yet to see another tool that integrates these purposes to deliberately focus collaboration on the growth of individual learners, rather than on the development of either a more discrete product or a more general paradigm of knowledge exchange and construction.

Through its design, the PLP creates a purposeful learning community that supports articulated standards and tangible evidence of their attainment, while supporting the particular needs of individual learners in their individual growth processes. In this sense, the primary outcome of the PLP is a continuous process of learning and renewal, embodied in the ongoing growth of the learner on whom the energies of the collaborating community are focused. By offering a tool set that places each learner at the center of an online community, the PLP focuses collaboration among mentors on the learner's personalized growth, reflection, and learning.

References

Barrett, H. (1998). What to consider when planning for electronic portfolios. *Learning & Leading with Technology*, 26(2), 6-13, October. Retrieved online August 23, 2003, from: *http://transition.alaska.edu/www/portfolios/LLTOct98.html.*

Bentley, T. (1999). Students empowered in Montpelier [Our Generation page]. *Times Argus Newspaper,* Barre-Montpelier, VT, December 8.

Carroll, T. (2000). If we didn't have the schools we have today, would we create the schools we have today? Keynote speech at the *AACE/SITE Conference,* San Diego, CA.

Friedrichs, A. (2000). Continuous learning dialogues: An ethnography of personal learning plans' impact on four River High School learners. Unpublished doctoral dissertation, University of Vermont.

Friedrichs, A. & Gibson, D. (2003). Personalization and secondary school renewal. In J. DiMartino, J. Clarke & D. Wolf (eds.), *Personalized learning: Preparing high school students to create their futures,* pp. 41-68. Lanham, MD: Scarecrow Education.

Copyright © 2005, Idea Group Inc. Copying or distributing in print or electronic forms without written permission of Idea Group Inc. is prohibited.

Gibson, D. (1999). Mapping the dynamics of change: A complexity theory analysis of innovation in five Vermont high schools. Unpublished doctoral dissertation, University of Vermont.

Gibson, D. (2000a). *Complexity theory as a leadership framework*. Montpelier, VT: Vermont Institute for Mathematics, Science, and Technology (VISMT). Retrieved online July 1, 2003, from: *http://www.vismt.org/pub/ComplexityandLeadership.pdf.*

Gibson, D. (2000b). Growing towards systemic change: Developing personal learning plans at Montpelier high school. In J. Clarke, B. Bossange, C, Erb, D.Gibson, B. Nelligan, C. Spencer, & M. Sullivan, M. (eds.), *Dynamics of change in high school teaching: A study of innovation in five Vermont professional development schools,* pp. 99-128. Providence, RI: Brown University.

Moffat, J. (1998). *The universal schoolhouse: Spiritual awakening through education*. Portland, ME: Calendar Island Publishers.

National Staff Development Council (NSDC). (2001). *National standards for online learning*. Oxford, OH: National Staff Development Council.

Newmann, F.M. & Wehlage, G.G. (1995). *Successful school restructuring: A report to the public and educators by the Center on Organization and Restructuring of Schools*. Washington, DC: American Federation of Teachers. (ERIC Document Reproduction Service No. ED387925).

Riel, M. & Polin, L. (2004). Communities as places of learning. In S. Barab, R. Kling, & J. Gray, (eds.), *Designing for virtual communities in the service of learning*. Cambridge, UK: Cambridge University Press.

Rogers, E. (1995). *Diffusion of innovations*. New York: The Free Press.

Schlager, M., Fusco, J., & Schank, P. (2002). Evolution of an on-line education community of practice. In K. A. Reninger & W. Shumar (eds.), *Building virtual communities: Learning and change in cyberspace*. NY: Cambridge University Press.

Sherry, L. (2000). The nature and purpose of online conversations: A brief synthesis of current research. *International Journal of Educational Telecommunications*, 6(1), 19-52.

Sherry, L. & Myers, K.M.M. (1998). The dynamics of collaborative design. *IEEE Transactions on Professional Communication*, 41(2), 123-139.

Sherry, L., Tavalin, F., & Billig, S.H. (2000). Good online conversation: Building on research to inform practice. *Journal of Interactive Learning Research*, 11(1), 85-127.

Stiggins, R.J. (1997). Student-centered classroom assessment. Upper Saddle River, NJ: Merrill, Prentice Hall.

Tavalin, F. (1998). *A guide to online critique*. Montpelier, VT: The WEB Project. Available at: The WEB Project, 58 Barre Street, Montpelier VT 05602, and at: *tavalin@sover.net.*

Tavalin, F. & Boke, N. (1998). *A guide to online discussion*. Montpelier, VT: The WEB Project. Available at: The WEB Project, 58 Barre Street, Montpelier VT 05602, and at: *tavalin@sover.net.*

Copyright © 2005, Idea Group Inc. Copying or distributing in print or electronic forms without written permission of Idea Group Inc. is prohibited.

Wiggins, G. (1989). Teaching to the (authentic) test. *Educational Leadership, 46:* 41-46.

Wilson, B., Ryder, M., McCahan, J., & Sherry, L. (1996). Cultural assimilation of the Internet: A case study. In M. Simonson (ed.), *Proceedings of selected research and development presentations.* Washington DC: Association for Educational Communications and Technology.

Copyright © 2005, Idea Group Inc. Copying or distributing in print or electronic forms without written permission of Idea Group Inc. is prohibited.

Chapter XI

Problems and Opportunities of Learning Together in a Virtual Learning Environment

Thanasis Daradoumis
Open University of Catalonia, Spain

Fatos Xhafa
Polytechnic University of Catalonia, Spain

Abstract

This chapter explores new ways of collaborative learning in a virtual learning environment based on our acquisition of knowledge from previous experience. We identify both the problems faced in real collaborative learning practices and the ways these problems can be overcome and turned into opportunities for more efficient learning. These issues concern pedagogical, organizational, and technical elements and constraints that influence the successful application of collaborative learning in distance education, such as efficient group formation, the nature of collaborative learning situations that promote peer interaction and learning, the student roles and tutor means of supervising and guiding the learning process, and an effective assessment of group work. The proposed methodology not only achieves better learning outcomes but also contributes to the tutor's professional development in a networked learning environment that facilitates social interaction among all participants while building on existing skills.

Copyright © 2005, Idea Group Inc. Copying or distributing in print or electronic forms without written permission of Idea Group Inc. is prohibited.

Introduction

One of the basic requirements for education today is to prepare learners for participation in an information society in which knowledge is the most critical resource for social and economic development. Moreover, contributed expertise and networked activities more and more characterize the emerging work environment. Elaborating, managing, and extending knowledge while productively collaborating with others and functioning within networks of experts will be essential for interactive and open organizations of the future.

Besides this generic objective, in the context of distance learning, one of the fundamental issues is to provide quality teaching and learning. This fact is even more imperative today when the rapid development of the information and communication technologies has initiated a shift away from conventional distance learning to networked learning. The result of these technological advancements has given rise to virtual learning environments or virtual campuses where the communicative process is crucial. In this line, the Open University of Catalonia (http://www.uoc.edu) has built a large and complex organizational virtual campus that provides an innovative pedagogic model for distance learning and teaching.

In this broad networked learning community infrastructure, our work seeks to investigate and facilitate learning and social interaction. In particular, we have started exploring the possibilities for new forms of learning and teaching by proposing the design of a methodology that promotes and encourages learning and collaboration through smaller communities of learners working together. Our involvement in this project has given rise to different methodological approaches to and practices of networked collaborative learning, depending on several factors such as: the nature of the experience (the type of virtual collaborative learning activities), the individual and group objectives, assessment issues, the tutor and student roles and commitment level, and the technology used for the implementation of the different practices.

In this chapter, we describe a methodological framework that uses existing technology, the Basic Support for Cooperative Work (BSCW) system (Bentley, Horstmann, & Trevor, 1997), and applies an innovative scenario for developing a Project-Based Collaborative Learning (PBCL) practice that is adequately embedded in a real practical educational context. In this context, we examine the conditions and methods that influence and enhance active learning through collaborative project development in shared workspaces, as well as some methods for triggering collaborative processes. Our approach brings new expectations and requires changes in attitudes and reward structures for both the learners and the teachers, such as new roles, different pedagogic and learning methods, and technological and training supports that enable learners to build up social structures, encourage learning, and develop critical thinking skills.

From a methodological point of view, this allowed us to identify that the lifecycle and progress of learning groups in a virtual environment goes across four critical processes (or phases) that require defining specifications that are quite different from those applied in individual learning in virtual environments. These phases are: *Group formation, consolidation, development* and *closing* (Daradoumis, Marquès, Guitert, Giménez, & Segret, 2001).

Copyright © 2005, Idea Group Inc. Copying or distributing in print or electronic forms without written permission of Idea Group Inc. is prohibited.

Our research relies on real case studies concerning collaborative learning and work in several undergraduate courses in a virtual learning environment. Based on the implementation and analysis of such educational practices with clearly differentiated goals, contents, and methodologies, our research enabled us to explore several possibilities that are related to the main objectives we set above.

We also take into account various recent research approaches in Computer-Supported Collaborative Learning and Work (CSCL and CSCW). On the one hand, there exist a few well-developed approaches that concern group formation and groupware's life (Pipek & Wulf, 1999; Supnithi, Inaba, Ikeda, Toyoda, & Mizoguchi, 1999); however, several of the goals and aspects mentioned above remain unclear.

On the other hand, some other research has focused more on the development of ways of observing and assessing collaborative knowledge building (Baker, deVries, & Lund, 1999; Greif 1998; Koenemann, Carroll, Shaffer, Rosson, & Abrams, 1999; Krange, Fjuk, Larsen, & Ludvigsen, 2002; La Marca, Keith Edwards, Dourish, Lamping, Smith, & Thornton, 1999; Soller, Wiebe, & Lesgold, 2002).

Our research goes further through an analysis and evaluation of the different collaborative learning situations, which allowed us to draw interesting insights about the structure and function of effective peer groups. In general, our research showed that a satisfactory culmination of the above four processes constitutes an important factor for the success of a collaborative learning experience. More specifically, our analysis indicated that an adequate realization of the four processes sets the rules and the conditions that should hold in order to create an appropriate context that favors quality of learning in a group and helps the learners to receive maximum educational benefits.

The chapter is organized as follows: First, as regards *group formation and consolidation*, we explore the different processes involved in constructing effective virtual collaborative learning groups, especially why, when, and how these processes affect group formation and to which degree they guarantee the creation of well-functioning and successful learning groups. A student can benefit from collaborative learning only if he/ she participates in supportive learning teams. For this reason, our research interest is to aid and provide the means for the configuration of learning groups that are appropriate for different learning situations. A key issue in this process is to make the educational function and structure of collaborative learning groups clear by identifying and making explicit both the individual and group learning and social goals, as well as the relationships, interaction processes, and roles that determine the nature and idiosyncrasy of the group.

Then, as regards *group development and closing*, we identify and address a variety of issues that have been raised in real collaborative learning practices and require a thorough analysis and study in order to reveal both the problems faced when working in-group and the ways these problems can be overcome and turned into opportunities for more efficient learning. These issues concern the various kinds of pedagogical, organizational, and technical elements and constraints that take part in and influence the successful application of collaborative learning in distance education.

The result of a three-year experience (1999-2002) in PBCL with a variety of virtual learning groups led us to propose a new methodology design for PBCL that is currently implemented, experienced, and tested and is starting to yield much better outcomes.

Copyright © 2005, Idea Group Inc. Copying or distributing in print or electronic forms without written permission of Idea Group Inc. is prohibited.

In this chapter, we show that the design and realization of our methodological framework, which models the collaborative learning process, has to take into account at least the following four key issues that are related to the above four group processes:

- An efficient group structure, organization, and planning that can be beneficial to all group members (group formation and consolidation processes).

- The nature of the collaborative learning situations and tasks that promote peer interaction and learning (group development process).

- The specific roles and the means the tutor has to take in supervising and guiding the learning process of the students (group formation, consolidation, and development processes).

- The realization of an effective assessment of group work (group closing process).

Our work shows a particular interest in investigating ways to process all the information that is available both in the global shared workspaces and in the specific group spaces to allow for an efficient group and individual tracking, awareness, and assessment. The proposed methodology not only achieves better learning outcomes for students but also contributes to the tutor's own professional development in a networked learning environment, where special emphasis is given to promoting technology's possibilities to facilitate social interaction between tutor and students and among students while building on existing skills.

Group Formation and Consolidation: Towards an Efficient Group Structure and Organization for Virtual Collaborative Learning

In this section, we face the question of how to achieve an efficient group structure and organization that supports and promotes effective virtual collaborative learning. In particular, we explore the factors that influence and promote the creation of a group and, more particularly, the processes that take place and govern and condition the group construction.

The composition of effective peer groups for collaborative learning becomes even more important in our case, since these groups have to be constructed from "scratch"; this means that their members do not know each other, and they also have to carry out both the acquaintance process and the final group formation at a distance, in a virtual shared workspace. For this reason, we regard group formation and consolidation as dynamic collaborative processes, guided by the setting of the following two goals that motivated and conducted our research:

Copyright © 2005, Idea Group Inc. Copying or distributing in print or electronic forms without written permission of Idea Group Inc. is prohibited.

1.   Explore the different processes involved in constructing effective virtual collaborative learning groups, especially why, when, and how these processes affect group formation and to what degree they guarantee the creation of well-functioning and successful learning groups (*formation phase*).

2.   Make the educational function and structure of collaborative learning groups clear, by identifying and making explicit both the individual and the group learning and social goals, as well as the relationships, interaction processes, and roles that determine the nature and idiosyncrasy of the group (*consolidation phase*).

As a matter of fact, on the one hand, the ultimate aim of a learning group is to achieve the common learning goal of carrying out a learning activity successfully and, on the other, to pursue a private benefit for its members by promoting learning and enabling better learning outcomes. This premise has to be taken into account when forming a learning group.

Processes Involved in the Group Formation Phase

To implement the group formation phase, we propose a four-step scheme that consists of well-defined processes that are carried out in a virtual shared workspace and whose purpose is to engage students in activities that lead to the creation of well-functioning learning groups. Figure 1 shows the general approach followed, while Table 1 presents the details of each process involved.

Given a 15-week lecture period for a semester course, the group formation phase, which is basically carried out asynchronously, may take on average about eight days to be

Figure 1: The group formation phase.

Copyright © 2005, Idea Group Inc. Copying or distributing in print or electronic forms without written permission of Idea Group Inc. is prohibited.

Table 1: Description of the group formation processes (average duration: 8 days).

	When	Why	How
Initiation	Initial action of the approach (duration: 2 days)	To initiate students into the new experience of virtual collaboration and to enable them to understand the notion and function of collaborative learning groups.	All students collaborate together to resolve a specific case study about what they need to know and do in order to construct effective virtual collaborative learning groups.
Introduction	Second action of the approach (duration: 1 day)	To provide both one's own and other relevant information in order to enhance a deeper knowledge of each other and to promote a better interaction.	Students should work out a personal report with important information, such as personal data, expertise level, work pace, available working time, temporal coincidence, goals, and attitudes toward collaborative learning, social aspects of collaboration, and previous experience in groupware.
Negotiation	Third action of the approach (duration: 4 days)	To form a learning group that satisfies both individual and group goals or to search for an open group that better fits one's personal goals and needs.	Each student initiates a negotiation process, either with individual candidate members whose characteristics match his/her own in order to form a reliable and effective learning group, or with a possible open group in order to become an active member of it.
Group Proposal	Final action of the approach (duration: 1 day)	To inform and ask the tutor to approve the definite formation of a learning group and the initiation of the next collaborative phase.	A member of the recently constructed learning group informs the tutor about the group's constituent members, facilitates each member's data, and asks for its final approval.

completed. In that case, Table 1 indicates the suggested timeframes for each process. Our experience with PBCL practices during the period 1999-2002 (six semester courses) involved about 250 students in total, that is, more than 60 learning groups, each group consisting of four members.

A culture for collaboration must be based on relationships characterized by trust, motivation, encouragement, mutual support, and openness. Since people are in general reluctant to be engaged in a shared experience when they do not know the other parties, our approach proves to be effective, since it gives students the time and the opportunity to meet, interact in an informal networked setting, begin to develop relationships, and evaluate and learn each other's interests and intentions well enough to figure out the most adequate collaborators with which to form a group. Taking time to build this "social capital" at the beginning of the collaborative practice increases the effectiveness of the team later on. Even more, it was proved that it encourages some less able or less enthusiastic students to join good groups, a fact that could be very hard for them otherwise.

Experience with group formation in different learning situations revealed more benefits if this phase is carried out adequately, that is, according to the methodology and guidelines proposed. This fact increases the probability of achieving well-functioning and successful collaborative learning groups. These benefits are summarized as follows:

Copyright © 2005, Idea Group Inc. Copying or distributing in print or electronic forms without written permission of Idea Group Inc. is prohibited.

1. The *initiation* and *negotiation* processes allow students to increase their understanding of virtual collaboration and get a feeling of both possible benefits and problems that are intrinsic to it.

2. The *introduction* process encourages students to set clear individual goals and expectations while fomenting openness and honesty as regards their capabilities, skills, and attitudes to collaboration.

3. This initial level of collaborative expertise gives students both the motive and the feeling of confidence and readiness to tackle subsequent collaborative phases better.

4. Students become familiar with a collaborative learning technology before they start real work.

More specifically, the formation phase is best accomplished if it is carried out sequentially in a bottom-up order (as shown in Figure 1). Doing so, the experience acquired in one process serves to set up the basis for a more effective participation and completion of the next process. In that sense, the tutor plays an important role in supervising, guiding, and motivating students through the whole process. Taking into account that group formation is carried out asynchronously, the tutor's role is considered crucial.

First, the tutor should monitor the virtual shared workspace by observing and checking that each student is participating and contributing correctly to each process, and intervene when he/she detects low participation or misleading use of the workspace.

Second, the tutor should especially monitor the negotiation process by orientating and supporting those students who seem not to find their way, as well as organizing and restructuring the workspace itself, since the interaction load can be very heavy, and this may make it difficult for the students to locate adequate candidate members or an incomplete group they want to join.

Finally, the tutor should approve all the groups formed, identify individual members left without a group, and assign them to a group as adequately as possible. The latter constitutes a delicate and difficult decision. Experience has shown that groups whose formation was not based on students' own initiative tended to fail or have poor outcomes. For this reason, it is very important that the tutor looks at the curriculum of these students very carefully and opts for the best match among them.

The basic criterion on which the tutor has to base his or her decision regarding how to arrange students in possible well-functioning groups is the degree of commitment shown by each student. This can be inferred by looking at the students' contributions and interaction with others in the shared workspace during the first three processes. A careful examination of a student's contributions is needed in order to find out important elements that can lead to the identification of the correct student profile and commitment and, subsequently, to an adequate group placement decision. These elements include the student's intentions, possibilities, needs, expectations, desires, motivation, responsibilities, availability, and skills. Doing so, the tutor is able to identify those students who show weak commitment and most probably are going to drop out the course, which in fact happens in most cases.

Copyright © 2005, Idea Group Inc. Copying or distributing in print or electronic forms without written permission of Idea Group Inc. is prohibited.

Figure 2: The group formation phase implemented in the BSCW system.

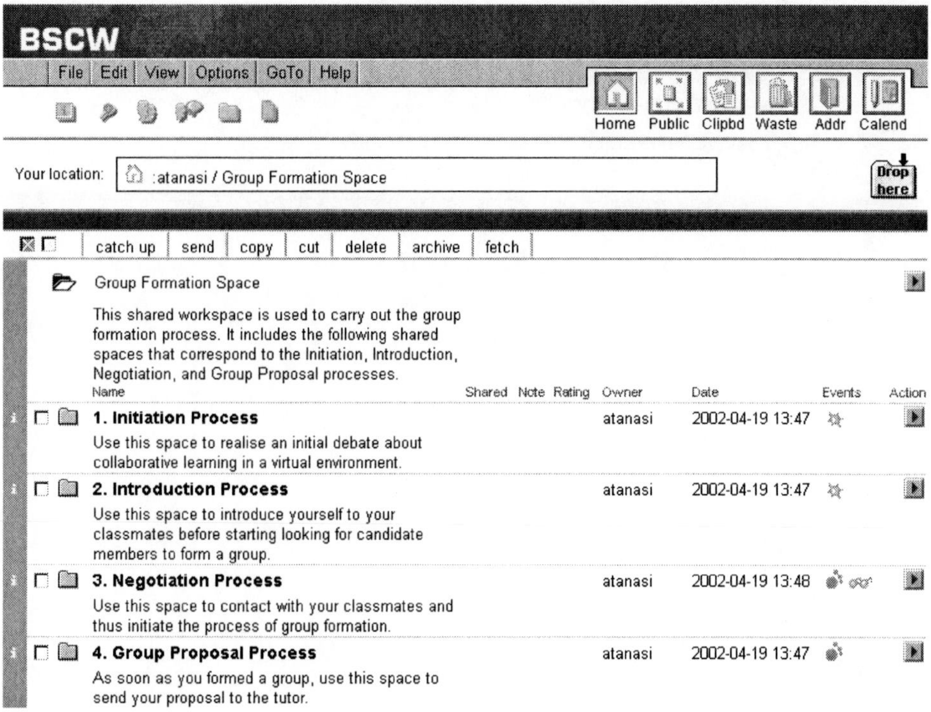

The completion of this approach results in the formation of groups that consist of members with the strongest possible degree of commitment, which satisfies the necessary requirements needed for efficient collaborative work, learning, and achievement of the course common and individual objectives.

Figure 2 shows the basic implementation of the four group formation processes on the BSCW system. Each process is represented by a shared workspace, which provides all the necessary means and functionalities for carrying out the sequence of actions that make up the process.

For the sake of an example, Figure 3 shows part of a negotiation process, which clearly illustrates that different types of actions may take place, depending on the participants' intentions, needs, and wishes.

Copyright © 2005, Idea Group Inc. Copying or distributing in print or electronic forms without written permission of Idea Group Inc. is prohibited.

Figure 3: Part of a negotiation process carried out on the BSCW system.

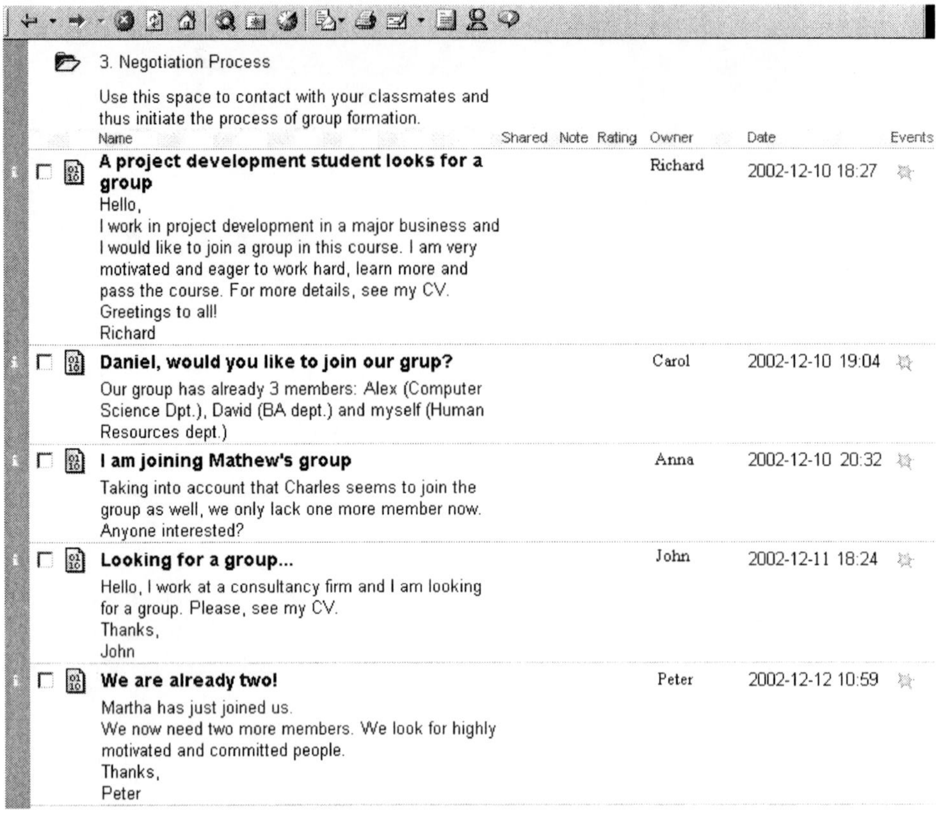

Consolidating Effective Virtual Learning Groups

When the group formation phase is over, the students can still take a week to consolidate their group. We consider this phase important in order to achieve effective and successful learning groups. In fact, each group goes through this process with a basic objective: to make the educational function and structure of the group clear.

To that end, each group has to work out and establish the group regulations — the rules and the conditions that should apply and that will conduct the working methodology and goals of the whole group, as well as the goals and relationships of the individual group members during the development phase. Most importantly, the group regulations should be both specific and flexible so that they build up an appropriate context that favors quality of learning in group and helps the learners to receive maximum educational benefits.

At this stage, our analysis of student interactions identified several elements or aspects that can help determine and delineate the educational function and structure of the group while, at the same time, are shown to influence, support, and enhance the group

Copyright © 2005, Idea Group Inc. Copying or distributing in print or electronic forms without written permission of Idea Group Inc. is prohibited.

development and learning process. In other words, this analysis indicates that these elements constitute the influencing factors that may yield effective (or ineffective) group learning experiences, depending on whether the group members are able not only to establish a clear identification of these elements but also to achieve the mutual acceptance or reinforcement of these elements at the end of the consolidation process. The ultimate aim of this approach is to build both a group and individual student model to promote meta-cognitive skills that help learners understand themselves, increase group performance, and improve learning. Subsequently, we turn to describe each of these elements or aspects:

- Individual and group learning and social goals

This concerns the identification and setting of each member's individual learning goals in addition to the common group learning and social goals to be accomplished. This goal specification depends on the collaborative activity to be performed, as well as on the possibilities, skills, time availability, and commitment of each member. A clear identification of the goals and the responsibilities of each member will result in elaborating an adequate working methodology, good planning and timing, and a fair and viable assignment and distribution of the constituent tasks to be performed. In addition, it will contribute to a more logical, intuitive, and structured organization of the group's shared workspace.

- Relationships among group members

Analysis of the student interactions, mainly during the group consolidation (but also in the formation) phase, revealed that several types of relationships are developed and sustained by group members. Such relations are: *confidence, commitment, responsibility, motivation, acquaintance, coordination, support, encouragement, openness,* and *equality in contribution, responsibility, and opportunity*. These relations characterize different types of group interaction and the roles that can be played by each member. Most importantly, they reveal important information that may help the tutor understand the internal functioning of the group, which makes it possible for him/her to check the appropriateness, capabilities, and viability of the group.

This consists of identifying possible weak points, which then allows the tutor to take needed remedial action. A learning group is considered to be viable if each member has reached a fairly deep degree of *acquaintance* with the other members' profiles. This is an important element that should occupy a prominent position in the group model.

Experience has shown that in order to achieve well-functioning groups, their members should know fairly well, and from the very beginning, essential features of the other members, such as their potential, skills, and time availability. For instance, knowing each other's skills allows the group to better determine how the whole group can benefit from the individual skills of its members, and thus cope better with the current learning situation.

Another important characteristic that governs group viability and success is *commitment*. A clear knowledge of each member's level of commitment to the common group goal transmits and creates confidence and security for all members. In case some members of

Copyright © 2005, Idea Group Inc. Copying or distributing in print or electronic forms without written permission of Idea Group Inc. is prohibited.

a group initially show weak commitment, promoting discussion between those members with members demonstrating a strong commitment can help increase the confidence level and produce better collaboration among all the members of the group. Thus, despite the long way the group has to cover during the development phase of the project in order to achieve the common goal, all its members feel confident that they will finally manage to complete it all together.

Moreover, when clear evidence in the shared workspace shows that the above characteristics are present in the group member relationships, the members are able to define a more accurate group model with clearer and differentiated roles, propose well-defined and more feasible goals, and draw a more robust planning and timing of the learning activities.

On the contrary, our experience proved that lack of the above elements in the group member relationships or only a weak presence of them in the shared workspace of the group is evidence of a rather problematic situation in the group. For instance, when members have loose or vague acquaintance of each other, they will not be well-informed and knowledgeable with regard to the involvement and skills of the other members. When their interactions are ambiguous and lead to unclear commitment, there is a high probability that this group will fail to achieve its objectives. In fact, it is demonstrated that further important relationship elements, such as motivation, responsibility, coordination, or support, are also missing or are not properly maintained among group members.

A Methodological Design for the Project Development Phase

The development phase constitutes a period of 12 weeks in which the real collaborative work and learning takes place in each group. In particular, each group is assigned a particular BSCW workspace, and the group members are engaged in collaborative work for implementing a project. To incite and promote collaborative interactions and make the collaborative project development and learning possible, the tutor provides the students with a specific methodology, which leads students through a guided process that involves achieving several learning objectives. This is reflected by a timing schedule that is suggested to the students, setting out the project phases and the learning objectives of each phase, thus creating an adequate context to collaborate with particular rules and tasks.

Subsequently, this context gives each group the potential to set its specific goals for accomplishment and to elaborate an appropriate planning and organization of the group workspace that best fits the members' goals and needs.

First, we describe the problems detected by the application of an initial methodology design used for collaborative project development, which are related to pedagogical, organizational, and technical constraints. Then, we show how this experience suggested a change to the original model and the design of a new methodological framework that best fits project-based collaborative work and learning (PBCL), provides more opportu-

Copyright © 2005, Idea Group Inc. Copying or distributing in print or electronic forms without written permission of Idea Group Inc. is prohibited.

nities and possibilities, and facilitates both learning and better outcomes. The function-ality and suitability of new model is also experienced and tested.

Limitations and Drawbacks in Modelling Collaborative Project Development

The project consists of developing a software product for a real-world problem. The project development consists of five well-differentiated phases: project specification and planning, design, implementation, testing, and final report. A detailed description of the purpose and content of each phase is beyond the scope of this chapter. For a detailed explanation, see Daradoumis, Xhafa, and Marquès (2002).

The initial model was based on a heavy application of collaborative work and learning at all project phases; that is, the development of each phase was entirely carried out collaboratively. The groups ran on a democratic basis with a coordinator at each phase (each group member had this responsibility at a particular phase). The role of the coordinator is intended not only to develop skills on leading a project, but also to establish the communication between the group and the tutor and to coordinate the collaboration among the group members. The experience of several semesters showed that, in fact, this is a valid model for realization of a project in a virtual group of four members. However, we have observed that the realization of all project tasks collaboratively tends to fail in groups with weak commitment. This fact resulted in several problems at pedagogical, organizational, and technical levels that we identify below.

At the pedagogical level, we observed that the extremely high degree of collaboration required to complete tasks and obtain deliveries among members was achieved, to some extent, at the expense of individual contributions. As a consequence, it was difficult for the tutor to do a proper tracking of the individual work, since it was delivered as a collaborative product. Moreover, we observed several situations in which the work volume was not correctly distributed and completed by all group members. The weak commitment of some implied an overload for other group members.

At the organizational and structural level, the high degree of collaborative interactions causes the generation of a big volume of information that has to be properly organized and managed. Information management can be very hectic, time-consuming, and coun-terproductive, thus hindering the group progress and learning. Moreover, this situation can affect the group structure itself, since it may provoke a member to abandon the group because of unexpected work overload. In this case, the tutor has to restructure the affected group, either introducing a new member to the group or reallocating its remaining members to other groups. In both cases, however, the entry of a new member in an existing group is most probably fraught with serious difficulties of adaptation and reorganization of the current group model. Finally, since the communication load is very heavy, the group members, and especially the coordinator, spend more time in managing interaction than contributing to real work and learning.

At the technical level, an increasing collaboration to solve the project tasks implies more frequent, or even intensive, interactions for decision making or conflict resolution, which are mostly synchronous. In that case, BSCW, being basically an asynchronous collabo-

Copyright © 2005, Idea Group Inc. Copying or distributing in print or electronic forms without written permission of Idea Group Inc. is prohibited.

rative tool, provides limited support to synchronous communication, since its functionalities are not sufficient to carry out complex online debates. As a consequence, the groups have to use extra tools to hold and record a discussion and then reflect the results of it on their shared workspace in BSCW. In addition, an intensive use of the BSCW system slows it down and remote access to it becomes tedious.

A Model for PBCL that Promotes Better Collaboration, Interaction, and Learning

To overcome the above problems, we decided to apply a new methodology for PBCL that is based on increasing the individual contribution of members and integrating it adequately to collaborative learning activities during the overall process. Research on cooperative learning indicates that cooperative learning methods are most likely to enhance learning outcomes if they combine group goals with individual accountability. Interestingly, as we will see below, a clear separation of concepts and responsibilities - individual versus collaborative - not only does not damage collaborative learning, but instead it gives rise to new learning situations.

The proposed methodology adapts PBCL to the learner's needs and goals by designing an adequate combination of individual and collaborative learning activities that are mostly based on asynchronous interactions among group members. The purpose of the model is to alleviate the communication and work load of the group members and focus on quality rather than quantity interactions.

To achieve this, the tutor designs the software project to be developed in four sub-systems -roughly of the same amount of work- and each group member is assigned to elaborate a subsystem within a phase. Since the project is going to be completed in five phases, the subsystems assigned to members are rotated so that each member is assigned a different subsystem during the first four phases. On the contrary, the last phase -in which a final report has to be elaborated- is completed collaboratively among members. More specifically, we separate each phase of the project into two subphases: the first one consists of assigning individual learning activities and concludes with an individual delivery, whereby the second one consists in setting up the collaborative activity of putting together the individual deliveries into a common delivery of the current phase.

Thus, for example, to complete the design phase, the members of the group first work individually to complete the design of the subsystem they are assigned -each student is assigned one subsystem out of four subsystems. Next, the students unify their individual designs into a common delivery -the design of the system as a whole. In fact, the separation of the work into individual and collective learning activities gives rise to new learning situations. In particular, on the one hand, even during the individual work, the members have to collaborate among themselves, as they need to resolve subsystem dependencies. As a matter of fact, they may make decisions themselves, as far as their subsystem is concerned, but, in most of the cases, they tend to consult the rest of the members in search for the best decision. On the other hand, while unifying the individual deliveries into a common delivery, each member has to revise the individual deliveries of the rest of the members, and all of them have to discuss and make decisions together on how to solve the expected problems of the individual deliveries. Moreover, the tutor

Copyright © 2005, Idea Group Inc. Copying or distributing in print or electronic forms without written permission of Idea Group Inc. is prohibited.

gives them an annotated report including the most important observations about their individual deliveries, which prompts the group members to discuss and sort them out as well.

In fact, our methodology introduces the notion of debate as a reinforcing element to collaboration and learning in the sense that group members should employ it construc-tively in order to unify their individual outcomes into a common product. In other words, this fact requires that members go through the individual work of others with a critical approach, while making suggestions for further improvement – all-in-all, these kinds of constructivist debate not only promote learning but also support and strengthen the group cohesion.

This scenario is repeated at each of the four phases of the project, and we have observed that these learning situations allow students to acquire a more accurate view of the project as a whole and a better understanding of the theoretical concepts of the software development cycle.

Our experience shows that in this way, the commitment of the members is considerably increased and, consequently, the outcome of the project is as well. In fact, the new methodology allows each member to keep track of the project development easier and carry out the project tasks assigned to him/her more effectively, "switching" to the collaboration mode when it is really necessary. This reduces anxiety, reinforces the group cohesion, and considerably diminishes the probability that a member abandons the group or that the group breaks up.

As a matter of fact, we experienced the new methodology in a virtual class of 60 students, in which 15 groups of four members were formed, 14 of them following the group formation approach. This experience showed that all 14 groups were kept alive during the whole project development process, and the great majority of them had excellent results. In other words, 55 students out of 60 continued and concluded their studies successfully, while the five students who dropped out did so at the beginning of the course for reasons that were alien to the process followed.

Finally, as for the tutor, this way of carrying out the project development collaboratively helps him/her identify in time those students that are faced with serious difficulties due to their lack of technical knowledge. In such a case, the tutor is able to give the students proper guidance and new insights to continue with the project more effectively, thus reducing the likelihood that they will drop out of the project.

Closing the PBCL Cycle

The closing phase of the PBCL practice includes the final assessment of the project outcome as well as of the collaborative learning itself, that is, the proper functioning of the learning group as such. This final learning situation is very interesting since students have to judge the whole project development process, the learning methodology, the organization aspects of the group, etc. This constitutes the culminating point of the PBCL that certainly gives students the confidence to face real-world project developments in their professional life.

Copyright © 2005, Idea Group Inc. Copying or distributing in print or electronic forms without written permission of Idea Group Inc. is prohibited.

As regards the previous model, it was very difficult for the tutor to assess correctly each individual member; it was hard to identify and evaluate both the quantity and the quality of individual work, since it was encapsulated by intensive group work that did not make clear who in the group did what. Thus, the tutor was not able to confirm whether individual goals were really achieved or learning was improved.

To that end, the new model provides a more effective assessment of collaborative learning, since individual work and learning can be completely identified and measured in each project phase, whereby the members' contributions can be easier traced and judged as to whether were being collaborative and supportive to group work.

Conclusions and Future Work

This chapter presented a complete model for project-based collaborative learning based on four basic phases: *group formation, consolidation, project development,* and *closing.* We identified and addressed a variety of issues that have been raised in real collaborative learning practices, and we showed how we were able to overcome important problems that made it difficult to achieve a fruitful collaboration. Our current collaborative learning model seems to provide more opportunities and real possibilities for tutors and learners to get involved in effective instruction and learning, respectively. Future work includes a more detailed analysis of all the above issues in order to explore and determine them at a much finer grain. We would like to study and estimate how this model influences several parameters that are related to the shared workspace, such as the communication load, the documentation load, the workspace organization, the group structure, and the information and interaction management. This will require statistical and/or data-mining analysis of real data we have already gathered from a considerable number of group workspaces. The ultimate aim of this analysis is the elaboration of taxonomy and a detailed description of the relationship between the different kinds of the above influencing factors. This will provide us a better understanding of group interaction and determine how to best support the collaborative learning process.

Acknowledgments

This work has been partially supported by the Spanish MCYT project TIC2002-04258-C03-03.

References

Baker, M., de Vries, E., & Lund, K. (1999). Designing computer-mediated epistemic interactions. In S.P. Lajoie & M. Vivet (Eds.), *Proceedings of the 9th International*

Copyright © 2005, Idea Group Inc. Copying or distributing in print or electronic forms without written permission of Idea Group Inc. is prohibited.

Conference on Artificial Intelligence in Education (AI-ED 99), pp. 139-146. Le Mans, France, July. Amsterdam: IOS Press.

Bentley R., Horstmann, T., & Trevor, J. (1997). The World Wide Web as enabling technology for CSCW: The case of BSCW. *Computer-Supported Cooperative Work: Special issue on CSCW and the Web, Vol. 6.* Boston, MA: Kluwer Academic Press.

Daradoumis, T., Xhafa, F., & Marquès, J.M. (2002). A methodological framework for project-based collaborative learning in a networked environment. *International Journal of Continuing Engineering Education and Life-long Learning.* Special Issue on Collaborative Learning in Networked Environments, 12 (5/6): 389-402.

Daradoumis T., Marquès, J.M., Guitert, M., Giménez, F., & Segret, R. (2001). Enabling novel methodologies to promote virtual collaborative study and learning in distance education. In *Proceedings of the 20th World Conference on Open Learning and Distance Education* (The Future of Learning – Learning for the Future: Shaping the Transition). Düsseldorf, Germany, 1-5 April.

Greif, I. (1998). Everyone is talking about knowledge management. In *Proceedings of ACM 1998 Conference on Computer-Supported Cooperative Work,* pp. 405-406. Seattle, WA: ACM Press.

Koenemann, J., Carroll, J. M., Shaffer, C. A., Rosson, M.B., & Abrams, M. (1999). Designing collaborative applications for classroom use: The LiNC Project. In A. Druin (ed.), *The design of children's technology,* Chapter 5, pp. 99-119. San Francisco, CA: Morgan Kaufmann.

Krange, I., Fjuk, A., Larsen, A., & Ludvigsen, S. (2002). Describing construction of knowledge through identification of collaboration patterns in 3-D learning environments. In G. Stahl (ed.), *Proceedings of the Computer Support for Collaborative Learning Conference, pp.* 7-11. January.

LaMarca, A., Keith Edwards, W., Dourish, P., Lamping, J., Smith, I., & Thornton, J. (1999). Taking the work out of workflow: Mechanisms for document-centered collaboration. In S. Bødker, M. Kyng, & K. Schmidt (eds.), *Proceedings of the 6th European Conference on Computer-Supported Cooperative Work,* pp.1-20. Copenhagen, Denmark, 12-16 September. Boston, MA: Kluwer Academic Publishers.

Pipek, V. & Wulf, V. (1999). A groupware's life. In S. Bødker, M. Kyng, & K. Schmidt (eds.), *Proceedings of the 6th European Conference on Computer-Supported Cooperative Work,* pp. 190-218. Copenhagen, Denmark, 12-16 September. Boston, MA: Kluwer Academic Publishers.

Soller, A., Wiebe, J., & Lesgold, A. (2002). A machine learning approach to assessing knowledge sharing during collaborative learning activities. In G. Stahl (ed.), *Proceedings of the Computer Support for Collaborative Learning Conference (CSCL '02),* pp. 128-137. 7-11 January.

Supnithi, T., Inaba, A., Ikeda, M., Toyoda, J., & Mizoguchi, R. (1999). Learning goal ontology supported by learning theories for opportunistic group formation. In S.P. Lajoie & M. Vivet (eds.), *Proceedings of the 9th International Conference on Artificial Intelligence in Education (AI-ED 99),* pp.67-74. Le Mans, France, August. Amsterdam: IOS Press.

Copyright © 2005, Idea Group Inc. Copying or distributing in print or electronic forms without written permission of Idea Group Inc. is prohibited.

Chapter XII

Web-Based Learning by Tele-Collaborative Production in Engineering Education

Amiram Moshaiov
Tel-Aviv University, Israel

Abstract

This chapter deals with the need and the potential for reforming design projects into web-based learning by tele-collaborative production in engineering education. An overview of related topics is provided, including the impact of computer-mediated communication (CMC) on engineering and engineering education, the role of social creativity and dominance of multidisciplinary thinking in modern engineering, assessing designers and the design process, and more. In addition to discussing the need and the potential for reforming engineering design projects, two major strategies for web-based learning by collaborative production in engineering education are discussed. It is concluded that short projects focusing on early design stages should be encouraged for the current assimilation of tele-collaboration, whereas long and complex design tasks may currently be better handled in a local framework.

Copyright © 2005, Idea Group Inc. Copying or distributing in print or electronic forms without written permission of Idea Group Inc. is prohibited.

Introduction

"Scientists discover the world that exists; engineers create the world that never was."

Theodore von Kármán

The new era of information and communication technology (ICT) involves both the need to train engineers for tele-collaboration, and the opportunities to change the traditional teaching and learning methodologies in engineering education. A flexible and effective combination of learning engineering principles and experiencing some design process is necessary for achieving a balance between theoretical knowledge and application skills. Design projects, assigned to small teams of engineering students, are traditionally used for this purpose, providing a mechanism for learning by collaborative production. Reforming such design projects into web-based learning by tele-collaborative production in engineering education (Tele-CPEE) is in its infant stage. However, considering such a reform is timely due to the increasing role of distributed design and manufacturing in modern engineering and globalization.

Despite major advancements in computer-mediated communication (CMC), computer-supported collaborative learning (CSCL), computer-supported cooperative work (CSCW), and collaborative engineering, there is accumulating evidence that Tele-CPEE is not simple to realize. Tele-CPEE means distributed projects in the collaborative learning process of engineering design and production. Tele-collaboration indicates that a substantial distance may exist between the partners involved, including not just the students but also the project planning partners. The physical gap complicates the cooperation and the infrastructure needed for Tele-CPEE, in comparison with those required for Local-CPEE (web-based learning by local-collaborative production in engineering education, such that collaboration takes place within an institute and/or its vicinity). As such, it requires ad hoc extended enterprising infrastructure, which might be more difficult to set up in comparison with that needed by Local-CPEE. At present, the practice of Tele-CPEE in educational settings is primarily driven by research interests. Its future may depend both on its merits and on the ability to overcome its inherent difficulties.

This chapter presents a pedagogical and engineering-practical approach in discussing the future of CMC-based collaborative production in engineering education. Three major questions are answered. The first question is a preliminary one and deals with the need to have design projects at all. The next resulting questions are: Should we tele-collaboratively produce in engineering education? and Should we focus on Local or on Tele-CPEE? In answering these questions, this chapter points at the difficulties of implementation and the possible benefits to the students. Based on the current extensive review and supported by our recent research efforts and experience, we propose here that short, well-planned projects focusing on early design stages should be encouraged for the current assimilation of Tele-CPEE. Long and complex design tasks may be better handled in a Local-CPEE framework before attempting to employ them by Tele-CPEE. It is noted that a profound assimilation of Local and Tele-CPEE will constitute a reform of design education. Finally, we list several related topics for future research and implementation.

Copyright © 2005, Idea Group Inc. Copying or distributing in print or electronic forms without written permission of Idea Group Inc. is prohibited.

Background

Preface

The influence of technology on society and, in particular, the emergence of high technology has brought engineering to the attention of society at-large. More than ever, the role of engineering in advancing the world is becoming apparent, together with its potential hazards to the environment and the future of civilization. Concerns about recent developments in biotechnology and genetic research, current tension regarding global terrorism, and the ease of availability of engineering know-how and materials provide a reminder to the ever- increasing importance that engineering education be accompanied with education for global responsibility. Burns (2000) suggested the need for a broad education for technology and its significance in relation to the greater power of technology to inflict harm. In an introductory letter in the Fall 2002 issue of the *MIT Spectrum*, MIT President Charles Vest made the following statement:

MIT has developed a plan to prepare students for life through an educational triad of academics, research, and community...Involvement with the broader community, meanwhile, helps to teach students communication skills, interpersonal and leadership skills, and critical thinking about societal issues...

The recent ThinkCycle project of the MIT Media Lab and the associated 2nd International Conference on Open Collaborative Design for Sustainable Innovation demonstrate the use of engineering design for community involvement activities led by MIT (details are available at http://www.thinkcycle.org/home). Moshaiov (2000a) has stressed the potential benefits of Tele-CPEE in relation to some issues of global responsibility and proposed a special student design competition. His suggestions are based on an original concept that he has termed TCRCT for Tele-Collaborative Robot Competition Team, and on the idea of including collaboration assessments as part of the scoring. In a later publication, Moshaiov (2000b) argued that high-school level competition may be superior to college-level in promoting assistance to developing countries. In this chapter, the focus of attention shifts from considering tele-collaboration in the context of engineering education for global responsibility to some general issues pertaining to both Local and Tele-CPEE and their potential realization in engineering education.

Impact of CMC on Engineering

Machines and, in particular, those developed during the industrial revolution were for the most part more powerful than human beings, but it was only with the availability of computers that they have started to posses some "intelligence." According to Sowa (1984), the term cybernetics was coined by Norbert Wiener in 1948 for "the entire field of control and communication theory, whether in the machine or in the animal." Sowa explained that cybernetics and artificial intelligence are closely related but have focused

Copyright © 2005, Idea Group Inc. Copying or distributing in print or electronic forms without written permission of Idea Group Inc. is prohibited.

on different levels of mental processing. Advancements in cybernetics and its realization resulted in serious discussions of the possibility that robots will become superior to humans (e.g., Minsky, 1994). Robots are used now to study nature and in particular for understanding intelligence, as explained in Pfeifer and Scheier (1999). Moreover, researchers are trying to imitate nature and automatically build artificial life forms (e.g., Lipson & Pollack, 2000). The practicality of the symbiosis between humans and computers is most notable when considering CSCW and CSCL as detailed below.

The widespread use of personal computers with their associated networks has, for some time, led to attempts at not only using these resources for distributed data processing and control of machines, but also for communication. A significant result of CMC is its influence on tele-collaboration abilities of humans. Terms, such as CSCW, CSCL, and "Groupware" have been introduced and accompany on-going applied research activities. Social scientists and computer and information scientists have worked together to create information systems more sensitive to human organization and needs (Bowker, Turner, Star, & Gasser, 1997). The sociotechnical movement proposed that changes in both technical and social systems were needed to obtain their right balance in the total system. A classical example of such a consideration is the impact of cognitive psychologists on technical design of Human-Computer Interaction (HCI). User-centered design of computer interfaces became the central philosophy (Rogers, 1997). In contrast to HCI that focuses on the individual user, CSCW and CSCL clearly concentrate on collaboration and therefore may "truly" manifest the "socio" element of the sociotechnical approach. The general process of tele-collaboration has many sociotechnical aspects that are detailed in the common literature on virtual (distributed) teams. The challenges faced by such teams are related to: performance control, communication, team building, cultural issues, cost, technology complexity, workflow, technical support, effort recognition, inclusion vs. isolation, and management resistance (Haywood, 1998). The significance of these aspects is that distributed teams may fail to accomplish their mission due to the many added problems created by distance and by the communication difficulties of tele-collaboration. CSCW methodologies and tools are constantly developed to assist both managers and members of distributed teams. These developments have certainly influenced current engineering practice as detailed below.

Engineers have used computers extensively to handle the complexity of their profession. Moshaiov (2000a) overviewed many issues concerning the use of computers in engineering with a special focus on collaborative engineering in modern era. Originally, computers were used for their data-processing capabilities in engineering analysis, design, and manufacturing. Early use of computer-based drawings improved the capabilities of engineers to present and convey engineering information. Yet, such drawings were discussed either face-to-face or by early means of distance communication. The traditional use of computers in engineering work has been changing due to the availability of computer networks and methods of CMC. Moreover, the availability of CMC is accompanied by other technological, economical, and political events that are shaping the post 2nd World War era. As a result, world industry is changing rapidly. Both small and large manufacturers have to respond rapidly to the market demands. The term JIT (just-in-time) has become a common concept among industrial engineers, expressing both the market demand and the need for restructuring industrial enterprises. The interconnected world economy and the need to stay competitive in JIT consumer-

Copyright © 2005, Idea Group Inc. Copying or distributing in print or electronic forms without written permission of Idea Group Inc. is prohibited.

manufacturer interrelations, together with the availability of ICT, created a situation in which working habits and structures are changing. Industries are increasingly forming joint design and manufacturing teams. These teams are not necessarily isolated within one company but are formed by cross-functional, multicompany, and multinational teams. As a result of the market pressures, and based on CMC, extended enterprises evolved from the concepts of agile competitors and virtual organization (Goldman, Nagel, & Preiss, 1995). The multinational, extended industrial enterprise has been described as the collection of all the suppliers, and the customer, in the total value-adding chain, with interaction among people being facilitated by extensive use of networked computer systems. The extended industrial enterprise, which straddles national borders and covers the globe, involves, among other factors, the key elements of tele-collaborative design and prototype development.

A fundamental issue in the evolution of virtual enterprises is the efficient utilization of distributed engineering resources that can be achieved by their integration via the international communication highways. This means that implementation of CMC for engineering design and production is not restricted within a given enterprise. The primary engineering resources, which are integrated by CMC, are the distributed human resources, the information systems, and the manufacturing facilities.

Computers can help reduce the labor needed by the industry. For example, the shipbuilding industry has traditionally been a labor-intensive industry. Computers were introduced to this industry as a means to help the designers. Traditional naval architects spent many hours learning how to manually draw the hydrodynamic shape of a ship. Both drawing and weight calculations involved tedious work, and a great deal of the education and engineering work efforts were devoted to them. Similarly, ship strength calculations were done manually. Due to the very complex nature of the structural details of ships, both the strength calculations and the detailed drawings of the ship structure were just as tedious and labor intensive as the shape drawing and weight calculations. Computers have changed ship design dramatically, and efforts to use them to assist manufacturing have been carried out in parallel. In the past, the European shipbuilding industry has suffered from the cost of labor and the inability to stay competitive to the extent that modern shipbuilding shifted to Japan and later on to Korea. Researchers and shipbuilders from Japan, Korea, USA, and Europe have made intensive efforts to automate the shipbuilding industry (e.g., Chryssostomidis, 1990). Recently, the world shipbuilding industry has started to realize that CMC can be used not just within a shipyard. The concept of global virtual enterprising of the world shipbuilding industry is currently being explored and reports are encouraging (e.g., Filling, Diggs, & Helgerson, 2000).

Using a recent review by Andreason (2001) of twenty years of international conferences on engineering design (ICED), several important observations are made. In the late 1990s, the number of papers on networking grew immensely. Another important trend is the increased number of papers during the late 1990s on coordination, cooperation, teams, and human resources in engineering design and product development. The review efforts of Andreason led to a conclusion that product development is now seen as a new profession with a focus on team competence. Moreover, ICT is setting the agenda for design practice, and efforts are being made to understand the human operator. Cybernetics, artificial intelligence, and ICT have narrowed the divide between engineering and non-exact sciences such as philosophy, psychology, and social science. Monarch,

Copyright © 2005, Idea Group Inc. Copying or distributing in print or electronic forms without written permission of Idea Group Inc. is prohibited.

Konda, Levy, Reich, and Ulrich (1997) discussed the crossing of the sociotechnical divide as related to engineering, and point out the significance of mapping the networks in the making. Moshaiov and Kaynak (2001) suggested a framework to use tele-collaboration based competitions within engineering education for studying a new field, which they have termed socio-mechatronics. According to their description,

The nature of information and communication technologies gives a unique opportunity to record and analyze the information flow that takes place during tele-collaboration including any discussions by the team members. Moreover, the whole process of tele-collaboration involves information systems and design tools that use some presentations of the engineers' knowledge. A suggested challenge of socio-mechatronics is therefore to examine how can failure be avoided and performance be improved. Strengthening the members' semantic communication by tailored groupware poses a special challenge. In doing so, we have to understand not just the team members' professional languages but also their sociological and psychological performance. The flow of information during tele-collaboration leaves a digital trace that can be used for overcoming the socio-mechatronic challenge. (p.43)

The research framework suggested by Moshaiov and Kaynak (2001) resulted in the COMEDI proposal involving industry, research institutes, and universities. Some research work related to the proposal involves the assessment of engineering groupwork and is presented in Moshaiov, Kaynak, Reich, Bar-El, Klunover, and Akin (2002). Universities with their research and teaching infrastructure provide a setting for research in groupwork as related to learning and working. The above discussion suggests that a strong research motivation exists to include tele-collaboration in the engineering education system.

Impact of CMC on Engineering Education

Engineers are educated to use computers, which became the major tool of supporting their work. Yet it appears that while the use of CMC in collaborative engineering is quite evident and growing, its use for collaborative learning of the engineering profession is lagging behind. Williams and Roberts (2002) claimed that, in comparison with primary and secondary education, "the full strength and weaknesses of CSCL in a tertiary environment have not been fully explored." Citing Moshaiov, et al. (2002),

Modern educators should be concerned with both tele-collaboration in education and education for tele-collaboration. Traditional higher education methods, in which a professor is giving a frontal lecture, tend to have minimal groupwork if at all. Nevertheless, many educators advocate the use of groupwork in learning, and point at the many possible virtues of experiencing teamwork. (p.1)

Copyright © 2005, Idea Group Inc. Copying or distributing in print or electronic forms without written permission of Idea Group Inc. is prohibited.

Brandon and Hollingshead (1999) provided an overview of issues involved in CSCL, as related to their descriptive model of CSCL theory and research activities. Their model involves *inputs* (social-behavioral, social cognitive, course-CSCW fit, student variables), *processes* (behavioral and cognitive), and *outcomes.* They view CMC and instructors as variables that influence the relations between inputs and processes. Collaborative learning isn't anything new. Small group discussions and study sessions are collaborative learning activities that have been used in education and training for decades. Jianhua and Akahori (2001) reviewed many cooperating learning methods. They claimed that most of these methods can be adopted for a web-based collaborative environment. They divided the various possible methods into two groups: learning through interpersonal communication, and learning through collaborative production. In general, the notion of a product in engineering education is wide, including joint reports on a subject, laboratory reports, design project reports, and design prototypes. Citing McMurrey and Dunlop (1999) in Williams and Roberts (2002), "McMurrey and Dunlop (1999) believe that strategies can be developed to transform the conventional study material to a form that would be optimal for CSCL." In reference to Tu (2000) and Soller, Wiebe, and Lesgold (2002), Moshaiov et al. (2002) suggested that:

An important element of web-based collaborative learning is the ability to organize the groupwork into a format, which can help monitoring and evaluating the tele-collaboration by the instructor. Careful planning of the working format and in-process and post-process analyses of the recorded information might help improving our understanding of the processes that take place during tele-collaboration and possibly serve to advance the methods of web-based learning by groupwork. Such studies are important due to the relative infancy of computer-mediated communication (CMC) in education and the understanding that CMC can both enhance and inhibit interaction (Tu, 2000)...In systems for future computer-mediated learning through collaborative production many aspects might be explored, evaluated, and perhaps monitored. These may include, for example, shared understanding (Soller et al., 2002), conflict resolution, social creativity, and group motivation. (p.1)

Understanding the needs and the inherent potential of CMC in engineering education is apparent from the willingness of universities to try distributed student projects starting almost two decades ago. Finholt, Sproull, and Kiesler (1990) claimed that the groups of students that frequently used CMC and, in particular, email, outperformed those that did not. Their study involved ad hoc software development teams, which consisted of senior students who participated in a required information systems course at CMU during the fall of 1986. The communication behavior and performance of seven teams were compared. It should be pointed out that the study was aimed at understanding working groups rather then learning groups (no explicit collaborative learning goals). All teams used both CMC and other means of communications (face-to-face, phone, hard-copy memos).

Some CMC-based multinational teaming in students' projects might be complicated to organize, yet such attempts are being carried out (e.g., Fadel, Lindeman, & Anderi, 2000; Henderson, de Pennington, Baxter, & Wells, 2000; Moshaiov, et al., 2002). The common

Copyright © 2005, Idea Group Inc. Copying or distributing in print or electronic forms without written permission of Idea Group Inc. is prohibited.

denominator of such efforts is that they are still in an ad hoc research stage and are sporadic. The realization of such Tele-CPEE endeavors requires careful planning and infrastructure. Reviewing such attempts, it becomes clear that while potentially possible, reforming design projects into web-based learning by tele-collaborative production in engineering education (Tele-CPEE) is in its infancy. This chapter provide some background and thoughts with the hope that it may help both in assessing possible alternatives and designing the computer-based tools that will be required for the realization. The following sections provide some understanding about the professional activities of engineers and the related current trends in global engineering.

Social Creativity in Modern Engineering

Horvath (2001) declared that "In spite of the fact that there have been efforts towards a kind of mechanization (computerization and/or automation) of engineering design, it remains one of the most human-related activities featuring intellectualism, creativity, and ingenuity." Engineers "create the world that never was" (von Kármán, *http://www.llnl.gov/ llnl/06news/Community/super_science_newsletter/*), either by making new things or by making old things better. The creativity process may be described as turning ideas into realities. During the creation process, engineers synthesize, solve problems, and innovate. According to http://www.eweek.org/, the elements of creativity in engineering and in other areas of life are: challenging, connecting, visualizing, collaborating, harmonizing, improvising, reorienting, and synthesizing. In other fields such as science and arts, creativity is also significant. For example, scientists create new knowledge about the world - "discover the world that exists" (von Kármán, *http://www.llnl.gov/llnl/ 06news/Community/super_science_newsletter/*).

To accomplish their work, engineers study both the natural and artificial worlds. They do use knowledge that has been created by scientists, but they also create ("discover") new knowledge about these worlds, as well as new methods to create such knowledge. Engineers create knowledge in accordance with their professional needs and, in particular, those that are related to the process of product development. In his contemporary survey of scientific research into engineering design, Horvath (2001) stated, "by generating knowledge about design and for design, discipline oriented (scientific) research is instrumental to the development of engineering design." For further discussion on the interplay between science and design, the reader is referred to Cross (1993) and to the books by Simon (1996), and by Hubka and Eder (1996). The need to balance between acquiring knowledge and acquiring skills becomes apparent from the above understanding of engineers activities. This need is also apparent from the history of engineering and science (see Burstal, 1965).

Artists are also creative but their emphasis is different. Artists commonly use some aesthetics to achieve their goals of raising thoughts and/or emotions. It appears that artists, in general, work primarily as individuals. This may be due to the large flexibility of goal definition experienced in art. Industrial design seems to be a mixture of engineering and arts where both practical and aesthetics considerations are taken into account. In contrast to artists and industrial designers, engineers are directed at the practicalities of developing useful things, and aesthetics becomes a secondary consideration. Due to

Copyright © 2005, Idea Group Inc. Copying or distributing in print or electronic forms without written permission of Idea Group Inc. is prohibited.

the complexity of modern engineering individual-based artifacts are scarce, and the pressure for collaboration and cooperation increases. Fischer (1999) used the term "social creativity" when addressing domain-oriented design environments. He explains that an individual human mind is limited, especially in the context of certain domains such as design and points at the lack of support for collaborative design. His discussion provides some insight to the changes needed to make the Web more supportive of social creativity. Communication is essential for social creativity, and CMC makes it possible to have distributed creativity and, at the same time, requires new communication skills. Referring to Hirsch, Thoben, and Hoheisel (1996), Moshaiov et al. (2002) stated that:

The expected practice of global distributed design and production requires some new skills and training. It has been recognized that engineers are not easily adopting themselves to new methods and tools that corporations are trying to assimilate (Hirsh et al., 1996). Engineers should be trained in the relevant technologies, become familiar with international cooperation and networking, and develop communication skills that withstand multi-cultural environment. (p.1)

The significance of communication in engineering education is also reflected in a recent study by Dannels (2002). The study illustrates the role of orality in the faculty and students' epistemologies and pedagogies with a focus on across-curriculum and in the disciplines communications. Citing from Dannels, "The 1995 report from the National Board of Engineering Education includes recommendations for a dramatic redesign of engineering curricula toward a more professional, socially oriented focus."

Being creative in a regular sense is not sufficient in modern engineering. Profits and survivability of products depend on engineers' ability to make creations in accordance with the market demands. This means that modern engineers should produce competitive products. Modern markets demand fast delivery of high quality products at competitive price. Consumer expectations and willingness to spend in a pursuit for better products are reflected in shorter life cycle of products. Fast consuming increases the significance of innovation and ingenuity in modern engineering. The complexity of modern engineering and the need to stay competitive require careful planning of engineering teams and the ability to dynamically re-arrange them according to the changing needs. It is therefore that social creativity in engineering design should be understood as an attempt to maximize human resources to fit into the right assembly of knowledge, skills, and behaviors. The right team is strongly influenced by the dominance of multidisciplinary thinking, as explained below.

In summary, engineers create both things and knowledge and also acquire useful knowledge for their creations. Their education curricula should maintain a balance as related to their mission. The engineering missions and knowledge have evolved into traditional disciplines, but, as explained below, some major changes occur, and the clear boundaries between the traditional disciplines are less apparent.

Copyright © 2005, Idea Group Inc. Copying or distributing in print or electronic forms without written permission of Idea Group Inc. is prohibited.

Dominance of Multidisciplinary Thinking in Modern Engineering

Professional activities of engineers encompass design, production, repair, and mainte-nance of artifacts, systems, and production lines. During their professional lifetimes, engineers may also become managers or salesmen based on their expertise and aptitude. The complexity of the knowledge and the skills that are required to perform engineering activities and the nature of modern systems mean that engineers must cooperate and work in multidisciplinary groups. According to a recent report, which has been prepared for the American Society of Mechanical Engineers (ASME), multidisciplinary ap-proaches promise to set both the context and the agenda for science and engineering in the twenty-first century (Heggy & Bowman, 2001). The report lists the following items as the drivers of multidisciplinary thinking in engineering: ICT, economic prosperity, globalization, diminishing lifetime of many products, corporate consolidation, and complexity and variety of products. Several significant implications arise from the dominance of multidisciplinary thinking. Among these are: the emergence of the multidisciplinary engineer, facilitation of novel research discoveries and transforma-tional learning, changes in engineering education, need for new skills, need for learning to learn, and uncertainty of career path. The significance of reforming or at least adopting current engineering education methodologies to better accommodate the dominance of multidisciplinary thinking becomes apparent. The recent emergence of educational programs in new interdisciplinary engineering fields such as mechatronics, which blends mechanical engineering with electronics and software engineering, reflects the ever-growing need for multidisciplinary engineers. In reforming engineering education, the new skills needed should be acknowledged. Heggy and Bowman (2001) stated that operating in multidisciplinary environment means that engineers should have broader intellectual perspective. Modern engineers need to comprehend and translate the assumptions of other fields. They should be able to manipulate information into knowledge and understand communication across disciplines. The perception and cognitive capacities needed by the individual interdisciplinary engineer, such as a mechatronic engineer, may differ from those needed by engineer who specializes in a particular discipline such as civil engineering. To successfully operate in several disciplines, the interdisciplinary engineers should not sacrifice their basic knowledge. Profound understanding of fundamentals will become even more important in a multidisciplinary environment (Heggy & Bowman, 2001). Programs' length, focus, and quality, as well as individual capabilities are likely to influence the balance between broad and deep understanding by new engineers. The need for both "broad" and "deep" understanding appears to be critical. Broad understanding is probably most important in evaluating different design alternatives; nevertheless, it is the deep and meticulous expert understanding of a particular discipline and subject that may be crucial for making a product competitive and avoiding design mistakes. Interdisciplinary programs are likely to produce engineers who may be well equiped to handle multidisciplinary engineering tasks, but market forces suggest that the help of experts is unaviodable.

Given the dominance of multidisciplinary thinking and the understanding that an individual can not fully posses the spectrum of knowledge and skills required for

Copyright © 2005, Idea Group Inc. Copying or distributing in print or electronic forms without written permission of Idea Group Inc. is prohibited.

competitive engineering, it is reasonable to assume that the need for engineering collaboration will increase. Recently, Moshaiov and Kaynak (2001) examined the commonly accepted definition of mechatronics in view of the need and possibility to tele-collaborate. According to their statement:

Mechatronics is commonly defined as the synergistic integration of mechanical engineering with electronics and intelligent computer control in the design and manufacture of products and processes. The above definition of mechatronics clearly shows that mechatronics requires teams of specialists from different fields. It may be misleading to consider the word synergy, in the above definition, as only a synergistic integration of knowledge from different engineering fields. In fact, engineering work involves not just knowledge synergy but also the synergy of the team members' efforts to work as one team rather than individuals. Teamwork is essential for the engineering process at large and in particular for mechatronics to be successful. (p.41)

Justifying the consideration of socio-mechatronics, Moshaiov and Kaynak (2001) added:

Mechatronic teams, and especially virtual (distributed) teams, create a need for a consideration of a blend of aspects not just from mechanical, electrical, and software engineering, but also from sociology, psychology and pedagogy (p.41).

Acknowledging that individual-based engineering constitutes a small part in the history of engineering, Moshaiov and Kaynak (2001) further addressed the mechatronic engineering community:

The reader may wonder what is new in our visionary message and why should we consider the new term of socio-mechatronics at the present. The critic may argue that engineering work has always involved teamwork hence no new consideration is needed. Furthermore, the opponent may refer us to the current engineering practice where a large portion of engineering work is supervised and/or advised by managers, industrial engineers, industrial psychologists, and other individuals that take into consideration the socio-aspects of teamwork. We claim that the traditional practice is at least insufficient to cope with today's engineering challenge and we call for a consideration of socio-mechatronics by the mechatronic engineering community. (p.41)

The above demonstrates that multidisciplinary thinking in engineering does not stop at the mixture of traditional engineering disciplines but may combine also non-exact sciences.

Copyright © 2005, Idea Group Inc. Copying or distributing in print or electronic forms without written permission of Idea Group Inc. is prohibited.

Assessing Designers and the Design Process

In modern computer-based tutoring systems, student models are used to evaluate the user and make intelligent decisions regarding which exercise to give to the student, when to interrupt, what level of explanation to give, etc. (e.g., Zhou & Evens, 1999). Models to support collaborative learning are less common (Jermann, Soller, & Muehlenbrock, 2001). Designer models are almost non-existent, and computational assessments of collaborative design process in an education setting is in its infancy. Some initial overview of performance assessment as related to tele-collaborative design can be found in Moshaiov, et al. (2002). Here some general background on assessments and their role in industry is given. Evaluation, reorganization, and monitoring procedures are becoming more essential for the success of organizations in the increasing competitive nature of the global marketplace. A drastic approach to the quality of organizations is the Total Quality Management (TQM) methodology. It is essentially a concept of changing organizations for improving quality. The changes may occur at different levels, including human services, organizational culture, and the organization's decision-making processes and power bases.

TQM includes the evaluation of humans in the organization. Books in social psychology, such as by Davis (1969), discuss variables that influence group performance. Such variables may include: group size, group composition, group cohesiveness, and norms. Evaluating team performance and individuals within teams requires some substantiated methodologies. Such methods do exist but should be tailored to specific domains (e.g., Brannick, Salas, & Prince, 1997). Distributed groups may face more difficulties than face-to-face groups, as discussed in Haywood (1998). Yet, the digital records and CMC tools may help monitor their work. Methodologies and tools are constantly being developed to assist both managers and members of distributed teams. A major issue in web-based groupwork is the collaboration by the distributed partners. Several important questions to be raised with respect to tele-collaboration are:

- Can (and how) the members of a distributed design team be chosen effectively, based on understanding of their fitness for tele-collaboration?

- Can (and how) the distributed group members cooperate effectively and reach their goal?

- Can (and how) the difficulties associated with such a process be overcome?

- Can (and how) the process be designed to become effective?

- Can (and how) the process be monitored?

The TQM advocators and others have put a large emphasis on giving the customers a major voice. Evaluating customer needs and satisfaction is part of modern engineering and constitutes part of a larger effort of quality control. Quality control has in fact become a profession by itself with the ultimate goal of improving the product quality. Quality and customer satisfaction would not be achieved without a careful and professional design

Copyright © 2005, Idea Group Inc. Copying or distributing in print or electronic forms without written permission of Idea Group Inc. is prohibited.

process. Engineering design is a key element in the entire modern engineering process. The designers should put together their knowledge and find creative solutions to a multiobjective and constrain problem. Moreover, the chosen solution should be competitive and robust to the dynamic nature of the modern market (Fowlkes & Creveling, 1995).

The modern design process requires evaluation and monitoring procedures. These are essential with the increasing competitive nature of the global marketplace. Professional engineers complement the design process with design evaluations at the different design stages. These are termed Critical Design Reviews (CDRs). It should be pointed out that a design is evaluated in many respects, including aspects with remote, indirect, and direct effect on the customers. An early station of the "design evaluation" is located in the pre-design stage of the requirements. Understanding the requirements of the customers and establishing a company policy is a complicated procedure that may dramatically affect the entire process. Once the requirements are understood, several concepts are commonly developed and evaluated. The selected concept, which will be transferred to the preliminary and detailed design stages, is hoped to be better than the one selected by the competitors. The entire design and production process is commonly characterized by some feedback and re-evaluations, and relies heavily on past experience.

A large effort is invested in trying to understand and predict the customer behavior. Customers' dissatisfaction may require a complete redesign and manufacturing of the product. It is therefore essential for the success of the design process that the outcomes of the design team will be constantly evaluated for their technical merits and commercial values.

Mixing Science, Engineering, and Fun in Current Design Education

The demand for engineers in the industrial world has increased dramatically with the information and communication revolution. Countries such as the USA and Japan are trying to raise public awareness to the significance of engineering and to attract young people to science and engineering (e.g., National Engineers Week, as portrayed at: http://www.eweek.org/). Creighton (2002) overviewed some recent developments and trends in reaching out to get pupils excited about engineering in the USA. According to her article, the campaign to bolster engineering education at the pre-college level is just beginning to provide models that might be adopted by individual school districts to produce more engineers and make society more technology-literate. Pre-college education programs and activities in engineering are spreading. Kolberg, Reich, and Levin (2003) describe in details such a high-school level program in Israel, which has been successfully introduced in non-technical schools. The program is planned around a project of designing and building robots. Robots and participating in design contests seem to be a natural way to stir the imagination of young people. For example, the program for Israeli high schools is affiliated with The Trinity College Fire Fighting Home Robot Contest (Hartford, CT, USA). Citing Moshaiov (2000a), "The FIRST Foundation (For Inspiration and Recognition of Science and Technology) conducts regional and national design competitions throughout the United States and it has already expanded to

Copyright © 2005, Idea Group Inc. Copying or distributing in print or electronic forms without written permission of Idea Group Inc. is prohibited.

Canada. It establishes a bridge between high schools, the industry, and universities." Design competitions may serve to support design education at all levels (preliminary, high school, undergraduate, graduate, industrial training). Furthermore, the contests bring together design educators, students, industry, and engineering societies. They promote a shared vision and create an "edutainment" (education and entertainment) environment, adding a fun element into the engineering education process (e.g., Asada, D'Andrea, Birk, Kitano, & Veloso, 2000).

For the most part, current engineering education at universities around the globe is carried out with some balance between theory and practice. Design courses, projects, and often contests are used as the elements for practicing engineering while learning. Details on design courses, projects, and often on contests can be found on many of the websites of engineering programs, and it is beyond the scope of this chapter to review them all. Observing a substantial list of such websites and based on personal knowledge, it is safe to say that there are countless ways in which design education is carried out. For example, in some prestigious engineering schools, most notably MIT, student participation in design contests is actively encouraged. In fact, the MIT 2.007 Design Course integrates a contest within the course syllabus (dating back to the early 70s).

Most engineering societies provide some details about design contests that they sponsor or acknowledge. The tremendous variety of design competitions and projects that are available to students reflects, to a large part, the actual variety of issues that are of interest to current industries. Given that the use of tele-collaboration by engineers is spreading, it should be expected to have an influence on design education in the coming future. Similarly, the trend to start educating for design at pre-college education levels may also have some effect on the future of engineering education in the long run.

Summary

The above overview of topics does not constitute a complete literature review, yet it reveals the magnitude and variety of issues involved when dealing with collaborative production in engineering education and, in particular, with tele-collaboration. Industrial needs are linked with education needs, and a balance between theory and experience in engineering education has always been practiced. Modern engineering and CMC requires a disscusion for better understanding and planning of modern engineering education. Some basic questions concerning collaborative production in engineering education are raised and discussed below.

Discussion

Should We Have Design Projects at All?

The time framework of engineering programs may vary according to national and/or institute preferences. Yet, regardless of the program extent, there is always difficulty

Copyright © 2005, Idea Group Inc. Copying or distributing in print or electronic forms without written permission of Idea Group Inc. is prohibited.

when trying to reform. From personal experience, changing the undergraduate mechanical engineering program at Tele-Aviv University (TAU) took six years of extensive debate and a similar consecutive period for changing the name and graduate program of the author's department. Moreover, faculty members may dislike changes for different reasons, some of which may be egocentric (see, for example, "Effect on academic stuff" in Williams and Roberts, 2002). Consequently, any substantial program changes need strong justifications and careful planning. The need for such debates and planning is constantly increasing, given the dynamics of modern engineering.

It can always be argued that design projects should not be included in the engineering education system and that they should belong to on-the-job training. This statement may look like rhetoric given the general consensus that teaching design and having design projects as part of the engineering education is a must. This understanding can be easily verified by searching a representative list of websites of engineering programs, and is reflected in the background of this chapter (subsection on Mixing Science, Engineering, and Fun in Current Design Education). Yet, we should treat the above argument seriously. The unsatisfied critic may suggest that no general knowledge is acquired during a design project, and therefore they should be abolished. In fact, any project concentrates on a particular problem, hence the student acquires particular knowledge that may end up irrelevant to the student's actual area of work. In this respect, some engineering schools promote a stronger link with industries, resulting in contracts with the students and on-the-job training while still formally in school. This improves the likelihood of making the design project "no-waist" of time, as the student may end up having a life career in the hosting company. The critic may add that such an arrangement is not common. Furthermore, given the extreme value of time, the suggestion to abolish design projects should be taken seriously.

Selecting general and fundamental knowledge that may prove to be valuable to the career of the engineer is certainly an important consideration. The days of the Industrial Revolution in Britain, which involved men who "were for the most part uneducated in scientific knowledge," are gone, and scientific knowledge plays an ever-increasing role in advancing engineering. Moreover, given the competitive nature of modern engineering, there is no room for a compromise on knowledge, and professionalism is a must for survival of industries and their engineers. This is especially true with the dominance of multidisciplinary thinking in engineering, which puts pressure on curricula planners. Considering, for example, mechanical engineering education, we note that possible related disciplines include electronics, chemistry, material science, biology, computer science, industrial engineering, aeronautical engineering, civil engineering, and more. We can easily pick some general course of the fundamentals of related disciplines with the hope of significantly improving the ability of mechanical engineering students to cope with the dominancy of multidisciplinary thinking. Using the same argument, we should note that learning about design is a process of acquiring general knowledge about design, and it is somewhat difficult to properly acquire it without experiencing it while learning the fundamentals.

General knowledge on design can be extremely valuable in modern engineering. Consider the extreme case of setting up of a start-up by young inexperienced engineers who possess a great idea. In contrast to large companies in which an infrastructure for design and design procedures exists, the members of the start-up may need to rely on their

Copyright © 2005, Idea Group Inc. Copying or distributing in print or electronic forms without written permission of Idea Group Inc. is prohibited.

understanding of the design process and product realization to avoid many possible pitfalls. Design projects facilitate the development of many important general engineering skills, such as leadership, organization, teamwork, and the ability to compromise in a group decision process when time, resources, and knowledge constraints are faced and conflicts must be resolved. These abilities are extremely important in start-ups, which are likely to have low structuring of management to the extent of horizontal management, but also in any engineering career due to modern market pressures. Moreover, knowledge is readily available with current ICT.

In the future, this will probably be even more profound given potential advancements such as the expected semantic Web. It is therefore arguable that the focus of future education should be more on "learning to learn" and acquiring engineering skills, rather than on knowledge acquired during undergraduate studies. It is realistic to assume that individuals with different skills and knowledge will be dynamically assembled into an extended team via ICT. Their success will strongly depend on their abilities to work together, rather than on their level of individual knowledge. Any individual with a great deal of knowledge will be an asset but not a must. Any individual who has poor collaboration skills will be a burden. This conclusion is supported by the observations made in the background subsections on Social Creativity in Modern Engineering and on Dominance of Multidisciplinary Thinking in Modern Engineering.

Design courses, design projects, and contests can contribute to "learning to learn" and acquiring engineering skills. They present an opportunity for the student to put the information gathered during studies to practical use and to try to comprehend to what extent it is valuable to the problem he is trying to solve. It is perhaps for the first time that the link is not so trivial (in comparison with regular homework). Moreover, for design, additional knowledge is likely to be needed forcing the student to learn on his own. The recommendations of The National Board of Engineering Education, as stated in Dannels (2002), for a dramatic redesign of engineering curricula toward a more professional, socially oriented focus, can be met by keeping and even increasing the role of design projects in the curricula. Collaborative production can certainly encourage communication in the education process.

In summary, a flexible and effective combination of learning engineering principles and experiencing some design process is necessary to achieve a balance between theoretical knowledge and application skills, including professional communication skills. Design projects assigned to small teams of engineering students are traditionally used for this purpose, providing a mechanism for learning by collaborative production. Reforming such design projects into web-based learning by tele-collaborative production in engineering education (Tele-CPEE) is in its infancy; hence, many relevant issues and questions can be raised.

Should We Tele-Collaboratively Produce in Engineering Education?

The current understanding is that ICT has the potential to strongly impact education-at-large, spanning the entire life of the potential learners. In several educational areas,

Copyright © 2005, Idea Group Inc. Copying or distributing in print or electronic forms without written permission of Idea Group Inc. is prohibited.

CMC-based tools have reached some maturity. Possible reformation of engineering education should therefore be addressed, and the question of including learning by tele-collaborative production is raised. It should be acknowledged that institutes and individuals may find it difficult to change, even if technology exists and some possible merits may be realized.

Computers have been used for many years in engineering education for several applications including, for example, computer-aided design and analysis. Most commonly, software tools that were used in industry were adapted for such purposes. It is reasonable to assume that engineering programs will not oppose CMC because of the increasing evidence that CMC is used in daily engineering work, as detailed in the background subsection on Impact of CMC on Engineering. Yet, the main question should not be "Will CMC be used in engineering education?" but, rather, "Should CMC be used at all?"

Supported by all the favoring arguments presented on the above question on design projects, and the current industry trends to form extended enterprises, it appears reasonable to conclude that universities should strongly promote tele-collaboration. Moreover, modern competition does not stop at the engineering world. Universities are facing competition as also, and the distance education capabilities that are spreading by CMC add a new dimension to competition in the education arena. Traditional universities may find that they have to explore tele-collaboration to stay competitive. Technical education is usually linked to industry, and many of the higher education entities are involved with industry-motivated R&D. When considering current sporadic activities of students' tele-collaboration, as referred to in the section on Impact of CMC on Engineering Education, it becomes apparent that most of it is research motivated. Research motivations are two-fold, including education and engineering, and a combination of the two. These "non-pedagogical" motivations appear quite legitimate due to the fact that engineering education is linked with the engineering profession. Currently, the engineering educational community is in a transition stage exploring and making decisions about reforms. It may take a while before we fully comprehend the capabilities and limits of introducing Tele-CPEE in engineering. Meanwhile, students may benefit in taking an active role in experimenting with tele-collaboration as motivated by the industry. It may open doors for them. They can get out of school with an extra edge of knowledge on tele-collaboration that can make a difference. Yet some caution should be taken. The uncertainties and risks involved in tele-collaboration in the educational system, at present, is high. Students may be trapped due to the enthusiasm of universities and industries to do research. Students should not be exploited, and careful planning must therefore take place. For example, the edutainment part (see subsection on Mixing Science, Engineering, and Fun in Current Design Education) may be more difficult to achieve with reduced face-to-face meetings. Tele-socializing alone may be unsatisfactory to students, and they may prefer not to get involved. Moreover, the entire tele-collaboration process may be uncomfortable for some of them. The possible difficulties and risks are well documented in the current literature on CMC in education, and some relevant references can be found in the subsection on Impact of CMC on Engineering Education. Moshaiov et al. (2002) provided a list of 11 major difficulties that had to be overcome during their bi-national design project. These can be categorized as managerial, pedagogical, initiation, coordination, and resources difficulties. Given the expected difficulties, and in view of the arguments against the inclusion of design projects at-large,

Copyright © 2005, Idea Group Inc. Copying or distributing in print or electronic forms without written permission of Idea Group Inc. is prohibited.

as discussed above on the general first question, investing in CMC-based CPEE is questionable. It is somewhat difficult to weigh the pedagogical and practical benefits vs. the pitfalls and to compare them with current practice. It can be easily argued that the current design projects should not be replaced, and that tele-collaboration has no justifiable added value. It can also be claimed that on-the-job training for tele-collaboration should take place and that the industry is better equipped for that. It is probably the lack of clear resolution of the conflicting forces that requires further studies of the use of CMC-based CPEE. Perhaps future development of tailored CSCL tools for design, which may better support learning than the existing collaborative design tools (CSCW), will make the situation clearer. Having some assimilation strategy seems important due to the inherent merits of tele-collaboration in engineering and the research and development forces that may push its use in engineering education systems. Such a strategy is discussed below.

Should We Focus on Local or on Tele-CPEE?

The variability of realizing design projects and competitions in engineering education, as described in the background subsection on Mixing Science, Engineering, and Fun in Current Design Education, is both beneficial and problematic. This is especially true when trying to design the infrastructure and software tools. In contrast to collaborative learning in general educational settings, the engineering design setting may involve many possible categories as related to potential partners from non-educational organizations, namely, industries. Moreover, in certain situations, additional stakeholders, namely, engineering societies, may come into play as explained below. Possible basic categories include university-university, university-industry, and university-high school partnerships. A coalition partnership is also possible in which there are more than two collaborating parties. In Tele-CPEE at least two of the parties are remotely located and substantially use the Web for the tele-collaboration. The above categories should be understood in relation to possible modes and phases of the collaboration, including planning, learning-by-production, and assessing. In other words, during the planning and assessing phases, partners may include, for example, two universities and one industry, whereas in the learning phase, the partners may involve primarily only the two universities. Similar categories have been suggested in Moshaiov (2000a) with respect to specially designed competition for tele-collaboration. For projects that may involve such competitions, additional players may come from engineering societies, especially during the planning and assessing stages. Local-CPEE means that partners are located within an institute and its vicinity, and therefore they may easily communicate face-to-face when necessary. In contrast to Tele-CPEE, experiencing Local-CPEE means enjoying both worlds – CMC and face-to-face. Therefore, it appears that much of the difficulties that may be faced by the students can be reduced by Local-CPEE. Moreover, planning partners of Local-CPEE may enjoy the same situation and can therefore better handle the project especially in critical situations.

Planning and assessing phases will commonly involve professors and engineers, but the inclusion of students may also be considered. This is probably easier to implement in Local-CPEE than in Tele-CPEE. While peer-to-peer assessments are common in collabo-

Copyright © 2005, Idea Group Inc. Copying or distributing in print or electronic forms without written permission of Idea Group Inc. is prohibited.

ration, the inclusion of students in the project planning may require some explanation. Planning, among other things, means setting the design requirements. Commonly, design requirements are negotiable in engineering design, and designers tend to re-think them in accordance with the development of their problem understanding while trying to find a solution. This means that if we want to simulate carefully real-life situations, we should consider letting the students get involved in re-planning the project as part of the project. Inclusion of students in the entire process may also be justified under the concept of participatory design, where the project planning and execution is considered as a pedagogical product and the students are the customers. Such an approach seems reasonable considering that student satisfaction is essential for successful assimilation of new methodologies.

The availability of commercial tools for professional collaboration can accelerate the ability of both universities and industries to get organized for Tele-CPEE. Nevertheless, caution should be taken, and careful planning may require different strategies for Tele-CPEE vs. Local-CPEE. The extent of such projects with respect to these two frameworks is discussed in the following section.

Solutions and Recommendations

Formally, the process of product development has several stages. There are several ways to model the process, and the following discussion is based on a common process description with the understanding that several iterations between the stages are quite common during the industrial realization of product development. Excluding the stage of revealing the need for the design, the main stages are setting requirements and understanding the requirements, conceptual design, preliminary design, detailed design, and manufacturing. In current practice, the extent of students' projects may vary from short projects focusing on the earlier design stages to full projects that include developing a prototype. In view of the expected difficulties in experiencing Tele-CPEE, it appears reasonable to suggest that short projects will be used in a Tele-CPEE framework. Short projects, such as described in Moshaiov et al. (2002) that focus on the early stages of the design process, provide a way to avoid or at least reduce the complexities of planning and executing tele-collaboration. This means increasing student satisfaction and reducing the likelihood of student frustration. Moreover, in such short projects the focus is on social creativity, which occurs in the conceptual design stage, and not on the tedious detailed design and manufacturing stages that are commonly split among the students.

Long and complex design tasks, which contain detailed design and manufacturing of a prototype, are not simple to realize in educational settings that involve more than one institute and require tele-collaboration. This is especially true when the development of a prototype is included. In contrast to the virtual product created during the design stages, which can be shared by the remote partners, the prototype produced during the development stage can't be shared. Prototype development in Tele-CPEE framework means harsh splitting of the tasks with reduced pedagogical benefits and increased

Copyright © 2005, Idea Group Inc. Copying or distributing in print or electronic forms without written permission of Idea Group Inc. is prohibited.

difficulties. Such long projects may therefore be better handled in a Local-CPEE framework before attempting to employ Tele-CPEE.

Future Trends and Research

The trend to introduce pupils to engineering at the primary and secondary levels and the possibility that most of the future in-coming college students will have previous design experience may reach a level requiring higher engineering education to be revised accordingly. In such a case, the previous design experience of in-coming students may affect decision makers in the higher engineering educational systems when discussing the necessity and extent of future design projects.

Reforming design education for tele-collaboration may be accelerated due to the availability of tools for tele-collaborative design, which are developed for the industry. Research should be carried out on the added pedagogical benefits when considering CMC-based CPEE from the student point of view. To what degree industrial tools will be adequate for the educational setting should be explored. The potential of specially developed tools for Tele-CPEE and their design should also be investigated. Another important ongoing issue is the strengthening of the team members' semantic communication by tailored groupware. From a cybernetic point of view, "closing the loop" using users and process models requires problem definition. Is it education for collaboration in the industry, or education for individual understanding of the design process? Do we focus on education for achieving better design, or for achieving better designers? To what extent does a correlation exists between team assembly, team performance (during the process), and the product quality? — an important open question that will require not just definitions of metrics, but also extensive studies to answer. To what extent will we be able to analyze communication during design and use it to improve the design process and outcome? It is noted that a mixed text and image analysis may be required to approach some of these questions, reflecting the mixed communication used by engineers.

Conclusion

World industry is changing rapidly. Both small and large manufacturers have to respond quickly to market demands. Organizations are increasingly forming joint design and manufacturing teams that are not necessarily isolated within one company but are formed by cross-functional, multicompany, and multinational teams. The extended enterprise involves, among other factors, the key elements of tele-collaborative design and prototype development. Globalization and the associated trend towards an intensely competitive and rapidly changing business and engineering environment have accelerated the development of novel collaborative design tools and methodologies. These changes should be accompanied by a discussion, such as given here, on the possible

Copyright © 2005, Idea Group Inc. Copying or distributing in print or electronic forms without written permission of Idea Group Inc. is prohibited.

revision of the relevant educational and training issues, particularly education for engineering tele-collaboration. Engineers should be trained in the relevant technologies, become familiar with international cooperation and networking, and develop communication skills that withstand multicultural and multidisciplinary environments. Design projects assigned to distributed team of engineering students provide a mechanism for learning by tele-collaborative production.

An extensive review of the different aspects of CMC in engineering is given in this chapter. Several fundamental questions have been raised on the necessity of design projects and tele-collaboration in engineering education. At present, it appears unclear to what a degree the efforts needed are worth the added pedagogical benefits when considering CMC-based CPEE from the student point of view. Careful planning is recommended and a distinction should be made between two major frameworks: Local-CPEE and Tele-CPEE. Finally, it is concluded that short well-planned projects may better suit the latter.

Acknowledgments

The author of this chapter would like to express his appreciation to the Design Society for acknowledging the COMEDI proposal, and to his colleagues and friends, including Okyay Kaynak, Yoram Reich, Alex Duffy, and Georgi Demirovski, for their encouragement.

References

Andreason, M.M. (2001). The contribution of design research to industry: Reflections on 20 years of ICED conferences. *Proceedings of the International Conference on Engineering Design, ICED 01*, p. 3-10, Glasgow, UK.

Asada, M., D'Andrea, R., Birk, A., Kitano, H., & Veloso, M. (2000). Robotics in Edutainment. In *Proceedings of the IEEE Int. Conf. on Robotics and Automation*, pp. 795-800. San Francisco, CA.

Bowker, G.C., Turner, W., Star, S.L., & Gasser, L. (1997). Introduction. In G.C. Bowker, S.L. Star, W. Turner, & L. Gasser (eds.), *Social science, technical systems, and cooperative work*, pp. xi-xxiii. Hillsdale, NJ: Lawrence Erlbaum Associates, Inc.

Brandon, D.P. & Hollingshead, A.B. (1999). Collaborative learning and computer-supported groups. *Commnication Education*, 48: 109-126.

Brannick, M.T., Salas, E., & Prince, C. (eds.) (1997). *Team performance assessment and measurement: Theory, research, and applications*. Hillsdale, NJ: Lawrence Erlbaum Associates, Inc.

Copyright © 2005, Idea Group Inc. Copying or distributing in print or electronic forms without written permission of Idea Group Inc. is prohibited.

Burns, J.D. (2000). Learning about technology in society: Developing liberating literacy. In J. Ziman (ed.), *Technological innovation as an evolving process,* pp. 299-311. Cambridge, UK: Cambridge University Press.

Burstal, A.F. (1965). *A history of mechanical engineering.* Cambridge, MA: MIT Press.

Chryssostomidis, C. (ed.) (1990*). Automation in the design and manufacture of large marine systems.* New York: Hemisphere Publishing Corporation.

Creighton, L. (2002). The ABCs of engineering. *Prism* (online periodical), 12(3). Available at: *http://www.asee.org/prism.*

Cross, N. (1993). Science and design methodology: A review. *Research in Engineering Design,* 5: 63-69,

Dannels, D.P. (2002). Communication across the curriculum and in the disciplines: Speaking in Engineering. *Communication Education,* 51(3): 254-268.

Davis, J.H. (1969). Group performance. Reading, MA: Addison-Wesley Publishing Company.

Fadel, G.M., Lindeman, U., & Anderi, R. (2000). Multi-national around the clock collaborative senior design project, The 2000 ASME Curriculum Innovation Award Honorable Mention. Available at: *www.ASME.org/educate/cia.*

Filling, C., Diggs, G., & Helgerson, D. (2000). MAAST Pilot: A prototype of collaborative environments. *J. of Ship Production,* 16(1): 12-26.

Finholt, T., Sproull, L., & Kiesler, S. (1990). Communication and performance in ad hoc task groups. In J. Galegher, R.E. Kraut, & C. Egido (eds.), *Intellectual teamwork: Social and technological foundations of cooperative work,* pp. 291-325. Hillsdale, NJ: Lawrence Erlbaum Associates.

Fischer, G. (1999). Domain-oriented design environments: Supporting individual and social creativity. In J. Gero & M.L. Maher (eds.), *Proceedings of Computational Models of Creative Design IV,* pp. 83-111, Key Centre of Design Computing and Cognition, University of Sydney, Australia.

Fowlkes, W.Y. & Creveling, C.M. (1995). *Engineering methods for robust product design.* Reading, MA: Addison-Wesley.

Goldman, S., Nagel, R., & Preiss, K. (1995). *Agile competitors and virtual organizations: Strategies for enriching the customer.* New York: Van Nostrand Reinhold.

Haywood, M. (1998). *Managing virtual teams: Practical techniques for high-technology project managers.* Boston, MA: Artech House.

Heggy, P.H. & Bowman, M.C.S. (2001). New dimensions in multidisciplinary thinking: Issues, trends and implications for mechanical engineers and ASME International. In Global Foresight Associates, *Report of the ASME International Committee on Issues Identification,* New York: ASME.

Henderson, M., de Pennington A., Baxter J., & Wells V. (2000). The global engineering design team (GEDT): Transatlantic team-based design for undergraduates, *2000 ASEE Annual Conference and Exposition,* St Louis, USA.

Hirsch, B.E., Thoben, K.-D., & Hoheisel, J. (1996). The change of production management paradigms and the consequences for the education of engineers. In F.-L. Krause,

Copyright © 2005, Idea Group Inc. Copying or distributing in print or electronic forms without written permission of Idea Group Inc. is prohibited.

K. Preiss, M. Shpitalni, & A. Shtub, *Proceedings of the Computer-Integrated Extended Manufacturing Enterprise (CIEME)*, Haifa and Herzlia, The Agility Forum.

Horvath, I. (2001). A contemporary survey of scientific research into engineering design. In *Proceedings of the International Conference on Engineering Design, ICED 01* Glasgow, UK.

Hubka, V. & Eder, W.E. (1996). *Design science*. London: Springer-Verlag.

Jermann, P., Soller, A., & Muehlenbrock, M. (2001). From mirroring to guiding: A review of state of the art technology for supporting collaborative learning. In *Proceedings of the 1st European Conference on CSCL*, pp. 324-331. Maastricht, Netherlands.

Jianhua, Z. & Akahori, K. (2001). Web-based collaborative learning methods and strategies in higher education. *ITHET 2001*, July 4-6, Kumamoto, Japan.

Kolberg, E., Reich, Y., & Levin, I. (2003). Project-based high school mechatronics course. *International Journal of Engineering Education, 19*(4): 557-562.

Lipson, H. & Pollack J. B. (2000). Automatic design and manufacture of artificial life forms. *Nature*, 406: 974-978.

McMurray, D.W. & Dunlop, M., E. (1999). The collaborative aspects of online learning: A pilot study. Available at: *http://ultibase.rmit.edu.au/Articles/online/mcmurry1.htm.*

Minsky, M. (1994). Will robots inherit the earth? *Scientific American*, 271(4): 109-113.

Monarch, I.A., Konda, S.L., Levy, S.N., Reich, Y., & Ulrich, C. (1997). Mapping sociotechnical networks in the making. In G. C. Bowker, S.L. Star, W. Turner, & L. Gasser (eds.), *Social science, technical systems, and cooperative work,* pp. 331-354. Hillsdale, NJ: Lawrence Erlbaum Associates, Inc.

Moshaiov A. (2000a). Information technology, virtual reality, and robot competitions in the promotion of culture of peace and assistance to developing countries. In *Proceedings of the Conference on Information Technology based Higher Education and Training (ITHET2000)*, pp. 212-217, Istanbul, Turkey.

Moshaiov, A. (2000b). Toward the implementation of the TCRCT concept for the promotion of culture of peace and assistance to developing countries. In *Proceedings of EPAC 2000 - The 21st Century Education and Training in Automation and Control*, pp. 9-14 Skopje, Macedonia.

Moshaiov, A. & Kaynak, O. (2001). Socio-mechatronics in the era of information technology. Plenary Paper in *the Proceedings of the 1st International Conference on Information Technology in Mechatronics (ITM01)*, pp. 40-45, Istanbul, Turkey.

Moshaiov, A., Kaynak O., Reich Y., Bar-El, S., Klunover, A., & Akin, L. (2002). Evaluating groupwork in tele-collaboration – Application to socio-mechatronics. In *Proceedings of the 3rd International Conference on Information Technology-Based Higher Education and Training (ITHET2002)*, Budapest, Hungary.

Pfeifer, R. & Scheier, C. (1999). *Understanding intelligence*. Cambridge, MA: MIT Press.

Rogers, Y. (1997). Reconfiguring the social scientist: Shifting from telling designers what to do to getting more involved. In G.C. Bowker, S.L. Star, W. Turner, & L. Gasser

Copyright © 2005, Idea Group Inc. Copying or distributing in print or electronic forms without written permission of Idea Group Inc. is prohibited.

(eds.), *Social science, technical systems, and cooperative work beyond the great divide*, pp. 57-77. Hillsdale, NJ: Lawrence Erlbaum Associates, Inc.

Simon, H. (1996). *The science of the artificial*. Cambridge, MA: MIT Press.

Soller, A., Wiebe, J., & Lesgold, A. (2002). A machine learning approach to assessing knowledge sharing during collaborative learning activities. In *Proceedings of the Computer-Support for Collaborative Learning (CSCL 2002)*, pp.128-137, Boulder, Colorado.

Sowa, J. F. (1984). *Conceptual structures: Information processing in mind and machine*. Reading, MA: Addison-Wesley Publishing Company.

Tu, C.-H. (2000). Critical examination of factors affecting interaction on CMC. *J. of Network and Computer Applications, 23*: 39-58.

Vest, C. (2002). Letter from the president. *MIT Spectrum*, Fall Issue [online]. Available at: *http://web.mit.edu/giving/spectrum/fall02/letter-from-the-president.html*.

von Kármán. Available at: *http://www.llnl.gov/llnl/06news/Community/super_science_newsletter/*.

Williams, S. & Roberts, S.T. (2002) Computer supported collaborative learning: Strengths and weaknesses. In *Proceedings of the ICCE 2002*, pp. 328-331, 3-6 December, Auckland.

Zhou, Y. & Evens, M. W. (1999). A practical student model in intelligent tutoring system, In *Proceedings of the 11th IEEE International Conference on Tools with Artificial Intelligence*, pp. 13-18, 9-11 November, Chicago, IL.

Ziman, J. (ed.) (2000). *Technological innovation as an evolving process*. Cambridge, UK: Cambridge University Press.

Copyright © 2005, Idea Group Inc. Copying or distributing in print or electronic forms without written permission of Idea Group Inc. is prohibited.

Chapter XIII

Relational Online Collaborative Learning Model

Antonio Santos Moreno
Universidad de las Américas Puebla, México

Abstract

This chapter describes an instructional online collaborative learning model that addresses the phenomenon from a systemic human relations and interaction perspective. Its main purpose is to aid students in their social building of knowledge when learning in a CSCL environment. The model argues that knowledge building in a networked environment is affected by the communication conflicts that naturally arise in human relationships. Thus, the model is basically proposing a way to attend to these communication conflicts. In this line, it proposes a set of instructional strategies to develop the student's meta-communication abilities. The concepts and instructional suggestions presented here are intended to have a heuristic value and are hoped to serve as a frame of reference to: 1) understand the complex human patterns of relationships that naturally develop when learning in a CSCL environment, and 2) suggest some basic pedagogical strategies to the instructional designer to develop sound online networked environments.

Introduction

Computer-Supported Collaborative Learning (CSCL) has been an important emerging research paradigm for the field of educational technology for almost a decade. Through these years, most of the researchers working in this area have been optimistic about its

Copyright © 2005, Idea Group Inc. Copying or distributing in print or electronic forms without written permission of Idea Group Inc. is prohibited.

benefits to education (see Harasim, Calvert, & Groeneboer, 1997; Jonassen et al., 1997; Stahl, 2002b; Willis, 1994). The main purpose of the CSCL field has been to offer innovative instructional strategies to avoid the low-level type of learning related to Internet correspondence courses (Comeaux, Huber, Kasprzak, & Nixon, 1998).

Research results have been promising; the studies made in the area usually show positive outcomes. For example, research has been done to prove that students develop their higher order mental abilities when they learn collaboratively online (Archer, Garrison, Anderson and Rourke, 2001; Arnseth, Ludvigsen, Wasson and Mørch, 2001; Bonk and Reynolds, 1997; Wang, Tzeng, & Chen, 2000); other authors have emphasized the collaborative aspects of online communication when students learn asynchronously as opposed to face-to-face learning (Ellis, 2001; Hiltz, 1998; Napierkowski, 2001); others have focused their research questions on administrative variables like the different group compositions, which usually show higher achievement for heterogeneous groups (Bernard and Lundgren-Cayrol, 2001; Lee and Chen, 2000; Nagai, Okabe, Nagata and Akahori, 2000).

Despite these results, Lipponen (2002) has raised some very interesting issues questioning CSCL research; he states that there is no agreement about the concept of collaboration and that "...there exists little research on how students participate in networked mediated collaboration" (p. 75). Specifically, it is not yet clear how students build knowledge collaboratively in asynchronous communication activities.

According to his comments, on the one hand, it is possible that most of the positive results shown so far could just be confirming the collaborative learning's effects on achievement, whether students are learning or not in a CSCL environment, a result fully documented in the face-to-face collaborative learning literature made available by well-known authors like Johnson and Johnson and Slavin.

In addition, judging from the literature reviewed in this chapter, it appears that not many research projects have considered the importance of understanding knowledge as something that is located in the group as a result of the activities done in the group; that is, knowledge is in the community rather than the individual, as studies done under the social constructivism perspective would suggest (see Imel, 1991; Resnick, Levine,& Teasley, 1991)[1].

Consequently, the group's shared cognitive processes related with the social building of knowledge in a technologically supported environment are not entirely understood. In order to contribute to this understanding, the present chapter describes an instructional online collaborative learning model that addresses the phenomenon from a systemic human relations and interaction perspective. Its main purpose is to aid students in their social building of knowledge or in their "relational cognition," a concept that this chapter suggests, that means that cognition is within human interaction when learning in a CSCL-networked environment. To achieve its purpose, the model incorporates the idea that the students' relational cognition is affected by the communication conflicts that naturally arise in human relationships. Thus, the model is basically proposing a way to attend to these communication conflicts.

For this model, learning happens inside a social environment where learners experience a complex set of different types of relationships when working together to accomplish a common task. Thus, when students build knowledge collaboratively, they are basically

Copyright © 2005, Idea Group Inc. Copying or distributing in print or electronic forms without written permission of Idea Group Inc. is prohibited.

engaging in communication-learning and relational-cognition circuits, as well as the conflicts that naturally arise within them, which affect the quality of their learning. For example, when a distance-education student learns in a group, he might feel that his opinions are not being respected by the rest of the group's members and, for this reason, he may decide to withdraw from the learning task instead of communicating his feelings to others.

Therefore, with the intention of addressing these communication and relational conflicts, the model proposed in this chapter is based on literature done in the fields of CSCL, collaborative learning, constructivism, and systemic communication.

The model's essence is based on the proposition that—to solve the communication and relational conflicts described above-the instructional designer should look for more profound and meaningful changes in the student, and that this is achieved when students develop their meta-abilities, that is, 1) the ability to meta-communicate, or communicating about how we communicate; 2) meta-learning, or learning to learn; and 3) meta-cognition, or thinking about how we think and about how we feel.

The model presented in this chapter is basically focused on developing the learners' meta-communication abilities. To this end, it includes a set of instructional strategies; it is expected that future versions of the model will also incorporate metacognitive and learning to learn strategies.

The instructional strategies to promote the students' meta-communication abilities, because of their "meta" nature, are called second-level type strategies by the author. I propose that all learning experiences should have a second level where students can develop their meta-abilities.

The instructional strategies to develop meta-communication abilities presented here are based on the five axioms of human communication developed by Paul Watzlawick and his colleagues (1967) at the Palo Alto Mental Research Institute. Their research results were considered relevant to understanding the online learning phenomenon because they address issues such as the differences and conflicts between digital and analogical communication-an important concern when learning online because most of the communication is done in writing (digital), but this writing, at the same time, communicates feelings and emotions (analogical).

Although this relational model is still in its initial stage, the concepts and instructional suggestions presented here are intended to have a heuristic value and are hoped to serve as a frame of reference to: 1) facilitate understanding of the complex human patterns of relationships that naturally develop when a virtual community interacts during a collaborative learning process; and 2) based on those understandings, suggest some basic pedagogical strategies to the instructional designer to develop sound online collaborative-learning environments.

The chapter starts with a general view of the theoretical background of the area of CSCL, of the field of collaborative learning, and of the constructivist epistemology to form a frame of reference to situate the Relational Online Collaborative Learning Model presented here. There is a special focus on the social and cultural lines of constructivism due to the relational nature of the model and to introduce and explain the concept of relational cognition; both serve as the starting point for presenting and discussing the set of concepts that provide a base for the model. Then, Watzlawick´s Five Axioms of

Copyright © 2005, Idea Group Inc. Copying or distributing in print or electronic forms without written permission of Idea Group Inc. is prohibited.

Figure 1: Justification for the model.

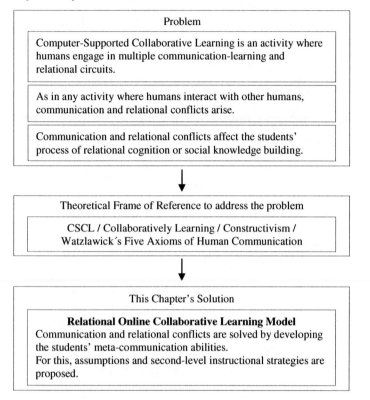

Human Communication are described, which serve as the basis for introducing the model's assumptions and the proposed instructional strategies (See the model's justification in Figure 1).

Theoretical Background

Computer-Supported Collaborative Learning

As said at the beginning of the chapter, CSCL has been part of the educational technology field for almost ten years; it combines the areas of Cooperative Learning with Computer-Assisted Learning (Graves & Klawe, 2002). Liponnen (2002) affirms that it is a field of study "…focused on how collaborative learning supported by technology can enhance peer interaction and work in groups, and how collaboration and technology facilitate sharing and distributing of knowledge and expertise among community members" (p. 72).

Gerry Stahl (2002a), states that 1) collaborative knowledge building, 2) group and individual perspectives, 3) mediation by artifacts (linguistic, cognitive, cultural, physi-

Copyright © 2005, Idea Group Inc. Copying or distributing in print or electronic forms without written permission of Idea Group Inc. is prohibited.

cal, and digital), and 4) interaction analysis are four themes that should be interrelated to form the theoretical frame of reference for the field of CSCL. The four have been studied independently in other areas of research, but putting them together gives a more coherent perspective to CSCL and shows how rich and complex it is.

In terms of the Stahl (2002a) proposition, the model described in the present chapter mainly addresses Themes 1 and 4; that is, it is suggesting how learners can increase their abilities for knowledge building through analyzing and improving the human communication relationships that naturally develop when humans interact. The model is especially focused on addressing the communication difficulties that predictably arise when humans learn together, such as misunderstandings, competition, and relationship conflicts- problems that are seldom considered in the collaborative-learning literature.

Collaborative Learning

Collaborative learning has been present in the educational discussion for more than two decades. It has grown as an important learning model and as a rather big set of instructional strategies. Research results have been mostly positive regarding the use of collaborative strategies inside the classroom. The belief is that students benefit academically and socially when working together to achieve a common goal. Some leaders in collaborative learning research like Johnson and Johnson (1982, 1990), Johnson, Johnson, and Holubec (1994a, 1994b); Slavin (1990, 1995) and Sharan (1990, 1994) have made important contributions in this area; for example, the concept of positive interdependence, which stresses the synergy that is developed due to the interactions among the participants by having learning outcomes beneficial to both the individual and the other group members. Johnson and Johnson (1994a) also introduce the notion that:

Placing socially unskilled individuals in a group and telling them to cooperate does not guarantee that they will be able to do so effectively. Skills such as leadership, decision making, trust-building, communication, and conflict management must be taught just as purposefully and precisely as academic skills. (p. 6).

Johnson and Johnson (1994a) have identified three theoretical perspectives on collaborative learning research:

1) Social Interdependence Theory, which sees the group as a system where the social structure affects the individuals' interactions;

2) Cognitive Development Theory, mostly related to the works of Piaget and his works in genetic epistemology. Piaget assumed that the process of cognition is an interaction between heredity and environment (Driscoll, 2000), and that individuals, when interacting, encounter cognitive conflicts that have to be solved, this being the intrinsic motivation to exchange information; and

3) Behavioral Learning Theory, which basically stresses the importance of extrinsic group reinforcers and rewards on learning and the analysis of conduct between group members.

Copyright © 2005, Idea Group Inc. Copying or distributing in print or electronic forms without written permission of Idea Group Inc. is prohibited.

The model discussed in this chapter would mostly fall under the social interdependence viewpoint because it stresses the relational aspects of collaborative learning. However, it also has strong influence from the cognitive perspective because it uses the posture of the constructivist epistemology, not so much from the biological aspects, but from the social approach.

In general, research contributions in this area are solid and are beneficial to the design and application of collaborative-learning strategies. However, most of those suggested for group learning focus mainly on structural or administrative variables (Duffy & Cunningham, 1996), such as student distribution in terms of gender, abilities, etc.; ways to ensure that all members work the same; ways to physically arrange the classroom to facilitate group work; the assigning of roles among members; the importance of specifying desired behaviors; ideas for monitoring and evaluating, etc. These strategies are, without a doubt, very useful to the teacher who wants to establish sound collaborative-learning processes, but they are not sufficient in themselves due to their structural basis and can be characterized as first-level strategies. The strategies proposed in this model are going to be called second-level strategies because of their different or "meta" nature, namely, communicating about how we communicate, and because they will be applied for the meta-development of the students learning in a group. Some of the strategies recommended by the literature could fall into this second- level category; for example, among the five essential elements recommended by Johnson and Johnson (1994a, 1994b) for designing good student cooperation is Group Processing, where members discuss how they are working in the group and how they are achieving their learning tasks. This is an example of a second-level strategy because students are in fact making a meta-analysis of their activities as members of a group.

The Relational Online Collaborative Learning Model, then, is aiming at the development of students by fostering their cognitive resources when interacting with the environment and with other human beings, that is, when they engage in what can be called a relational process of cognition. The next paragraph will explain this concept and its links to the constructivist epistemology.

Constructivism

Although there are several approaches to constructivism, what is most relevant is that this epistemology is an important philosophical turn. First, instead of believing in the existence of a reality outside the subject, constructivists assume that we invent or construct it through experience and, most important, that we do it with others. This apprehension of what and how we know puts in a new light to the way we understand knowledge and, of course, learning. For example, if the learners build their own unique interpretations of the world, then most of the responsibility (ability to respond) of the learning process falls on the learner. Moreover, as this is at the same time a social process, then cognition is merged within it and we must "...analyze the ways in which people jointly construct knowledge under particular conditions of social purpose and interaction" (Resnick, 1991). Thus, we can talk of a relational process of cognition or, to be more specific, my cognition, your cognition, and the relational cognition. A relational process of cognition will occur every time a group of students interacts collaboratively to pursue an educational goal.

Copyright © 2005, Idea Group Inc. Copying or distributing in print or electronic forms without written permission of Idea Group Inc. is prohibited.

Conceptually, relational cognition is almost synonymous to what Resnick, Levine and Teasley (1991) would refer to as socially shared cognition. However, we prefer to talk about a relational cognition to emphasize that the interaction or sharing includes not just mental, but behavioral, communicative, emotional, and even spiritual aspects. This concept of relational cognition is at the very essence of the Relational Online Collaborative Learning Model.

The Relational Online Collaborative Learning Model

In addition to the concept of relational cognition, the model will be described by first introducing the concepts of communication-learning circuits, second-level instructional strategy, and meta-development, which will serve as the general base for the suggested instructional strategies. These will be presented as emerging from Watzlawick's five axioms of communication (Watzlawick et al., 1967) and from a set of assumptions.

Communication-Learning Circuits

The model views the CSCL experience as a group of learners within a virtual social environment engaged in relational cognition processes mediated by communication technologies with the aim of completing an educational task (See Figure 2).

When thinking in terms of relational cognition processes, we can see the experience of learning online in a different light; this allows us to identify many circular patterns of human relationships that we just ignored before, regarding them as irrelevant, because we basically stressed the linear part of the process, i.e., teaching. The model sees the online learning experience as a system where all actors such as teachers, students, teacher's assistants, etc., interact within a social environment influencing each other all the time. These mutually influential relations could be done in many ways: speaking, writing, with gestures, and using all kinds of signs and media. In other words, communication circuits are established, as opposed to the more traditional sender-receiver dualistic model of communication. This circuit perspective is a key concept of the model; it is taken from the idea of Dewey (1896) that the total act precedes any discrimination of stimuli and response, very much in tune with systems theory. From this perspective, then, it is not important to ask who originated the process of communication, just that it happened, and that the whole process is composed of a myriad of circuits, which, using Krauss and Fussell's (1991) concepts, form a shared communicative environment with the main purpose of generating social knowledge.

In this model, we are interested in the multiple communication circuits that are naturally created when pursuing a learning endeavor with others, so we can call them communication-learning circuits. These circuits happen inside a social environment where learners experience a complex set of different types of relationships when working together to accomplish a common task. The idea is to understand the nature of these

Copyright © 2005, Idea Group Inc. Copying or distributing in print or electronic forms without written permission of Idea Group Inc. is prohibited.

Figure 2: The relational online collaborating model.

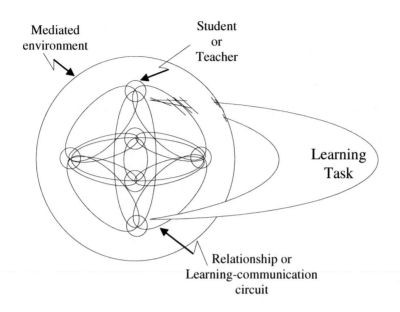

relationships and, from there, suggest pedagogical strategies to facilitate collaborative online learning.

In a CSCL environment, all of the communication circuits are mediated using all types of technologies, thus we must add the role of media to the formulation of this model. For example, the communication of gestures would be difficult in a learning-communication circuit using email or telephone in an audio conference. In this line, Rocheleau and Santos (2002) discuss the technological and cognitive aspects of media, both important when students are learning online. The technological aspect affects the decisions made about which technology could be used to support the process of teaching and learning. However, due to the nature of the Relational Model, the cognitive aspect of media is more relevant to it; that is, media can be used to support cognitive processes in such a way that they can become cognitive partners with the learners, instead of just serving them as mere devices to deliver the traditional ways of teaching and learning, i.e., unidirectional transport of information. From this viewpoint, media, then, forms an integral part of the learning environment acting as cognitive tools (Jonassen, 1996) aiding the process of relational cognition.

Second-Level Instructional Strategies

The instructional strategies proposed here are intended to be used for an online collaborative learning experience having the following basic structure: first, there is an explicit learning task (See Figure 3), which acts as the leader of the whole process, and second, there is a group of persons (teachers and learners) who interact with the purpose of attaining that task[2].

Copyright © 2005, Idea Group Inc. Copying or distributing in print or electronic forms without written permission of Idea Group Inc. is prohibited.

Figure 3: The learning microcosm.

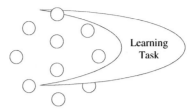

This structure is presented here in a seemingly simplistic way. Obviously it is not; rather, it is a complex microcosm where learners and teachers share more than just the goal of completing the task. In fact, they construct knowledge and their identities as persons (Lipponen, 2002) within a complex web of relational cognition circuits.

The model introduces the notion that there are two different levels of instructional strategies (IS). Both would have the basic purpose of changing the learning status of a student from one state to another, but first-level strategies move the learner from one point to another, while the second-level instructional strategies seek a change in change of position, or meta-change (Watzlavick, Weakland, & Fisch, 1974). For example, in motion, first level would be a mere a change in position, while a change in change of motion would be acceleration. Thus, first-level instructional strategies help move the learner from one learning state to another (a higher one, hopefully) while second-level strategies intend to accelerate the change by producing more profound and meaningful changes in the student. The perfect example of the second type would be strategies to develop metacognitive abilities in the learner. An example of a first-level IS would be "When learning collaboratively, the teacher should always evaluate individual and group work." While a second-level one would be "Make sure to have discussions where your students communicate among themselves about how they communicated during the time they were doing the learning task." All the instructional strategies presented in this model are at the second-level because there are already many first-level strategies, and, as we said, second- level instructional strategies produce more significant changes in the students. This notion of second-level instructional strategies is a new idea presented in this chapter and, therefore, it still needs research evidence to be supported.

Two Levels of Instruction

We do not mean that first-level IS are not important or that they should not be employed by an instructional designer. The general advice for the designer would be to include a sort of second-level of instruction, where second-level instructional strategies are applied so that students can reflect on how they learn, how they communicate, etc. The strategies suggested are presented so that, when applied, the probability increases that the students' learning will have a positive change. However, the speculation of this model is that learning cannot be qualitatively changed when using only first-level strategies, such as "Make sure that students' teams are formed with no more than five members." or "Assign collaborative and individual tasks to make sure that every student ends up working as hard as everybody else."

Copyright © 2005, Idea Group Inc. Copying or distributing in print or electronic forms without written permission of Idea Group Inc. is prohibited.

The model identifies three categories of second-level strategies corresponding to the following types of human abilities: 1) Meta-Communication, or communicating about how we communicate; 2) Meta-Learning, or learning to learn; and 3) Meta-Cognition, or thinking about how we think and about how we feel (See Figure 4). Because the last two have been thoroughly discussed in the literature and, due to the relational nature of the model, most of the discussion presented here is dedicated to Meta-Communication. However, the three are intricately related, i.e., when developing one, the others are also developed. Future versions of this model will incorporate the three categories.

Meta-Development

Due to their "meta" quality, we can say that the main purpose of the second-level type of strategies is to allow the meta-development of students (See Figure 4). One of the assumptions here is that, by meta-developing, students would develop the type of abilities that would allow them to surpass the lack of non-verbal communication that mostly characterizes online learning. We humans, when interacting with our fellow companions, use a lot of nonverbal types of communication; we use gestures, move our hands, our eyes, etc., to send all types of messages. However, in a computer-supported environment most of the communication is based on written language, which is not enough to form a complete human interaction system.

As said before, the Relational Online Collaborative Learning Model stresses the importance of having a second-level of instruction in any type of online course where second-level instructional strategies are applied. This is important in any type of learning experience, but is even more important in an online experience because most of the communication is done verbally.

In this line, then, the general advice for the designer would be to have two layers of instruction, with both happening all the time during the whole course (See Figure 5). In one, all the content instruction and the first-level strategies for collaborative learning (or some other pedagogy) can be applied. And in the other, the second-level type of strategies would be applied, during which students can learn more about themselves and how they learn and behave in a group as they reflect on how they learn, how they

Figure 4: Meta-development.

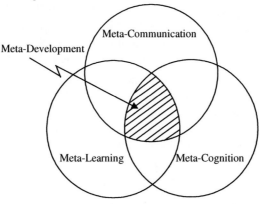

Copyright © 2005, Idea Group Inc. Copying or distributing in print or electronic forms without written permission of Idea Group Inc. is prohibited.

Figure 5: Two levels of instruction.

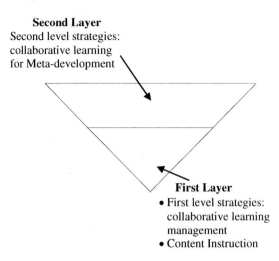

Second Layer
Second level strategies:
collaborative learning
for Meta-development

First Layer
• First level strategies:
  collaborative learning
  management
• Content Instruction

communicate, how they think, etc. For this model, this type of self-knowledge is as valid as that related with the course content.

In Figure 5, an upside-down triangle was used to suggest that second-level strategies should be used much more often than first-level ones when learning online. This proposal will obviously decrease the time available for the teaching of the content matter. However, the premise here is that students will in fact be gaining, because they would be developing the kind of meta-abilities that would enhance their capacities as students and as human beings in general.

Instructional Strategies

As we have discussed, the learners' communication-learning circuits are at the essence of this model; therefore, we have adapted parts of the theory of human communication developed by Gregory Bateson, Paul Watzlawick, and their colleagues at the Palo Alto Mental Research Institute (Watzlawick, Bavelas, & Jackson, 1967),[3] to understand their nature and dynamics and, from them, suggest a series of instructional strategies.

First, each axiom is presented and explained; then, from it, a group of assumptions are suggested to serve as the basis for the instructional strategies that are finally given for the designer of an online collaborative-learning environment.

The assumptions and the strategies recommended here are intended to have a heuristic value due to the initial stage of this model. The reader will notice that many of the assumptions and strategies that can be applied to more than one axiom are separated for clarification. But, it is possible to apply a strategy to solve more than one relationship problem, pertaining to one of the axioms, at a time.

Copyright © 2005, Idea Group Inc. Copying or distributing in print or electronic forms without written permission of Idea Group Inc. is prohibited.

Axiom 1: The Impossibility of Not Communicating

In the interaction with others, one cannot not behave and, if all behaviors inform, then one cannot not communicate. This axiom extends the concept of interaction between all actors in the system to include all possible types of behaviors, and still characterizes them as communication circuits even if one of them decides to stay as a mere observer. For example, by not participating in a certain chat discussion, a student would still be influencing the group. She or he would still be sending a message, perhaps that they are experiencing some type of confusion within themselves, with the logistics of the course, or with the content. From this idea, it follows, then, that even when some mutual miscommunication or misunderstanding occurs, we can still call that interaction a learning circuit. Therefore, for our model, all types of interactions have the same learning value. That is, even the interactions that are not (at least apparently) directed towards the learning goal would still be adding elements to achieve it.

Seemingly, this posture of accepting all communication circuits as part of the learning experience hinders the management of an online course because there would be too many to have under the control of the teacher. This is not so, because what we are in fact saying is that it is the social system as a whole—students, teacher and resources—that is aiming for the goal and is following the path to learning; thus, most of those interactions are experienced and regulated by students themselves. On the one hand, this posture adds uncertainty to the system, but, on the other, it also presupposes a teacher with plenty of confidence in the ability to respond to their students, i.e., their responsibility. For the designer of an online collaborative-learning environment, this means that it is important to build a learning space where confusion and errors are welcome and never punished.

Assumptions that emerge from this axiom include:

- All behavior is relevant to learning.

- All types of interactions have the same learning value.

- The more students are conscious of their behaviors and what they mean to them and to the system, the greater the quality of the collaboration.

- The more students trust the system, the greater the quality of the collaboration.

- Relationship problems are best solved by the students themselves.

Some strategies for teachers in an online collaborative-learning environment:

- Follow any behavior with care, because it can become an important indicator of the success or failure of the whole group. For example, if a student working on a certain team decides to drop out of the learning experience, inquire into the situation by sending him and the rest of his team an email investigating what could be the matter. This systemic view is important because his action might affect not just the work of the team but that of the whole course. The amount of behavior to take care of in a distance course could be staggering, thus it would be rather difficult to follow

Copyright © 2005, Idea Group Inc. Copying or distributing in print or electronic forms without written permission of Idea Group Inc. is prohibited.

this strategy for a single teacher. However, it is common for an online course to have several teacher assistants; one could be dedicated to this task during the whole course.

- Have sessions where students reflect on their behaviors. At the beginning of the course, have a chat session where all students openly state the different ways they usually behave when working in a team. This experience can be done in many different ways, for example, using a questionnaire so that they can identify their own style of working in a group. Or else develop a case to which they can relate and then have a chat session where all openly discuss how they learn in a group. The basic intention is to aid the students in gaining consciousness about how they behave when they learn collaboratively. In this way, they can realize, for example, that they always like to be the leader in a group, or that they tend to wait before talking, or that they feel more comfortable when somebody else does the leading.

- It is important not to have the second-level experiences directly tied to the grade of the student. The idea is that students not feel judged in any way and that they feel free to express feelings and information about themselves. A message stating this could be sent to the students at the beginning of the course to establish a sort of contract, and it can be reinforced during the course duration. Students would only believe in and understand trust by experiencing it intensely during the process of learning.

Axiom 2: The Content and Relationship Levels of Communication

In any type of interaction there are two levels of communication. At one level, communicants convey the content; at the other, they say something about that content. This second level, or the meta-communication level, is affected by the type of relationship that the communicants have or are establishing. For example, when a team of five students is interacting via email to solve a problem collaboratively, one of them might decide (without communicating it to the others) to go to the library and start researching some data for the problem, but when the others find out they get mad at him or her. These students are agreeing at the content level, i.e., it is a good thing to research data to solve a problem, but disagreeing at the relationship level, i.e., do not do it without informing the others. In this group, a rule had been established at the relationship level (but not openly).

Unfortunately, most people are not aware of the two levels, thus, the disagreement within the team of students will affect their learning unless the process is clarified for them or they are taught how to do it. Here what is desirable is not that students agree at both levels, but that they "agree to disagree"; that is, they understand and accept that others can have different viewpoints.

The above example illustrates how a group of people always develops rules of relationship from the start. Most of the time, these rules are not openly established; rather, they develop from each member's past experience and from the combination of these in the new relationship. Most, if not all, of the communication of content between the members of this group is affected by these rules of relationship. Thus, a lot of relationship troubles can emerge if they do not come to the open so that they can be accepted or changed, i.e.,

Copyright © 2005, Idea Group Inc. Copying or distributing in print or electronic forms without written permission of Idea Group Inc. is prohibited.

negotiated. When students are negotiating their rules of relationship as a group, we say they are meta-communicating. At this second level they meta-communicate information about the type of message that is being sent and how it should be taken.

The relationship part, then, clarifies the content part of communication. For example, compare the two following messages that a teacher might send by email to his students: 1) "Please make sure to finish your essays before the established deadline," and 2) "Finish your essays before the established deadline!" Both messages have more or less the same content but show two different types of relationship. It is rather easy for communicants to confound the two levels; it might happen that some students, when receiving the second message from their teacher would feel offended because they expect a teacher to always be kind when communicating with them.

The designer should include, as part of the course, learning experiences where students develop the ability to identify the two levels of communication and how to solve the common mistake of confounding them. That is, teach them how to communicate about their communication, or, in other words, how to meta-communicate.

Also, this axiom of communication can be applied to the subject matter itself; for example, two pieces of content like "a" and "b" are differently qualified by a "+" connector put between them than by a "-". So, the content that the student decodes should be designed including indicators of relationship as well. That is, they should be told how to relate with and how to process the information.

Assumptions that emerge from this axiom include:

- The relationship level qualifies the digital part of the communication.

- When learning in groups, members always develop rules of relation from the start.

- The nature (stated in rules) of the relationship among actors affects the way the content is learned.

- The more clearly the rules of relation among actors of a course are stated and negotiated, the fewer relationship difficulties between the group members and, thus, the better relational cognition processes.

- The earlier the rules of relationship are stated, the better.

- There are rules for the human communication and also for the human-subject matter relationship.

- The more explicit the rules of relation are between students and subject matter, the less ambiguous it is for them.

Some strategies for instructors in an online collaborative-learning environment:

- Students should gain consciousness about the difference between content and relationship. If it is the first time they are encountering these concepts, give them a learning experience where they can understand the difference. This can be done

Copyright © 2005, Idea Group Inc. Copying or distributing in print or electronic forms without written permission of Idea Group Inc. is prohibited.

in many ways. For example, it can be as simple as giving them some reading materials to study this difference, or, better yet, designing a constructivist learning environment where they can experience the difference. For example, send all the class a piece on the Mexican revolution and another on the Russian revolution (or any other type of content). To half of the class, also send another piece of information where a writer says how similar both are, and, send the other half a different piece stating how different the revolutions are. Then, let all students discuss the subject matter in a chat session without telling them the differences. Following the discussion, make them reflect on how differently one half of the class related to the content as compared to the other half. In this way they can identify the differences between content and relationship.

- Teach students how to meta-communicate. Form discussion groups regarding some of the content of your course, and, at certain time intervals, stop the discussion and ask them to communicate about how they are communicating. Make them realize that at the meta-communicative level it is almost impossible to tell lies.

- At the beginning of the course have a chat session where all students discuss what they see as the rules of relationship that would be followed during the class; that is, among students and other students, with the teacher and assistants, and even between teacher and his or her assistants. An example of this type of rule would be: "The teacher, not the students, is who decides the pace and direction of the whole course," or vice versa. Consequently, all the actors have the opportunity to negotiate the rules of relationship.

- If it comes to your attention that a certain group is having relationship troubles, have its members discuss how they have been communicating. Ask them to clearly identify the errors they have made regarding their relationships. Also, ask them to openly clarify and negotiate their rules of relationship.

- Design the content so that it includes information of the relationship level. If the content is available online in written form (or some other way), always include information in how students could relate (there could be more than one way) with that subject matter. Clearly separate the relationship information from the subject matter, so students can soon learn how to use it as an aid for their learning.

Axiom 3: The Punctuation of the Sequence of Events

As we have said, the learning system is composed of a myriad of communication-learning circuits, where, due to the systemic perspective, one cannot really say who started or finished a circuit. However, each person introduces a personal punctuation of the sequence of events that he or she experiences during a certain learning process. By punctuation, we mean that each person believes that the communication started at a certain point in time, and another might think it started at a different one. These diverse personal punctuations could hinder learning because each person believes that his or her punctuation is the right one.

For example, working in a team, a student decides to take a more passive attitude because he feels that the rest of the group makes all the decisions regarding the learning task

Copyright © 2005, Idea Group Inc. Copying or distributing in print or electronic forms without written permission of Idea Group Inc. is prohibited.

without considering his point of view. The group, on the other hand, decides to ignore him because it feels that he is too passive. In this example, the two different punctuations are patent: the student is passive because he sees the group's action of ignoring him as the cause of his passivity, and the team punctuates the process as if his action is the start or cause of its response of ignoring him. Who is right? It does not matter. The fact that the discrepancy exists is what matters and that they need to meta-communicate to solve their differences.

Another example- it is common in a group that some student, due to his or her personality, ends up doing most of the task's workload. He feels badly about it and thinks that he does not really like collaborative learning, but he feels obligated to make the effort because he thinks that the rest of the team is lazy. Now, the rest of the team members think that they let him do the entire job because, from the start, he disqualified most of their ideas. The difference of punctuation is clear: on the one hand, the student thinks he does the whole job because the rest are lazy and, on the other, the team thinks they let him do it because he would not let them cooperate. Again, neither of the two punctuations is right or wrong, but students need to become aware of the difference. Through meta-communicating this team could get to the root of the relationship problem. That is, through it they could realize that what is happening is that the student wanting to do the whole work has no confidence in others and that for them it is very comfortable to let somebody else do the job.

For this axiom, designers could plan learning experiences similar to the ones for Axiom three. However, instead of focusing on the content-relationship aspects of the team's relationship, he or she should focus on opening spaces for students to understand that it is possible to have different punctuations for any set of communication circuits. Also, when differences do happen and go unnoticed, the way to solve them is through meta-communication.

Assumptions that emerge from this axiom include:

- When learning in groups, members always develop their own punctuations of the process.

- Through meta-communication students identify their different punctuations and also the roots of those differences.

- The clearer the punctuations are, the more efficient the group's relational cognition processes are and, thus, their social learning.

- New differences in punctuation can emerge at any time during the learning process.

Some strategies for instructors in an online collaborative-learning environment:

- Make clear to students that the learning environment is a space designed in such a way that there is always freedom of expression. Express openly that any type of feeling and belief can be ventilated without any type of punishment.

Copyright © 2005, Idea Group Inc. Copying or distributing in print or electronic forms without written permission of Idea Group Inc. is prohibited.

- Design a learning experience where the students can understand the importance of punctuation in the team's relationships. This can be done in different ways, but an excellent one would be to take advantage of the situation when somebody expresses that he feels rejected by the group or when the team states that they have developed negative feelings toward a certain person because she is not doing her part of the work. Open a debate session with that team of students to discuss the circumstances.

- Dedicate an area of the course's Web page where students working in a team can go and chat by themselves to meta-communicate. Once students understand the concept and problems of punctuation in a communication, the best way is for them to solve their own relationship problems. After a while, whenever they encounter a relationship difficulty, they will tend to look for the origins of their problems using the communication abilities they are developing.

Axiom 4: Digital and Analogic Communication

We humans can communicate in two ways. For example, if one wants to express something about a cat, we could use a picture, a drawing of a cat, point to a real one, or else we could use an arbitrary sign that would represent it, like the word "cat." The former type is called analogic communication because the picture is analogous to the cat; and the latter is called digital because the sign "cat" does not in any way resemble the real animal. Another example of digital communication is the digit 7, which needs a semantic convention for us to understand it. When we speak using any language, we are always using both types of communication. Digitally, we make use of the signs and all the syntactic and semantic conventions of that particular idiom. And analogically we make use of body language, gestures, different voice intonations, etc., to send a different part of the message at a higher level of communication. In this sense, the analogical qualifies the digital part of the communication. This is done in any type of human communication. It is clearer in oral communication, but it is present when reading or writing as well. Have you read a good novel more than once because every time you "read" new ideas in it?

Writing is still the dominant mode of communicating in an online learning environment. Thus, apparently only digital communication is happening, but this would not be true. Imagine an example where the teacher sends the message, "Finish your essays before the established deadline, or face the consequences!" A lot of analogical communication is present here because it is with this type of communication that we form the relationship with others. Both types are present in any human communication. But paradoxical relationships can be established if both types contradict each other. Imagine that the same teacher had written as part of the course Web page, "I am a flexible person and always willing to hear from you." Students and teachers are always translating all the messages they receive from digital to analogical, something that could be very complex and full of possible errors.

Again, students should be aware of the two different types of communication, so that they can develop the abilities to use both as decoders or coders of all the messages sent during the learning process.

Copyright © 2005, Idea Group Inc. Copying or distributing in print or electronic forms without written permission of Idea Group Inc. is prohibited.

Here, the virtual context in which students interact to learn is also part of the analogical aspect of communication. Thus, a lot of thought should be given by the designer to the way the whole course is going to be structured, to the interfaces, to the metaphor used as the central organizing theme, etc.

Assumptions that emerge from this axiom:

- Both digital and analogical communications are necessary to decode a certain message.

- The analogical qualifies the digital part of the communication.

- Students learning online receive most of their information in digital form; however, they are always trying to translate those digital messages into analogical communication.

- The less contextual information, the greater the probability of error when translating from digital to analogical.

Some strategies for instructors in an online collaborative-learning environment:

- Learn how to include analogical information. The ability to do it can be learned, especially in the written communication mode.

- Always include analogical information. This can be done in several ways: develop the course material using plenty of graphics; use video depicting people in action whenever possible; or carefully develop written information following the principles of rhetoric.

- Teach students to understand what they read. In this way, they learn how to translate digital information to analogical efficiently.

- Teach students to write. That is, teach them how to build solid arguments through writing following the principles of logic. In this way, they learn how to encode analogical information in writing.

- Embed every learning task in a real-life context. Thus, always include and define the problem's context in great detail.

- Make sure students have enough information about a particular culture because "the activity of problem solving is influenced by the culture in which it is embedded" (Driscoll, 2000, pp 238).

- Always semantically define all the concepts used in the content. Have a dictionary or glossary always ready online, or use the ones available on Internet. It would be better if an "intelligent glossary" were used, that is, one that presents the definitions relevant to the particular piece of content a student is reading.

Copyright © 2005, Idea Group Inc. Copying or distributing in print or electronic forms without written permission of Idea Group Inc. is prohibited.

Axiom 5: Symmetrical and Complementary Interaction

In general, interactions between persons tend to change progressively during a social experience, including group learning. This movement can be done in two ways: 1) symmetrically, where everybody is considered at an equal level and thus a very competitive environment develops, or 2) complementary, where there are superior and inferior roles that define the relationship. Both types develop naturally during the length of a course, mainly due to the particular characteristics of the members of a certain group.

These two types of dynamic interactions can be used to define the relationship between all actors in a group engaged in collaborative learning (including teacher and assistants). Neither is better than the other. What is important is that neither should predominate; in a learning environment with sufficient freedom of expression, both would happen naturally during the length and breadth of a course or a learning task. The designer should accept that, at some times, the relationship between teacher and students or among students is complementary and at other times symmetrical. This is so because differences and equality should both be meta-goals of any course.

On the one hand, all communicants must accept that there is a hierarchy, a natural order, where the teacher is at a superior level because he or she knows something the student does not, for example, the general structure of the course. But they must also accept that, at other times, the teacher could also be at an inferior level; for example, in solving a problem, a certain student could have gained a great deal of knowledge regarding a certain topic and thus, know more than the teacher. Unfortunately, the words "superior" and "inferior" have a meaning attached to them that does not truly describe what is meant here; we use them not to denote "good" or "bad," but just to indicate differences in the hierarchy of the relationship.

These concepts of symmetry and complementarity are also related to the concept of leadership in a learning team. Traditionally, a student would take the leadership of the group mainly due to her or his personality; that is, some persons tend to take control of any situation from the start of any relationship, while others prefer to follow. The roots of these behaviors are, most certainly, situated in the individuals' early childhood experiences. Without judging if "to lead" or "to follow" is good or bad behavior, the designer should accept that each student has a learned way of behaving when interacting with others.

In this sense, when two or more leading personalities collide in a group, it is very probable that a symmetric type of relationship will develop in that group. The problem is that if these group members remain symmetrical all the time then one could expect that competition between them would quickly escalate to the sky. This can be a very destructive learning environment. On the other hand, if only followers end up forming a group, then a sort of negative symmetry also develops because symmetrical interaction is characterized by not having differences. Although in a different way, this type of symmetry also promotes a destructive environment.

Now, from the complementary perspective, if a leader type of student teams up with a follower then a complementary relationship is established; that is, the leader would be in a superior level and the follower in an inferior one. Apparently, this type of relationship would be more productive than the symmetrical one, but, in fact, it is not, because it results in a rigid type of relationship where the amount of group synergy is very low.

Copyright © 2005, Idea Group Inc. Copying or distributing in print or electronic forms without written permission of Idea Group Inc. is prohibited.

Thus, a productive collaborative learning environment needs to have all types of students and, therefore, both kinds of relationships. The ideal situation is where students realize what type of persons they are, and, from there, the team has a rolling leadership among all members; that is, all students take the role of leader or follower at one time or another. In this sense, leadership and "followership" are both equally important.

Assumptions that emerge from this axiom:

- All relationships are either symmetrical or complementary.

- A productive CSCL environment has both symmetrical and complementary relationships.

- The possible ways a group could function is infinite because the possible types of students are also infinite.

- Freedom of expression ensures that both types of relationships would happen.

- Each student has a learned way of behaving when interacting with others.

- The more predominant a type of relationship is, the less synergy the group would have, and thus, the less productivity.

- The more predominant a type of relationship is, the more destructive the group is.

- Leadership and "followership" are equally important.

- Meta-communication helps students gain consciousness of the ways they prefer to interact with others.

Some strategies for instructors in an online collaborative-learning environment:

- Design a learning experience where students discover what type of personality they have when learning in a team. For example, you can make a game where you randomly assign different roles to members of a certain team. One could act as the leader, another as a follower, another as having trouble at home, another who is plain lazy, etc. Give them a learning task and ask them to act the role that was assigned to them. Then, have a reflection session where they realize how the different types of students affect the group's work. Next, ask each one to reflect on the way he or she usually behaves when learning in a team. Ask them to send everybody in the class an email stating it. In this way each student is expected to start using different ways of relating after having more consciousness of their own ways of behaving in a group.

- The course should have a space where students can realize that, as humans, they are all the same and thus must respect each other when having different points of view. The course can include learning experiences where students are asked to engage in discussions with others where at one time they defend one position and, at another, they defend the opposite.

Copyright © 2005, Idea Group Inc. Copying or distributing in print or electronic forms without written permission of Idea Group Inc. is prohibited.

- When selecting members to form a team, do it with the whole class. Let students express their preferences, but make them aware of the importance or the concept of rolling leadership. Here you can make use of a first-level strategy and ask them to keep some type of diary where they leave a record of how they are exchanging their roles.

- After students have had the opportunity of working in different teams, ask them to reflect on and compare all of them in terms of their productivity. In this way they will reach conclusions about what makes a team develop synergy and the different ways a group can increase productivity.

- From the start of the course, inform students that you and your team of assistants are ready to hear any comment, complaint, or praise. Remind them of this fact several times during the course. It can even be a banner on the course Web page. This is a way of starting to develop an environment where freedom of expression prevails. Never punish a student in any way for being honest.

Conclusion

The Relational Online Collaborative Learning Model intends to offer a means of providing a humanistic educational experience to students in a computer-supported cooperative-learning environment. The model holds that a promising approach is to emphasize the relational aspect of the human experience of learning collaboratively online. Its basic proposal is to increase the meta-communication between all actors in the teaching and learning process and to consider the student as a human being responsible and capable of interpreting and constructing his or her own knowledge. In this sense, its main goal is to meta-develop students; that is, to let them build up their potentialities to accomplish a full life and not merely accumulate inert knowledge.

Pedagogically, the model accentuates the importance of having a second level of instruction in any type of course where second-level strategies are applied to the meta-development of students. This is important in any kind of learning experience, but is even more significant in an online experience because most of the communication is done digitally.

Digital communication is a rather recent human invention and, no doubt, a very important one, but it is not enough. This is mainly because we humans, when interacting with our fellow companions, prefer to use the analogic type of communication—we use gestures, move our hands, our eyes, etc., to send all types of messages. Nature communicates analogically. The model described here offers a method to include analogical communication as well in an online course.

CSCL offers many promises for our world; however, we face the challenge of widening it so that it adds to our good human qualities. One possible way to achieve this could be to forget the fatuous modern search for inventing a virtual exact image of reality and, instead, use our intelligence to build online worlds where students can associate with technology in innovative ways.

Copyright © 2005, Idea Group Inc. Copying or distributing in print or electronic forms without written permission of Idea Group Inc. is prohibited.

In conclusion, the Relational Online Collaborative Learning Model can serve as a base to develop learning environments where the goals of education are fulfilled; that is, where the persons develop with others to become better human beings.

References

Archer, W., Garrison, D.R., Anderson, T. & Rourke, L. (2001). *A framework for analyzing critical thinking in computer conferences.* Paper presented at the 2001 Euro-CSCL. Retrieved online June 5, 2003, from: *http://www.mmi.unimaas.nl/euro-cscl/Papers/6.doc.*

Arnseth, H.C., Ludvigsen, S., Wasson, B., & Mørch, A. (2001). *Collaboration and problem solving in distributed collaborative learning.* Paper presented at the 2001 Euro-CSCL. Retrieved online June 5, 2003, from: *http://www.mmi.unimaas.nl/euro-cscl/Papers/8.doc.*

Bernard, R. M. & Lundgren-Cayrol, K. (2001). Computer conferencing: An environment for collaborative project-based learning in distance education. *Educational Research and Evaluation: An International Journal on Theory and Practice*, 7(2-3): 241-261.

Bonk, C. J. & Reynolds, T. H. (1997). Learner-centered web instruction for higher-order thinking, teamwork, and apprenticeship. In B.H. Khan (ed.), *Web-based instruction*, pp. 167-178. Englewood Cliffs, NJ: Educational Technology Publications.

Brush, T.A. (1997). The effects on student achievement and attitudes when using integrated learning systems with cooperative pairs. *Educational Technology Research and Development*, 45(1): 51-64.

Comeaux, P., Huber, R., Kasprzak, J., & Nixon, M. (1998). Collaborative learning in web-based instruction. *Proceedings of Web-Net 98 World Conference of the WWW, Intranet and Internet.* Orlando FL. (ERIC Document Reproduction Services No. ED427693).

Dewey, J. (1896). The reflex arc concept in psychology. In J. A. Boydston, (Ed.), *John Dewey: The early works* (Vol. 5, pp. 96-109). Carbondale: Southern Illinois University Press.

Driscoll, M.P. (2000). *Psychology of learning for instruction.* Needham Heights, MA: Allyn & Bacon.

Duffy, T.M. & Cunningham, D.J. (1996). Constructivism: Implications for the design and delivery of instruction. In D. Jonassen, (ed.), *Handbook of research for educational communications and technology,* pp. 186-187. New York: Macmillan.

Ellis, A. (2001). *Student-centered collaborative learning via face-to-face and asynchronous online communication: What's the difference?* (ERIC Document Reproduction Services No. ED467933).

Garrison, J. (1998). Toward a pragmatic social constructivism. In M. Larochelle, N. Bednarz, & J. Garrison,(eds.), *Constructivism and education,* pp. 43 - 60. New York: Cambridge University Press.

Copyright © 2005, Idea Group Inc. Copying or distributing in print or electronic forms without written permission of Idea Group Inc. is prohibited.

Graves, D. & Klawe, M. (1997). *Supporting learners in a remote CSCL environment: The importance of task and communication.* Retrieved May 25, 2003, from University of British Columbia, Computer Science, EGEMS, Electronic Games for Education in Math and Science Web site: *http://www.cs.ubc.ca/nest/egems/.*

Harasim, L., Calvert, T., & Groeneboer, C. (1997). Virtual U: A web-based system to support collaborative learning. In B. H. Khan (ed.), *Web-based instruction,* pp. 149-158. Englewood Cliffs, NJ: Educational Technology Publications.

Hiltz, S. R. (1998). *Collaborative learning in asynchronous learning networks: Building learning communities.* (ERIC Document Reproduction Services No.ED427705).

Imel, S. (1991). *Collaborative learning in adult education.* ERIC Digest No. 113. (ERIC Document Reproduction Service No. ED334469).

Iowa Department of Education (1989). *A guide to developing higher order thinking across the curriculum.* Des Moines, IA: Department of Education. (ERIC Document Reproduction Service No. ED 306 550).

Johnson, D.W. & Johnson, R.T. (1982). The effects of cooperative and individualistic instruction on handicapped and non-handicapped students. *Journal of Social Psychology,* 118: 257-268.

Johnson, D.W. & Johnson, R.T, (1990). Cooperative learning and achievement. In S. Sharan (ed.), *Cooperative learning: Theory and practice.* New York: Praeger.

Johnson, D.W., Johnson, R.T., & Holubec, E. J. (1994a). *Cooperative learning in the classroom.* Alexandria, VA: Association for Supervision and Curriculum Development.

Johnson, D.W., Johnson, R.T., & Holubec, E. J. (1994b). *The new circles of learning: Cooperation in the classroom and school.* Alexandria VA: Association for Supervision and Curriculum Development.

Jonassen, D. H. (1996). *Computers in the classroom: Mindtools for critical thinking.* Englewood Cliffs, NJ: Prentice Hall.

Jonassen, D. H., Dyer, D., Peters, K., Robinson, T., Harvey, D., King, M, & Lougher, P. (1997). Cognitive flexibility hypertexts on the Web: Engaging learners in meaning making. In B. H. Khan (ed.), *Web-based instruction,* pp. 119-133. Englewood Cliffs, NJ: Educational Technology Publications.

Krauss, R.M. & Fussell, S.R. (1991). Constructing shared communicative Environments. In L.B. Resnick, J. M. Levine & S.D. Teasley (eds.), *Socially shared cognition,* pp. 172 - 200. Washington, DC.: American Psychological Association.

Lee, Y. & Chen, N. (2000, November 21-24). Group composition methods for cooperative learning in web-based instructional systems. Paper presented at the *ICCE/ICCAI 2000.* Taipei, Taiwan. (ERIC Document Reproduction Services No. ED454852).

Lipponen, L. (2002). Exploring foundations for computer-supported collaborative learning. In G. Stahl (ed.), *Proceedings of CSCL 2002, Computer support for collaborative learning: Foundations for a CSCL community,* pp. 72-81. Hillsdale, NJ: Lawrence Erlbaum Associates, Inc.

Nagai, M., Okabe, Y., Nagata, J., & Akahori, K. (2000, November 21-24). A study on the effectiveness of web-based collaborative learning system on school mathematics:

Copyright © 2005, Idea Group Inc. Copying or distributing in print or electronic forms without written permission of Idea Group Inc. is prohibited.

Through a practice of three junior high schools. Paper presented at the *ICCE/ ICCAI 2000. Taipei, Taiwan. (ERIC Document Reproduction Services No. ED454852)*.

Napierkowski, H. (2001). *Collaborative learning and sense of audience in two computer-mediated discourse communities.* (ERIC Document Reproduction Services No. ED456482).

Resnick, L.B. (1991). Shared cognition: Thinking as social practice. In L.B. Resnick, J.M. Levine & S.D. Teasley (eds.), *Socially shared cognition,* pp. 1-20. Washington, DC: American Psychological Association.

Resnick, L.B., Levine J.M., & Teasley, S.D. (eds.). (1991). *Socially shared cognition.* Washington, DC: American Psychological Association.

Rocheleau, J. & Santos, A. (2002). Socio-constructivist tele-learning environments: core foundations for future development. Unpublished manuscript.

Sharan, S. (1990). *Cooperative learning.* New York: Praeger.

Sharan, S. (ed.). (1994). *Handbook of cooperative learning methods.* New York: Greenwood Publishing.

Slavin, R.E. (1990). Learning together. *The American School Board Journal, 177*(8), 22-23.

Slavin, R.E. (1995). *Cooperative Learning: Theory, Research, and Practice.* Boston: Allyn and Bacon.

Stahl, G. (2002a). *Contributions to a theoretical framework for CSCL.* In G. Stahl (ed.), *Proceedings of CSCL 2002: Computer support for collaborative learning: Foundations for a CSCL community,* pp. 62-71. Hillsdale, NJ: Lawrence Erlbaum Associates, Inc.

Stahl, G. (ed.). (2002b). *Proceedings of CSCL 2002: Computer support for collaborative learning: Foundations for a CSCL community.* Hillsdale, NJ: Lawrence Erlbaum Associates, Inc.

Wang, W., Tzeng, Y., & Chen Y. (2000, November 21-24). A comparative study of applying Internet on cooperative and traditional learning. Paper presented at the *ICCE/ ICCAI 2000.* Taipei, Taiwan. (ERIC Document Reproduction Services No. ED454852).

Watzlavick, P. (1981). La realidad inventada, ¿cómo sabemos lo que creemos saber? [The Invented Reality, how we know what we believe we know?] Barcelona: Editorial Gedisa.

Watzlavick, P., Weakland, J.H., & Fisch, R. (1974). *Principles of problem formation and problem resolution.* New York: W. W. Norton & Company, Inc.

Willis, B. (ed.). (1994). *Distance education strategies and tools.* Englewood Cliffs, NJ: Educational Technology Publications.

Copyright © 2005, Idea Group Inc. Copying or distributing in print or electronic forms without written permission of Idea Group Inc. is prohibited.

Endnotes

[1]   It is important to notice that the *Proceedings of CSCL 2002* clearly show the intent to readdress research in this line.

[2]   According to the model of Complex Thinking developed by the Iowa Department of Education (1989), there are three major types of learning tasks: 1) problem solving, 2) designing, and 3) decision making. Learners make use of different mixtures of critical, creative, and basic thinking abilities when pursuing any of the three.

[3]   Although most of the work done by this research group was done several decades ago, their views tie in very well with the constructivist assumptions. In fact, Watzlawick has written extensively about it; for example, see *The Invented Reality, How we know what we believe we know* edited by Watzlawick in 1984 New York: Norton.

Copyright © 2005, Idea Group Inc. Copying or distributing in print or electronic forms without written permission of Idea Group Inc. is prohibited.

Chapter XIV

Online, Offline and In-Between:
Analyzing Mediated-Action Among American and Russian Students in a Global Online Class

Aditya Johri
Stanford University, USA

Abstract

Online collaborative learning is a situated activity that occurs in complex settings. This study proposes a sociocultural frame for theorizing, analyzing, and designing online collaborative- learning environments. The specific focus of this study is: learning as situated activity, activity theory as a theoretical lens, activity system as an analytical framework, and activity-guided design as a design framework for online learning environments. Using data gathered from a naturalistic investigation of a global online collaborative-learning site, this study reveals how these lenses and frameworks can be applied practically. The study also identifies the importance of design iterations for learning environments.

Copyright © 2005, Idea Group Inc. Copying or distributing in print or electronic forms without written permission of Idea Group Inc. is prohibited.

Introduction

In 1992, Salomon (1992, p.62) had this to say about the design and analysis of effective CSCL, "Given a reasonable of minimum of technological capability, the success or failure of cooperative learning is accounted for by entirely different and far more complex factors." Four years later, Salomon and Perkins (1996) made two more observations:

First, computers in and of themselves do very little to aid learning...[a]lthough it may make the enterprise more efficient and more fun. [L]earning depends crucially on the exact character of activities that learners engage in with a program, the kinds of tasks they try to accomplish, and the kinds of intellectual and social activity they become involved in, in interaction with that which computing affords. [S]econd, it has also become evident that no single task or activity, wondrous as it may be, affects learning in any profound and lasting manner in and of itself. Rather, it is the whole culture of a learning environment, with or without computers, that can affect learning in important ways. (p.113)

In the decade since Salomon made his first observation (1992), there has been a tremendous growth in computing technology and its implementation and use in educational settings. Computer-Supported Collaborative learning (CSCL) has been hailed as an emerging paradigm of instructional technology (Koschmann, 1996), and there is a profusion of literature related to CSCL and online/distance learning (Bonk & King, 1998; CSCL, 1997; EuroSCSL, 2001; Hoadley & Roschelle, 1999; Stahl, 2002). A close examination of this literature reveals that to a large extent the studies have focused solely on the technology and have paid little or no attention to the context in which the technology was implemented.

Online collaborative learning settings are places of complex interactions and outcomes, and I believe that sociocultural theories of learning, particularly Activity Theory (Engeström, 1987; Leont'ev, 1978; Vygotsky, 1978), can be a valuable theoretical lens to study such settings. Moreover, Activity System can be used as an analytical tool to analyze the setting (Cole & Engeström, 1993), and Activity-Guided Design can be used as a framework to design such environments. A common thread running through this chapter is that of mediated-action or activity. As this concept is discussed in detail later, I'll just give a quick introduction here. The primary concept is that cognition takes place as people are engaged in an activity that has a purpose and an object. The activity is mediated by artifacts that they use to act on the object to reach a desired outcome. As Pea (1993) explains,

While it is people who are in activity, artifacts commonly provide resources for its guidance and augmentation. The design of artifacts, both historically by others and opportunistically in the midst of one's activity, can advance that activity by shaping what are possible and what are necessary elements of that activity. (p.50)

Copyright © 2005, Idea Group Inc. Copying or distributing in print or electronic forms without written permission of Idea Group Inc. is prohibited.

The cognition or intelligence required for and the outcome of this activity is distributed across the artifacts and is not the sole property of the individual. "When I say that intelligence is distributed, I mean that the resources that shape and enable activity are distributed in configuration across people, environments, and situations. In other words, intelligence is accomplished rather than possessed" (Pea, 1993, p.50). Before launching into discussion of learning, I'll try to explain two concepts that would appear frequently in my discussion: Online and Collaboration.

Online

By online I mean a setting that uses any or all of the following technologies for communication: discussion software, mailing list or listserv, email, instant messaging; and it has either all classes being held online and no face-to-face interaction among the participants; or it follows a hybrid model, i.e., a mix of face-to-face and online classes.

Collaboration

Collaboration in the context of this study has the following characteristics:

1.   Genuine interdependence - Collaboration is distinct from cooperation in that collaboration requires "genuine interdependence" among participants. Cooperation can exist when participants distribute their work and then bring it all together; in return they might not learn anything from one another (Salomon, 1992).

2.   Production of knowledge - Another feature of collaboration is the production of knowledge, rather than just its assimilation or distribution. What can individuals do together that they cannot do separately? This is also like the apprenticeship model in some sense since students are expected to learn how to participate in communities of learning, a necessary part of higher education or work place. There is an assumption that there will be some internalization of knowledge as well, and students will learn new things that they can use later (Bruffee, 1984).

3.   Self-construction of task - Participants construct their own tasks rather than working alone on instructor-assigned tasks or problems (Cranton, 1996).

4.   Construction of joint activity space - Participants should come to a common understanding of what their goal is, and this understanding should develop through their conversations with one another (Peters & Armstrong, 1998).

Copyright © 2005, Idea Group Inc. Copying or distributing in print or electronic forms without written permission of Idea Group Inc. is prohibited.

Learning as a Situated Activity: Sociocultural Perspective

Vygotsky (1978) proposed that all higher order psychological functions, including learning, emerge first on a social or interpersonal plane and then on an internal or intra-personal plane. Moreover, human activity is mediated through artifacts and man and artifact shape, and is shaped by social and physical environment (Cole, 1996). Sociocultural theorists have advocated the usefulness of studying learning as a collaborative practice and have emphasized the situated and social nature of learning (John-Steiner & Mahn, 1996; Scribner, 1997; Vygotsky, 1978; Wertsch, 1991). They argue that to evaluate and study learning it is important to analyze the role of context, especially communication and collaboration. Moreover, according to Wertsch (1991) a sociocultural perspective presupposes that action is mediated and is inseparable from its context. Furthermore, he states that the goal of a sociocultural approach to mind "is to explicate how human action is situated in cultural, historical, and institutional settings" (Wertsch, del Rio & Alvarez, 1995, p. 11). According to the sociocultural lens then, learning is seen as situated, a part of the activity, context, and culture in which it is developed and used (Brown, Collins, & Duguid, 1989) and "in which practice is not conceived of as independent of learning and in which meaning is not conceived of as separate from the practices and contexts in which they are developed" (Barab, Barnett, Yamagata-Lynch, Squire, & Keating, 1999, p.104).

From a methodological perspective, a sociocultural approach allows researchers to investigate complex environments in their natural settings using multiple modes of inquiry. Therefore, this approach is particularly well suited to studying online collaborative-learning environments (OCLE) since OCLE settings are created on a premise that there will be social interaction among several participants that will be mediated by some technological artifact.

Activity Theory: A Theorectical Lens

Activity Theory (AT) refers to a line of theory developed by Leont'ev, Vygotsky, Luria and other Russian psychologists at the beginning of the last century (Engeström, 1987; Leont'ev, 1978) and although Vygotsky himself never explicitly examined the concept of activity, he strongly influenced the development of activity theory (Wertsch, 1981). Activity theory sees learning as a situated and social activity and interlinks the individual and social levels (Kaptelinin, 1996; Nardi, 1996). The basic unit of analysis in activity theory is an activity, which includes a context, and activities are directed towards objects by the need to transform the object into an outcome. As Kuutti (1996) points out, activity theory is not a theory per se; rather, it is "a philosophical and cross-disciplinary framework for studying different forms of human practices as developmental processes, with both individual and social levels interlinked at the same time" (p. 25). Over the past decade, activity theory has found application in learning (Barab et al., 1999; Barab, Schatz, & Scheckler, in press), human-computer interaction (Kuutti, 1991; Nardi, 1996),

Copyright © 2005, Idea Group Inc. Copying or distributing in print or electronic forms without written permission of Idea Group Inc. is prohibited.

and work practices (Engeström & Middleton, 1996). From a methodological standpoint, AT accounts for cultural, institutional, and social settings, and therefore provides a holistic macro-analysis. It provides conceptual resources to capture elements of a complex setting, allows for a varied set of data collection techniques, and emphasizes the user's point of view (Nardi, 1996).

What is an Activity?[1]

Activity is the unit of analysis in activity theory and is composed of subject, object, actions, and operations. Leont'ev (1978) proposed a hierarchical structure of activity according to which activities are organized into three hierarchical levels: activities, actions, and operations. Activities are done to fulfill a motive; actions are goal-directed processes carried out to fulfill a motive; and operations are functional subunits of actions that are carried out automatically. He stressed that activity has a collective nature and that the relations between these three central components of an activity are mediated in a reciprocal way (Kuutti, 1996). According to Engeström (1987), activity "is the smallest and most simple unit that still preserves the essential unity and integral quality behind any human activity" (p. 81). In focusing on activity as the basic unit of analysis, emphasis is put on the cultural, institutional, and social settings in which these activities occur. One can thus argue that AT also provides the necessary conceptual resources for capturing essential elements of a complex setting. As Barab et al. (1999) explain: "When discussing activity, activity theorists are not simply concerned with 'doing' as a disembodied action, but are referring to 'doing in order to transform something,' with the focus on the contextualized activity of the system as a whole" (p.78).

Artifacts and Mediation

A key idea in activity theory is the notion of mediation by artifacts. Activity is mediated through the use of artifacts. Every activity has an object towards which the subject's action is directed, and artifacts are tools that the subject uses to complete that action. Wertsch (1991) proposes that mediated action is the key to understanding how human action is situated in context. A common reformulation of Vygotsky's mediational triangle is shown in Figure 1.

Figure 1: Mediational triangle (Cole & Engeström, 1993, p.5).

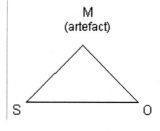

Copyright © 2005, Idea Group Inc. Copying or distributing in print or electronic forms without written permission of Idea Group Inc. is prohibited.

Figure 2: Activity system (Cole & Engeström, 1993, p.8).

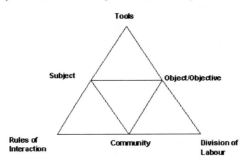

Activity System: An Analytical Tool

Engeström has established a simple structural model of the concept of activity and culturally mediated relationships within it (Engeström, 1987; Engeström & Cole, 1993). Engeström (2002) has replaced binary relationships by mediated relationships through the introduction of a third term that carries with it the cultural heritage of the situation, e.g., the relationship between the subject and object is mediated by a tool. In its simplest form, the model contains six elements and three mutual relationships. The relationship between subject and object is mediated by a *tool/artifact*; the relationship between subject and community is mediated by *rules*; and the relationship between community and object is mediated by *division of labor* (Figure 2).

Contradictions

Contradictions in AT signify a misfit within elements, between them, between different activities, or between different developmental phases of a single activity. According to activity theory, development occurs when contradictions are overcome (Engeström, 2002; Kuutti, 1996). In activity systems, this contradiction is renewed in "the clash between *individual actions and the total activity system*" (Engeström, 1987, p. 82, italics in original), and it has been suggested that these internal contradictions are what characterizes activity systems (Engeström, 1987; Leont'ev, 1978). Practically, contradictions help us recognize places of intervention and help improve a setting or a system.

Case Study: The Global Classroom Project

I'll now provide an example, from a recent study that I did, to analyze how activity takes place in an online learning environment. The focal premise of the study on which this analysis is based is that a technological system is situated within a complex environment

Copyright © 2005, Idea Group Inc. Copying or distributing in print or electronic forms without written permission of Idea Group Inc. is prohibited.

and that the productive use of that technology, or a lack thereof, is contingent upon the interaction among the different elements of that environment. The study investigated one such technology-supported learning environment-The Global Classroom Project (GCP). The Global Classroom Project is a web-based classroom that integrates online and face-to-face interactions to provide students from Russia and the U.S. a chance to engage in cross-cultural digital communication. The idea behind the GCP is that by engaging in cross-cultural communication students will learn about each other's culture first-hand from native students and also learn how to communicate with people from other cultures. I made a conscious decision not to focus on any one element of the GCP, especially the technology-WebBoard, but to try to look at all (or at least as many as possible) mediating factors and artifacts that could have influenced the learning environment. I believe that technology use is socially and culturally mediated; hence, to understand its use or misuse, one has to look at the context of technology use (Newstetter, 1998).

The Global Classroom Project[2]

The Global Classroom Project (GCP) provides an online distance-learning environment for students from the U.S. and Russia to collaborate on projects to produce text-based documents and/or digital artifacts such as websites or CD-ROMs. In addition to classes that are completely online, the GCP also offers face-to-face classes for students in their respective higher education institutions. The first pilot GCP class was offered in Spring 2000. Since then, a total of seven classes (both graduate and undergraduate level) have been offered over a three-year period. The purpose of the class is two-fold — to teach technical communication skills to the students (such as resume, proposal, and project report writing); and to teach them skills needed to work in a cross-cultural online environment. The learning philosophy behind the GCP is experiential learning, i.e., students learn best by personal experience that the instructors foster by providing them with a setting that emulates the workplace and brings up similar issues and problems.

The Technology

The GCP uses WebBoard, a web forums and chat software, as the platform for student interaction. WebBoard is a message board tool. WebBoard provides support for chat, graphics, archiving, and other technical features. According to its website, some of the leading uses for WebBoard are community building, technical support, online education, project collaboration, virtual meetings, and information management. In the GCP, WebBoard is used primarily as an asynchronous communication medium, to post messages and to exchange documents, usually as attachments. Communication is also supported by the use of email.

Interface of WebBoard

The WebBoard follows a predefined structure. The five main components of every WebBoard installation are:

Copyright © 2005, Idea Group Inc. Copying or distributing in print or electronic forms without written permission of Idea Group Inc. is prohibited.

- Boards/Forums: A board is the top level of the hierarchy in WebBoard. It is the name given to contain all of the Conferences, Topics, Messages, and Users for a particular instance of WebBoard. In the GCP, a board is created each semester for all the classes that are offered that semester.

- Conferences: They are the second level in the WebBoard hierarchy. Conferences contain topics. In the GCP, the instructors usually create the conferences. In a typical semester, the conferences may be Class Discussions, Group Discussions, Introductions, Welcome, Class Assignments, etc.

- Topics: They are the next level after Conferences. They are created by users and contain individual Messages. If a user posts a new Message that is not a reply to an existing Topic, it becomes a new Topic and is available for reply. Typical Topics in the GCP might be Thread Arrangement, Proposal Discussion, Project Discussion, etc.

- Messages: The final level in the hierarchy is Messages. Messages can be in the form of Reply to someone else's message, or they can be a New Post, in which case a new Topic will be created. Messages are also called Posts.

- Users: Users are members or people using the Board. There are different levels of users, from Administrators to Guests.

Students and Activity

The total number of students in the GCP class varies each semester and has ranged from 20 to 36 (American = 6 to 24, Russian = 9 to 30). The Russian students are typically graduate students enrolled in social sciences program, whereas the American students are either undergraduate and graduate and range from liberal arts to engineering majors. The major activity of the class is a group project to be submitted at the end of the semester. The groups consist of American and Russian students who are assigned an open-ended topic to research, write a proposal for their final project, and then work together to complete the project based on the proposal. The topics given to the students have ranged from "analysis of propaganda" to "comparison of online greeting cards." Several activities are given to the students that lead to the group project. They are asked to write a resume that is posted online and to come up with a list of annotated bibliographies that can be used for their project. They are also given a list of readings that are discussed electronically on the WebBoard and sometimes in the face-to-face classes.

Research Methodology

The study was ethnographic in nature, and data was collected using in-depth interviews, surveys, participant observation, analysis of online transcripts, and informal communication with participants. A total of 15 participants were interviewed. All the student interviews were face-to-face except one that was over the phone, and each interview lasted anywhere from 45 to 90 minutes. The primary subjects for the interview were American students. Furthermore, the researcher participated as a team member of a group

Copyright © 2005, Idea Group Inc. Copying or distributing in print or electronic forms without written permission of Idea Group Inc. is prohibited.

of six students for a period of eight weeks, took part in all the group activities, projects, and assignments, and also observed the class during that period. Other data-gathering methods included open-ended surveys and informal communication with students and the instructor. Detailed analysis of online WebBoard transcripts provided further data. Data was also gathered from the Russian instructor via email. The data used in this chapter are a subset of the larger data set and consist primarily of interviews and online transcripts.

Analysis

To look at how an activity is performed in an online collaborative-learning environment I've looked at a group of American and Russian students as it worked on a proposal for its final project. The students were supposed to choose a topic that they agreed upon, and they were given some guidelines to help them select a topic. Based on the assignment given to the students and their discussions up to this point, we can draw an activity system of their task to come up with a proposal. The components of the activity system would look something like Figure 3.

Next, let us look at a group, Group P, as it worked through this process. I've analyzed 60 messages sent over a period of five weeks between the American and Russian students and the instructors. The "Proposal Discussion" thread was started on September 29th, and the proposal was due on November 1st. The aim of this analysis is to highlight instances of contradictions or breakdowns[3] that were discovered as part of the analysis of the GCP as an activity system. The objective is also to contextually frame the breakdowns, to interpret them in a meaningful manner, and to reconstruct events as they might have actually occurred.

Figure 3: Ideal activity system for the Global Classroom Project.

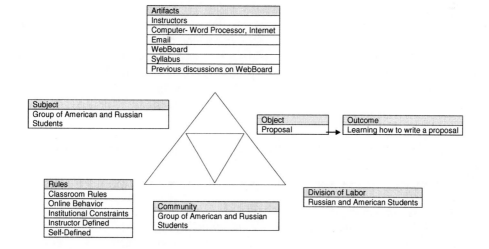

Copyright © 2005, Idea Group Inc. Copying or distributing in print or electronic forms without written permission of Idea Group Inc. is prohibited.

Figure 4: Number of messages posted per day.

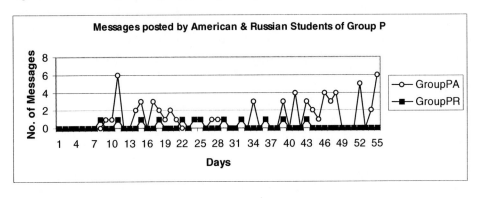

Why This Group?

To decide which group to analyze, I did a log analysis of one message thread across all of the four groups that worked on the project that semester and plotted the number of messages against the days to come up with the graph that displayed the number of messages across time for the "Proposal Discussion" thread. The graph for the group I've analyzed is shown in Figure 4.

After looking at the graphs and other qualitative characteristics, I decided to focus on Group P because it provided an adequate opportunity to explore a struggle between its members as they tried to come up with a topic for research. I believed this would help me to discern the points and reasons for breakdown among groups and in the GCP. Also, this group had the highest number of messages for the particular thread and time period, and therefore it provided more data. This group is by no means representative of all group discussions that took place between the students but is rather a unique case. The group consisted of four American students and three Russian students, which was typical of all groups that semester. The American group consisted of one graduate student and three undergraduate students. It had two female and two male members. The Russian group consisted entirely of female graduate students. All four members in the American group were from different majors: one was an Information, Design, and Technology graduate student, one was a Building Construction major, one was a Business major, and one was a Computer Science major. Two of the American group members were graduating seniors. The graduate student was appointed as the group leader by the instructor and was responsible for managing the group. The American classes met on Tuesdays and Thursdays, whereas the Russian students met on Tuesdays and Saturdays.

Broad Interaction Patterns: Some Visual and Numerical Data

Before I delve into in-depth analysis of the group, it would be helpful to look at some broad interaction patterns in the group. The network diagram (Figure 5) represents the group dynamics in terms of flow of messages. The arrows in the diagram represent

Copyright © 2005, Idea Group Inc. Copying or distributing in print or electronic forms without written permission of Idea Group Inc. is prohibited.

messages originating from a member of the group, both American and Russian, with the thickness of the arrows being proportional to the number of messages. It can be seen from the figure that most of the Russian messages were posted as a group, whereas American students posted individually, and most of the messages for the Americans came from the graduate student who was also the group leader.

Another important observation is that American students posted messages for other American students, whereas Russian students only posted messages for the American students. This means that American students were using the WebBoard to discuss a topic among them and to have a dialogue, whereas Russian students were using the WebBoard just to send messages to the American students.

Table 1 shows the number of messages per week for the students and the instructors. Some broad patterns that emerged are:

- The overall activity was highest in Week 2 and then tapered off for the next two weeks before picking up again in the fifth week. A closer analysis shows that this pattern was a result of the activity of the American students.

Figure 5: Network diagram for Group P.

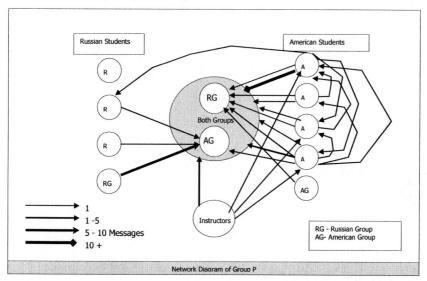

Table 1: Number of messages per week.

Week	Number of Messages			
	American Students	Russian Students	Instructors	Total
Week 1	6	2	3	11
Week 2	15	2	1	18
Week 3	4	3	0	7
Week 4	6	2	1	9
Week 5	11	3	1	15

Copyright © 2005, Idea Group Inc. Copying or distributing in print or electronic forms without written permission of Idea Group Inc. is prohibited.

- The Russian students were consistent with their postings and posted two or three posts every week. A look at the graph presented in Figure 4 would show that the Russian students also posted at a regular interval.

- The instructors posted very few messages within this thread, three in the first week and, at the most, one message in other weeks, although they met the students face-to-face.

Week 1 (Sept. 29 - Oct. 05)

During Week 1, 11 messages were exchanged in the "Proposal Discussion" thread, six written by American students, two by Russian students, and three by the instructors.[4] There were several contradictions that emerged as the activity unfolded over the week. To start with, a couple of American students were not able to follow the discussions since they were reading other threads and realized only later in the week that they had to follow the conversation in the "Proposal Discussion." The Russian students posted their message as a group, i.e., they signed off each message with the names of all the group members. The American students could never understand why the Russian students did this. A practical reason for this could be that the Russian students had limited access to computers and could only post during their class times. The effect of this behavior on the American students was greater than is apparent on the WebBoard discussions. The American students were disappointed and frustrated, and during an interview, one student commented that there was no incentive for her to post anything back since they only got back one post for every four posts they put up, and added that it should feel more like a conversation.

This brings us to another important distinction between the model of communication for American and Russian students. The American students look at electronic communication as conversation, an attitude they have no doubt acquired because of fast access speed and the use of Instant Messaging (IM) (the group reported that they had used IM during their brainstorming sessions, and all of them used it frequently). On the other hand, the Russian students used WebBoard more like traditional mail. Moreover, Russian students engaged in face-to-face group work since they had to meet during class to use the computers. On the other hand, the American students interacted only using electronic medium: WebBoard, emails, and Instant Messaging.

Week 2 (Oct. 06 - Oct. 12)

Week 2 had a total of 18 messages: 15 by American students, 2 by Russian students, and 1 by the instructors. A number of breakdowns occurred during this week. The Russian students were frustrated that the American students were not working together, and the American students were still frustrated with the lack of individual response from the Russian students. Within the American group, a division started based on the priority of the group members. The graduate student in the group was appointed the unofficial "leader" of the group by the instructor and was concerned more with logistics of the group work and the delivery of the final product compared to the topic at hand. Another

Copyright © 2005, Idea Group Inc. Copying or distributing in print or electronic forms without written permission of Idea Group Inc. is prohibited.

American student had become really frustrated with the whole exercise, and his comments show that he was used to making quick decisions and moving on (even if it meant that not everyone in the group could be happy), whereas here the decision-making process was taking a long time. After this message, he only posted three times during the next three weeks. But his comment leads us to something more significant that became apparent when I interviewed other students—the Engineering students did not work well together with Liberal Arts students and vice versa, since for the engineering students the class was required, whereas the Liberal Arts students pursued it because they were interested in the class. Therefore, the interest level and commitment of the students was different. Another difficulty arose for the American students when they tried to meet face-to-face. At least one student in the group had enrolled in the class precisely because he did not want to come on campus and wanted to participate electronically. Therefore, scheduling a face-to-face meeting became almost impossible. Another breakdown was the lack of knowledge of the American group about what the other American groups in the class were doing (there were three other groups) because the class met face-to-face infrequently. The Russian students in the group were concerned that they might interfere with what the other groups in the class were doing for their projects and therefore wanted to focus their topic based on this input. They had changed the context of their work from a group project to a class project.

Week 3 (Oct. 13th - Oct. 19th)

During Week 3, only seven messages were posted on the WebBoard: 4 messages by American students and 3 by the Russian students. The American students, frustrated by the lack of responses from the Russian students, only posted 4 messages during the week compared to 15 messages the week before. Also, the division in the American group was more apparent, with the graduate student desperately trying to divide the work between the group members and trying to get everything together. The graduate student also made an attempt to explain to the Russian students what Americans thought about collaboration and that they were deliberately making an attempt to include everyone in the discussion.

Week 4 (Oct. 20 - Oct. 26)

During Week 4, a total of 9 messages were posted: 6 by American students, 2 by Russian students, and 1 by the instructor. The American students removed the emphasis on "image of enemy," an idea forwarded by the Russian students from the proposal. The Russian students always took for granted that it would be the focus of their study, and the American students thought it was just one of the ideas forwarded by the Russian students that was open for discussion. Neither group talked about it specifically, and it was removed from the proposal. This left the Russian students in the dark, since they were no longer sure of the aim of the project.

Moreover, interaction among the different Russian groups in their class influenced their collaboration with the Americans. The American students did not really know what the other groups were researching other than what they could see on the WebBoard, as

Copyright © 2005, Idea Group Inc. Copying or distributing in print or electronic forms without written permission of Idea Group Inc. is prohibited.

expected of them by the Russian students. During this week, the American students replied as a group to the Russian students for the first time since the start of the discussion; however, it is important to note here that when the American group replied to the Russians students as a group, the post was signed off by three American students instead of all four. This suggests a breakdown among the American students in terms of group work.

Week 5 (Oct. 27 - Nov. 04)

A total of 15 messages were posted during the fifth week: 11 by American students, 3 by Russian students, and 1 by the instructor. The message by the instructor tried to please everyone, and it wasn't really clear on how the students should proceed. It failed to provide the direction that the students needed at that point. Something really interesting was happening at this point. The American graduate student ended up working on the proposal all by herself and was frustrated by the lack of response from the undergraduate students. So she decided to "scare the shit out of them" and purposely did not come to class the day the proposal was due.

In the last posts, the students mentioned that they should distribute the work and that the purpose of the distribution of work was not to limit collaboration but to move forward quickly as the deadline was approaching. Yet, it was obvious that the group work no longer required collaboration among the Russian and American students, since they had decided to split the work so that the American students worked on Art section, and the Russian students worked on News section, and then put it all together in the final paper.

Table 2 shows a list of contradictions identified from this analysis. Table 3 lists the total references to propaganda made by the students during the five weeks.

Table 2: List of contradictions for Group P.

No.	Contradiction	Element(s) of Activity System
1	Major of students in a group (Liberal Arts/Engineering)	Community/ Division of Labor
2	Means of communication (WebBoard/Email/IM/F2F)	Tool
3	Software (WebBoard/Email)	Tool
4	Structure of Task/Assignment (Open-ended/Closed-ended)	Tool/Object
5	Reason for taking the class (Required/Not Required)	Division of Labor/Community
6	Group Size (Small/Big)	Community/Division of Labor
7	Readings (Pertinent/Not useful)	Tool
8	Schedule (American/ Russian)	Rule/Tool
9	Interaction time (Small/Large)	Community/Division of Labor
10	Discussion on WebBoard (Project based/Personal)	Community/Object
11	Nature of classes (F-2-F/Online)	Rules
12	Discussion (F2F/Online)	Tool/Rules
13	Grading (Group/Individual)	Rules/Object/Tool
14	Communication Frequency (Frequency/Infrequent)	Rules
15	Communication Norms (Group email/Individual email)	Rules

Copyright © 2005, Idea Group Inc. Copying or distributing in print or electronic forms without written permission of Idea Group Inc. is prohibited.

Table 3: References to Propaganda.

Total References to Propaganda

1. Propaganda tools
2. Propaganda styles
3. Character of propaganda
4. Propaganda during
5. Type of propaganda
6. Use of propaganda
7. Transformation of propaganda
8. Area of propaganda
9. Propaganda through artwork
10. Progress in propaganda
11. Analysis of tools and content of propaganda
12. Technological metamorphosis of propaganda
13. History of propaganda
14. Development of propaganda
15. Means of propaganda
16. Attributes of propaganda
17. Change in propaganda
18. Evolution of propaganda
19. Categories of propaganda

Reflections: Learning to Collaborate and Collaborating to Learning

The American and Russian students were involved in two mutually co-existing activities: they were learning to collaborate using an online environment and simultaneously collaborating with one another to learn from each other. They had to work together to reach a decision about what they would do their project on, and also work on their communication and collaboration skills. This did not prove to be an easy task for them. They had to understand the *affordances* of the tools and artifacts available to them and use them in a meaningful manner. As has been reported in other studies, the students either failed to grasp the "affordances" of the learning environment, or they embraced them in ways that the designers of the environment had not foreseen (Halloran, Rogers, & Scaife, 2002; Holland & Reeves, 1996; Newstetter, 1998).

The Becoming of An Activity

Every activity is in a constant state of flux. A tool/artifact becomes the object, an object becomes the activity, and the activity changes, since an activity is only a sum total of its parts and if a part changes so does the activity. Therefore, an activity is always in the *becoming* rather than in the being. For instance, the definition or a common understanding of propaganda was seen as a tool at the start of the activity. As the activity progressed, it was apparent that the students had to come to an understanding of propaganda, so it became an object. Similarly, the instructions given by the instructors to the students were supposed to be a tool, but they also became an object, and the students tried to make sense of what the instructors were trying to say. Also, an activity may be composed of other activities and so it is more like a network of activities rather than a single activity. Halloran, Rogers, and Scaife (2002) have proposed the concept

Copyright © 2005, Idea Group Inc. Copying or distributing in print or electronic forms without written permission of Idea Group Inc. is prohibited.

of Activity Space to capture these dynamics, and Hyppönen (1998) has proposed the concept of Network of Activity.

Some Design Implications

Let us look at some implications for design that emerge from this analysis.

Nature of Tasks

The structure of the task has a profound impact on how the activity progresses. Let us look at two specific examples to understand the role of the structure of tasks in the GCP. In the first case, in Fall 2000, students in GCP were assigned a narrowly defined task where they had to compare Russian and American greeting cards on two websites. The project was to go to an online e-postcards site determined by the instructors and compare the Russian and American postcards. When I asked American students from this semester if they had any problems working with their Russian counterparts, they said they had none. From the transcripts on the WebBoard and from the interviews, it is evident that groups in this class had an easier time in completing their tasks as compared to other groups in some other classes.

On the other hand, Group P in the example above was tackling a task that had no boundaries. It was a true ill-structured task - "come up with an analytical report and a digital artifact," in an ill-structured domain - "propaganda," and ill-structured tasks in an ill-structured domain influenced collaboration and learning, and are closer to a real-world problem (Koschmann, Kelson, Feltovich, & Barrows, 1996). The groups in this class ran into various communication and collaboration problems. So what went wrong? Why did the group have so many problems? The biggest problem faced by the group was that the technology proved to be a hindrance in synthesizing the multiple perspectives forwarded by the group members:

1.   Less access to technology meant a communication lag that resulted in almost no feedback from the Russian students.
2.   Complex structure of the WebBoard led to decreased usability and resulted in students posting and reading the wrong thread.
3.   Students had different expectations of collaboration and communication, which are influenced by experience with technology.

This observation highlights a recurring tension that has profound implications for the design of online collaborative environments. If you design tasks that are open-ended, you have to make sure that tools available in the setting afford the communication and collaboration needed for the task; and if you design tasks that are too close-ended, collaborative-learning opportunities may be lost.

Copyright © 2005, Idea Group Inc. Copying or distributing in print or electronic forms without written permission of Idea Group Inc. is prohibited.

Scaffolding

A related issue is scaffolding. If things are not moving in the right direction, when should the instructors intervene and what should be the nature and level of scaffolding or intervention? Koschmann et al. (1996) summarize the requirements for instruction in an ill-structured domain with ill-structured activity:

[I]nstruction should facilitate adaptability in all these respects: It should build upon preexisting foundations, monitor for and encourage correction when misconceptions are identified, and foster the development of cognitive flexibility so that the learner's efforts toward learning have the greatest possible effect. (p. 91)

For instance, in my example from the GCP, there was little or no scaffolding provided by the instructors. Their intervention was either encouragement or logistical direction, but not help in bridging the misconceptions between the Russian and the American groups. The American students never found an answer to: How did their view of propaganda differ from the Russian students? Why did the Russian students reply as a group, and why did they want the American students to reply as a group too? The instructors were well aware of the problems encountered by the students, yet they didn't directly intervene because they believe in the teaching philosophy of "experiential learning" — the best way to learn about something is to experience it first-hand. They also believe that by going through the whole cycle of working on the project and by dealing with their problems, the students will be able to apply the knowledge and experienced gained in the real world if they face a similar problem later on. This may or may not work, and as can be seen from this example, scaffolding, especially about cross-cultural differences in the understanding of "propaganda," would have been an important lesson.

Technology

I believe there is an important lesson to be learned here in terms of how technology can influence collaboration in an ill-structured domain with an ill-structured task. The lesson is that mediation by technology might not always be useful in such a scenario and may actually obstruct interaction among students. Of course, on the surface the solution seems very simple - increase the access to computers for Russian students and all problems of communication and collaboration will go away. But that may not necessarily be the case. Through the interviews and through participant observation, I've realized that, in certain cases, face-to-face collaboration *may* be essential for open-ended and ill-structured tasks. The American students actually realized this by the end of the semester, and their face-to-face interaction increased substantially. One student reported having met for 15 hours straight with her group in order to get her work done and regretted that they did not meet face-to-face before. To alleviate communication and collaboration problems, a lot of the American student groups in the past have tried using instant messaging in addition to WebBoard and emails, and even though that helped, it did not eliminate the need for face-to-face meetings. The participation by Russian groups can

Copyright © 2005, Idea Group Inc. Copying or distributing in print or electronic forms without written permission of Idea Group Inc. is prohibited.

also serve as proof that face-to-face meetings lead to more productive collaboration as synchronous or asynchronous online communication.

Groups

For a group to work together on a task there needs to be what Salomon (1992) calls "genuine interdependence." Speaking from his personal experience, Salomon asserts that there is little success between collaborative teams in terms of pooling together their abilities, in terms of true collaboration, and in terms of learning outcomes. Cohen (1994) argues that when designing a task for cooperation, it is important to make sure that there is a reason for the group to interact:

One may give a group a task, but, unless there is some reason for the group to interact, students may well tackle the task as individual work. This is especially the case if each individual must turn out some kind of worksheet or report. This is also the case if the instructor divides the labor so that each person in the group does a different part of the task; the group has only to draw these pieces together in sequential fashion as a final product. The consequence of either of these patterns is that there is comparatively little interaction; people do not gain the benefits of using one another as resources, nor is there any basis for expecting the prosocial outcomes of cooperation. (p.11)

Since the tasks in the GCP are open-ended, the students themselves decide what role each of them will play and regularly divide the work among them based on their skill-sets. Invariably, the division was into a web designer, a researcher, and two writers. The engineering students took web designing, and the liberal arts students preferred writing. Neither learned much from the other and lost a valuable opportunity. In some instance, a single student ends up doing the majority of the work since the other students didn't finish their parts of the task.

Therefore, the way the instructors set up the problem, suggest procedures, and specify roles can do much to create interaction that is markedly superior to that produced by simply asking a group to reach consensus. The dilemma is that if teachers do not structure the level of interaction, they may well find that students stick to a most concrete mode of interaction, and if they structure the interaction too much, they may prevent the students from thinking for themselves and thus gaining the benefits of the interaction.

Scheduling and Logistical Factors

A number of factors not in the control of the instructor play a crucial part in an environment like the GCP. For instance, the schedule of classes, the class timings, the course number under which it is offered, and the length of interaction were some factors that were determined by the department through which the class was offered. This in turn determines the class size, the class composition, and to some extent the prior knowledge of the students coming into the class, and their expectations from the class. These factors

Copyright © 2005, Idea Group Inc. Copying or distributing in print or electronic forms without written permission of Idea Group Inc. is prohibited.

play a far more decisive role in combination than the instructors would prefer, but there is no way to control them. The only way to curb their influence is by design iterations - learn by experience how each factor influences the setting and then modify elements of the setting to make them work together (Miller, Trimbur, & Wilkes, 1994). Monitoring a discussion software can also prove to be a daunting task for the instructors once students start posting in different conferences and threads.

A Framework for Online Collaborative Learning: The Waterfall Model

The Global Classroom Project and most other web-based distance learning classes rest on a technology-driven supposition: computers will lead to communication; communication will lead to collaboration; and collaboration will lead to learning. Even though this is a simplistic interpretation, it can be extremely helpful in analyzing an online collaborative-learning environment. Using the Activity System as an analytical tool, we frame each step described above as an activity (Figure 6). As can be seen from Figure 6, the computer, which is an object in the first system, becomes the tool in the next activity system, and communication, which is the outcome of the first activity system, becomes the object in the second system leading to collaboration. In the succeeding activity system, communication is the tool, collaboration is the object, and the outcome is learning.

Figure 6: Waterfall model of online collaborative learning.

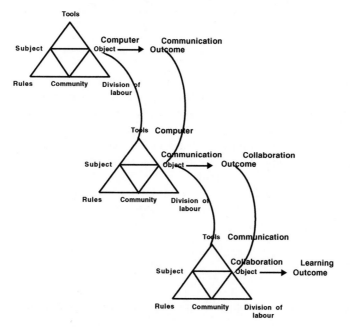

Copyright © 2005, Idea Group Inc. Copying or distributing in print or electronic forms without written permission of Idea Group Inc. is prohibited.

Theoretically, the implementation of the GCP follows this model closely but not entirely. In the next section, I've identified several factors that result in a breakdown in the process and their possible solutions.

Triangle One: Computer/Technology as the Object

Contradictions

In the GCP, the breakdown starts in the first activity system. The use of computer is mediated by access to technology. For the Russian students, this is a problem because they have limited access to computers. Restriction in access proves critical since it creates a communication lag between American and Russian students, which in turn restricts collaboration. In addition, the Russian students also have to overcome a language barrier since English is not their native tongue, which inhibits synchronous communication.

Possible Solutions

The easiest solution to suggest is to increase access to computers for the Russian students. It would also be helpful if American students were told beforehand that Russian students have restricted access to networked computers and that synchronous communication is not feasible due to low access and language barrier.

Triangle Two: Communication as the Object

Contradictions

The use of computer as a tool also has some inherent contradictions, the first of which is the use of WebBoard. The use of WebBoard creates a learning curve for both the American and Russian students. At the start of their projects, when the students are deciding upon a topic to investigate, using the WebBoard creates a lag in communication. A lot of American groups therefore supplement the use of WebBoard with face-to-face meetings. American students find this especially discomforting since they are so used to emails, and they don't see a reason for using WebBoard. Another problem with using WebBoard is that it does not lend itself well to all kinds of discussions. There are other technical and usability problems associated with the use of WebBoard that were discussed in a previous section.

Possible Solutions

One possible solution is to test other software for feasibility and usability for use in the

Copyright © 2005, Idea Group Inc. Copying or distributing in print or electronic forms without written permission of Idea Group Inc. is prohibited.

class. Training students on using WebBoard can also curtail problems associated with the usability of WebBoard.

Triangle Three: Collaboration as the Object

Contradictions

The use of communication as a tool for collaboration is influenced to a large extent by group size and the nature of the assignment. If the group size is small, communication and collaboration are easier. Also, an assignment that has been tailored to involve all the group members leads to a more fruitful collaboration. Since the goal is to learn through interaction, the more students interact with one another, the more opportunities there will be for learning. Collaboration is also influenced by differences in communication styles. For example, Russian students post as a group whereas the American students post individually. In a sense, the American students look at communication from a conversational point of view, something they have learned from using chat and IM. On the other hand, for most of the Russian students email is still an extension of normal/snail mail. This difference is also visible when you compare the posts of Russian and American students. The posts from the Russian students are invariably longer and more formal in writing style since they first discuss a topic among themselves and then post it. To compensate for their formal style, the Russian students use a lot of smileys and emoticons. In some classes, communication is also impeded by a difference in class schedules. For instance, one semester, the American students met on Tuesdays and Thursdays, whereas the Russian students met on Saturdays and Tuesdays. This was coupled with the fact that there is an eight-hour time difference between Russian University and American University.

Possible Solutions

Collaboration among students is determined by the nature of the assignments given to the class and upon the extent of communication required to complete the assignment. Since the activities in the GCP are typically open-ended and require a large amount of communication, collaboration usually suffers. Changing the nature of the activity can drastically change collaboration among students. If an activity requires limited communication between Russian and American students, which can be achieved given the current constraints, student satisfaction will increase. Giving individual assignments or specific breakdowns of group work among the members can enhance individual learning among students. Learning how to work with groups, especially with students from other cultures is the goal of the class. Readings that specifically discuss these aspects can be assigned to the students. Students can be given scenarios to work on where they can apply this knowledge - similar to case studies. After doing the case studies, when they interact with other students in their group-both in their respective countries and with students from the other country-there will be a greater chance for learning to take place.

Copyright © 2005, Idea Group Inc. Copying or distributing in print or electronic forms without written permission of Idea Group Inc. is prohibited.

Activity-Guided Design: A Framework for Design

Use of Activity Theory in design of educational technology and CSCL has been proposed and examined before[5]. Bellamy (1996) proposed that:

Activity Theory can inform our thinking about the process of designing educational technology to effect educational reform. In particular, through emphasis on activity, it becomes clear that technology cannot be designed in isolation of considerations of the community, the rules and the divisions of labor in which the technology will be placed. (p.127)

Bellamy (1996) also proposes three principles for the design of educational environment based on Vygotsky's work: authentic activities, construction, and collaboration. Barros and Verdejo (2000) show how activity theory can be used to model learning experiences and for designing software to support collaborative discourse (also see, Verdejo, Barros, & Rodriguez-Artacho, 2001). Gifford and Enyedy (1999) proposed the idea of Activity-Centered Design (ACD). They explain that:

Instead of placing either the teacher or the students at the center of the model, we propose that the focus should be to design activities that help learners develop the ability to carry out socially formulated, goal directed action through the use of mediating material and social structures. From this perspective both the social actors, and cultural tools are seen as resources that the students coordinate during activity. In the Activity-Centered Model, as students move through the activities they progress from being partial participants, heavily dependent on the material mediation of tools, to full participants, able to more flexibly use the cultural tools of the normative practice. (p.193)

Enyedy and Gifford propose the ACD as a framework for both the design and analysis of CSCL environments.

Although the theoretical principles underlying ACD and Activity-Guided Design (AGD) are largely the same, there are some significant differences in the framework I propose. In AGD, activity is not at the center of the framework but is the context for the overall design (Figure 7). As a matter of fact, no element is at the center, but they together make up the whole activity. As Nardi (1996a) explains:

Activity theory, then, proposes a very specific notion of context: the activity itself is the context. What takes place in an activity system composed of objects, actions, and operation, is the context. Context is constituted through the enactment of an activity involving people and artifacts. (p. 76)

Copyright © 2005, Idea Group Inc. Copying or distributing in print or electronic forms without written permission of Idea Group Inc. is prohibited.

 Therefore, when I talk about Activity-Guided Design, I'm thinking of an activity as the context or "collaborative contexts," as Hoadley (2002) calls them, i.e., "activities and cultural structures that support collaboration leading to learning." Second, I propose AGD as a framework for design *only* and not as a framework for analysis. I believe that the Activity System (Cole & Engeström, 1993) does a better job of analyzing an activity. I do not propose this framework as the only way or even the "right" way to design a learning environment but as an alternative to learner-centered (there is no one or "typical" learner) or knowledge-centered (there is no knowledge "there" but it is produced) design that can be especially useful for online collaborative-learning environments. The design of a task or assignment requires attention to the tools that will be used, the participants that will collaborate, and the outcomes of the task. The idea is to design an activity in the sense of cultural-historical activity or at least to make an attempt in that direction based on a model that can attempt to predict the outcomes. There will always be trade-offs in design (Pea, 1993), and iterative design of learning environments (Bruckman, 2002) and design experimentation (Brown, 1992; Hoadley 2002) can provide means to find the optimum solution.

The real test of the success of any educational technology starts once the technology is used in its natural setting and environmental factors start interacting with the technology. One obvious solution to implement the technology successfully would be to try to control as many factors as possible every time the technology is used. This is neither feasible nor desirable. The other alternative is to design for change and provide multiple affordances for students. In addition, it is essential to continually evaluate the environment after it is implemented and iterate to find the optimum solution. Moreover, as projects are scaled up to real-world context, factors that can affect a class may not always be predictable, and the pragmatic solution is to design for change, catalogue all possible influences, and improve upon them every semester. This case study of the GCP identifies the importance and need for iterative design of learning environments.

Figure 7: Activity-guided design framework.

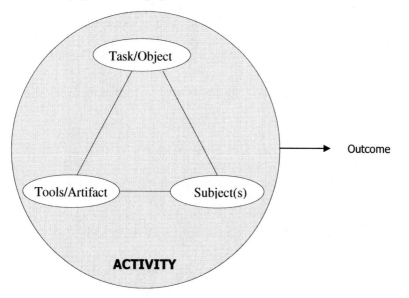

Copyright © 2005, Idea Group Inc. Copying or distributing in print or electronic forms without written permission of Idea Group Inc. is prohibited.

Conclusion

Within the realm of sociocultural theories of learning, I've identified Activity Theory as a theory that can be successfully applied to understand a complex learning environment and an Activity System to analyze it. I've also attempted to explain how the concept of activity can be used to design a learning environment. During my analysis, I've made a conscious effort to try to present evidence from both the micro and macro level of activity, therefore, the emphasis on the message-by-message interaction among the students on the one hand, and the Waterfall Model on the other; as McDermott (1993) proposes, "By institutional arrangements, we must consider everything from the most local level of the classroom to the more inclusive level of inequities throughout the political economy (preferably from both ends of the continuum at the same time)" (p.273). The Global Classroom Project is an outcome mediated by online activities such as emails and postings on WebBoard; offline activities such as face-to-face interaction and class discussions; and the interaction of online and offline activities—the in-between activities—emails that lead to face-to-face interaction or postings that extend class discussions; and, also things that are left unsaid or unacknowledged.

In a simplistic manner, several findings from the study can be identified: the affordance of the computer for communication may not be sufficient for ill-structured and open-ended tasks, and the affordance for communication needs to be supported by access to computers, user-friendly software, and by designing tasks that can be supported by the technology that is available. Groups that show a natural tendency to breakdown their tasks into easily manageable parts that can be supported by the current technology usually succeed in completing the tasks, and groups that fail to recognize the limit placed on collaboration by the technology are less successful at their tasks.

One topic that I've not talked about much is "what were the learning outcomes of the GCP?" It has been hard for me to identify specific learning outcomes in the study, as I started out with research questions that encouraged a contextual investigation and led me to explore factors that would lead to learning, namely, communication and collaboration, and failing which there can be little expectancy of learning outcomes. The use of computers (WebBoard), communication using the WebBoard, and the collaboration resulting from that communication are elements of the environment that got my attention as precursors to learning outcomes.

Salomon (1992) has differentiated between effects *of* technology and effects *with* technology. According to him:

Effects with *are the changes that take place while one is engaged in intellectual partnership with peers or with a computer tool, as, for example, is the case with the changed quality of problem solving that takes place when individuals work together in a team . . .and [E]ffects* of *are those more lasting changes that take place as a consequence of the intellectual partnership, as when computer-enhanced collaboration teaches students to ask more exact and explicit questions even when not using that system. (p.62)*

Copyright © 2005, Idea Group Inc. Copying or distributing in print or electronic forms without written permission of Idea Group Inc. is prohibited.

I believe both of them are essential if learning is to take place. Let us look briefly at some effects *of* and some effects *with* technology in the GCP.

Effects with technology are easy to identify: communication among students and instructors using WebBoard, discussions on readings on the WebBoard, exchange of documents among students, sharing of resources among students, e.g., URLs. Effects of technology are usually difficult to identify (Kolodner & Guzdial, 1996), but here are a few examples:

- Jason, who just graduated and is now working fulltime, says that he learned how to collaborate across time differences from the GCP. His work requires him to work on a project where a part of his team is on the West Coast. He has realized how important it is make sure that the other team gets his part of the work in time and is able to complete their work without any problem.

- Amy, another graduating student, says that she learned a critical lesson the hard way. She has realized that social interaction, especially upfront, is essential for productive group work later on.

- Cathy, who wants to be a high school math teacher after she graduates, believes that she has learned lessons in cross-cultural communication that will certainly help her in dealing with the diversity in her class.

- Many other students mentioned that they learned how to work in a group, although they learned it the hard way.

Changes in the Global Classroom Project

Over the years, several changes have been made to the GCP based on the feedback the instructors have received from the students and from their own experiences. The ratio of face-to-face classes has been increased. "Ice-Breaker" questions have been introduced at the start of online collaboration to increase social interaction. Students now have to sign a contract among themselves describing group responsibilities and promising to fulfill their roles. Student photos are put on the Web so that students can put a face to a name.

Since this study, two notable changes have been made: The assignments have a narrower focus so that they can be completed within the timeframe of the class, and pointed instructions are given to students on how to conduct research and the instructor meets privately with the students to help them in their research.

The findings further identify the benefits of continually evaluating an environment after it is implemented in a natural setting and of designing the learning environment flexibly. We have to think of a learning environment as an activity system, and the activity system as a distributed intelligence system. This has implications for both the analysis and design of a learning environment (Pea, 1993). During the analysis, we have to look for instances of intelligence that are distributed in the environment—in the artifacts, the students, and the rules. While designing the environment, we have to make sure that there is a process in place for the distributed intelligence to take place and for students to accumulate it.

Copyright © 2005, Idea Group Inc. Copying or distributing in print or electronic forms without written permission of Idea Group Inc. is prohibited.

Acknowledgments

This study was done as a part of my master's thesis at Georgia Institute of Technology under the supervision of TyAnna Herrington. I would like to thank Ty for providing me access to a rich research environment, and for her ideas, motivation and continued support. I would also like to thank Wendy Newstetter and Amy Bruckman, members of my thesis committee, for their help, support and guidance. I owe special debt to Wendy for introducing me to Activity Theory and for teaching me what it means to think and act like an ethnographer.

References

Barab, S., Barnett, M., Yamagata-Lynch, L., Squire, K., & Keating, T. (1999). Using activity theory to understand the contradictions characterizing a technology-rich introductory astronomy course. *AERA 1999*. In *Mind, Culture, and Activity, 9*(2): 76-107, 2002.

Barab, S. A., Schatz, S. & Scheckler, R. (in press). Using activity theory to conceptualize online community and using online community to conceptualize activity theory. *AERA*, Seattle, WA. April, 2001. To appear in *Mind, Culture, and Activity*.

Barros, B. & Verdejo, M.F. (2000). Analyzing student's interaction process for improving collaboration: The DEGREE approach. *International Journal for Artificial Intelligence in Education*. ISSN 1560-4292, Volume 11. Available at: *http://sensei.lsi. uned.es/~bbarros/papers/ijaied2000.pdf*.

Bellamy, R. E. (1996). Designing educational technology: Computer-mediated change. In B. Nardi, (ed.), *Context and consciousness: Activity theory and human-computer interaction*, pp.123-146. Cambridge, MA: MIT Press.

Bødker, S. (1996). Applying activity theory to video analysis: How to make sense of video data in HCI. In B. Nardi, (ed). *Context and consciousness: Activity theory and human-computer interaction*, pp.69-102. Cambridge, MA: MIT Press.

Bonk, C. J. & King, K. S. (eds.)(1998). *Electronic collaborators: Learner-centered technologies for literacy, apprenticeship, and discourse*. Mahwah, NJ: Lawrence Erlbaum Associates.

Brown, A. (1992). Design experiments: Theoretical and methodological challenges increasing complex interventions in classroom settings. *The Journal of the Learning Sciences, 2*(2):*141-178*.

Brown, J., Collins, A., & Duguid, P. (1989). Situated cognition and the culture of learning. *Educational Researcher*, 18: 32-42.

Bruckman, A (2002). Co-evolution of technological design and pedagogy in an online learning community. In S. Barab, R. Kling, & J. Gray (eds.), *Designing virtual communities in the service of learning*, pp.239-255. Cambridge, UK: Cambridge Available at: *http://www.cc.gatech.edu/~asb/papers/bruckman-co-evolution.pdf*.

Copyright © 2005, Idea Group Inc. Copying or distributing in print or electronic forms without written permission of Idea Group Inc. is prohibited.

Bruffee, K. (1984). Collaborative learning and the "conversation of mankind." *College English*, 46(7): 635-652.

Cohen, E. (1994). Restructuring the classroom: Conditions for productive small groups. R*eview of Educational Research*, 64:1-35.

Cole, M. (1996). *Cultural psychology: A once and future discipline*. Cambridge, MA: Belknap Press/Harvard University Press.

Cole, M. & Engeström, Y. (1993). A cultural-historical approach to distributed cognition. In G. Salomon (ed.), *Distributed cognitions, psychological and educational considerations*. Cambridge, UK: Cambridge University Press, pp. 1-46.

Cranton, P. (1996). Types of group learning. In S. Imel, (ed), *Learning in groups: Exploring fundamental principles, new uses, and emerging opportunities*. San Francisco, CA: Jossey-Bass.CSCL 1997: Proceedings available at: *http://www.oise.utoronto.ca/cscl/*.

Engeström, Y. (1987) *Learning by expanding: An activity-theoretical approach to developmental research*. Helsinki: Orienta-Konsultit Oy.

Engeström, Y. (2002). The activity system. Center for Activity Theory and Developmental Work Research. *University of Helsinki. Finland*. Available at: *http: www.edu.helsinki.fi/activity/6b.htm* (Retrieved 11/14/02).

Engeström, Y. & Middleton, D. (eds.) (1996). *Cognition and communication at work*. Cambridge, UK: Cambridge University Press.

EuroCSCL (2001). Proceedings available at: *http://www.mmi.unimaas.nl/eurocscl/*.

Gibson, J. J. (1979). *The ecological approach to visual perception*. Boston, MA: Houghton Mifflin.

Gifford, B. & Enyedy, N. (1999). Activity-centered design: Towards a theoretical frameworkfor CSCL. In *Proceedings of the Third International Conference on CSCL*, pp. 189-197. Available at: *http://www.gseis.ucla.edu/faculty/enyedy/pubs/Gifford&Enyedy_CSCL2000.pdf*.

Halloran, J., Rogers, Y., & Scaife, M. (2002). Avoiding groupware: Why students said 'no' toLotus Notes. In *Proceedings of the Fourth International Conference on CSCL*. Available at: *http://newmedia.colorado.edu/cscl/202.html* (Retrieved 11/14/02).

Hoadley, C. (2002). Creating context: Design-based research in creating and understanding CSCL. In *Proceedings of the Fourth International Conference on CSCL*.Available at: *http://newmedia.colorado.edu/cscl/230.html* (Retrieved: 11/14/02).

Hoadley, C. & Roschelle, J. (eds.) (1999). *Proceedings of the Third International Conference on CSCL*. Mahwah, NJ: Lawrence Erlbaum Associates Available at: *http://kn.cilt.org/cscl99/* (Retrieved 11/14/02).

Holland, D. & Reeves, J.R. (1996). Activity theory and the view from somewhere: Team perspectives on the intellectual work of programming. In B.A. Nardi (ed.), *Context and consciousness: Activity theory and human-computer interaction*. Cambridge, MA: MIT Press, pp. 257-282.

Copyright © 2005, Idea Group Inc. Copying or distributing in print or electronic forms without written permission of Idea Group Inc. is prohibited.

Hyppönen, H. (1998). Activity theory as a basis for design for all. *Presentation at 3rd TIDE Congress 23-25 June, 1998.* Available at: *http://www.stakes.fi/tidecong/213hyppo.htm* (Retrieved 11/14/02).

John-Steiner, V. & Mahn, H. (1996). Sociocultural approaches to learning and development: A Vygotskian framework. *Educational Psychologist,* 31: 191-206.

Kaptelinin, V. (1996). Computer-mediated activity: Functional organs in social and developmental contexts. In B.A. Nardi, (ed.) *Context and consciousness: Activity theory and human-computer interaction.* Cambridge, MA: MIT Press.

Kolodner, J. & Guzdial, M. (1996). Effects with and of CSCL: Tracking learning in a new paradigm. In T. Koschmann (ed.), *CSCL: Theory and practice of an emerging paradigm,* pp. 307-320. Mahwah, NJ: Lawrence Erlbaum Associates.

Koschmann, T. (1996). Paradigm shifts and instructional technology: An introduction. In T. Koschmann (ed.) *CSCL: Theory and practice of an emerging Paradigm,* pp.1-23. Mahwah, NJ: Lawrence Erlbaum Associates.

Koschmann, T., Kelson, A., Feltovich, P., & Barrows, H. (1996). Computer-supported problem-based learning: A principled approach to the use of computers in collaborative learning. In T. Koschmann (ed.), *CSCL: Theory and practice of an emerging paradigm,* pp.83-124. Mahwah, N.J: L. Erlbaum Associates.

Kuutti, K. (1991). Activity theory and its applications to information systems research and development. In H.E. Nissen, H.K. Klein, & R. Hirschheim (Eds.), *Information systems research: Contemporary approaches and emergent traditions,* (pp. 529-549). North Holland: Elvsevier Science Publishers B.V.

Kuutti, K. (1996). Activity theory as a potential framework for human-computer interaction research. In B.A. Nardi (ed.), *Context and consciousness: Activity theory and human-computer interaction.* Cambridge, MA: MIT Press, pp.17-44.

Leont'ev, A.N. (1978). *Activity, consciousness, personality.* Englewood Cliffs, NJ: Prentice Hall.

Miller, J., Trimbur, J., & Wilkes, J. (1994). Group dynamics: Understanding group success and failure in collaborative learning. In K. Boswort & S. J. Hamilton (eds.), *Collaborative learning: Underlying processes and effective techniques,* pp.33-58. San Francisco, CA: Jossey-Bass.

McDermott, R. (1993). The acquisition of a child by a learning disability. In S. Chaiklin, & J. Lave, (eds.), *Understanding practice: Perspectives on activity and context,* pp. 269-305. Cambridge, UK: Cambridge University Press.

Nardi, B.A. (ed.) (1996). *Context and consciousness: Activity theory and human-computer interaction.* Cambridge, MA: MIT Press.

Nardi, B. (1996a). Studying context: A comparison of activity theory, situated action models, and distributed cognition. In B.A. Nardi (ed.), *Context and consciousness: Activity theory and human-computer interaction.* Cambridge, MA: MIT Press, pp.69-102.

Newstetter, W. (1998). Of green monkeys and failed affordances: A case study of a mechanical engineering design course. *Research in Engineering Design* 10:118-128.

Copyright © 2005, Idea Group Inc. Copying or distributing in print or electronic forms without written permission of Idea Group Inc. is prohibited.

Pea, R.D. (1993). Practices of distributed intelligence and designs for education. In G. Salomon (ed.), *Distributed cognitions: Psychological and educational considerations.* Cambridge, UK: Cambridge University Press, pp.47-87.

Peters, J. & Armstrong, J. (1998). Collaborative learning: People laboring together to construct knowledge. In I.M. Saltiel, A. Sgroi, & R.G. Brockett (eds.), *The power and potential of collaborative learning partnerships*, pp. 75-85. San Francisco, CA: Jossey-Bass.

Salomon, G. (1992). What does the design of effective CSCL require and how do we study its effects? *SIGCUE OUTLOOK,* 21(3): 62-68, Spring, New York: ACM Press.

Salomon, G. & Perkins, D. (1996). Learning in Wonderland: What do computers really offer education? In S. Kerr (ed.), *Technology and the future of schooling*, pp. 111-130. Chicago, IL: University of Chicago Press.

Scribner, S. (1997). A sociocultural approach to the study of mind. In E. Tobach, R.J. Falmagen, M.B. Parlee, L.W.M. Martin, & A.S. Kapelman (eds.), *Mind and social practice: Selected writings of Sylvia Scribner,* pp.266-280. Cambridge, UK: Cambridge University Press.

Stahl, G. (ed.) (2002). Computer support for collaborative learning: Foundations for a CSCL Community. *Proceedings of CSCL 2002,* Jan. 7-11, Boulder, CO. Hillsdale, NJ. Lawrence Erlbaum Associates. Available at: *http://orgwis.gmd.de/~gerry/ publications/journals/cscl2002/CSCL 2002 Proceedings.pdf.*

Verdejo, M.F., Barros, B., & Rodríguez-Artacho, M. (2001). A proposal to support the design of experimental learning activities. In *Proceedings of the European Conference on Collaborative Learning '2001*, pp. 633-640. ISBN 90-5681-097-9. Maastrich McLuhan Institute

Vygotsky, L.S. (1978). *Mind in society: The development of higher psychological processes.* Cambridge, MA: Harvard University Press.

Wertsch, J.V. (ed.) (1981). *The concept of activity in Soviet psychology.* Armonk, NY: M.E. Sharpe.

Wertsch, J.V. (1991). *Voices of the mind: A sociocultural approach to mediated action.* Cambridge, MA: Harvard University Press.

Wertsch, J.V., del Rio, P. & Alvarez, A. (1995). Sociocultural studies: history, action and mediation. In J.V. Wertsch, P. del Rio & A. Alvarez (Eds.), *Sociocultural studies of mind* (pp. 1-34). NY: Cambridge University Press.

Endnotes

[1]   The term Activity, when used in this chapter, has this specific meaning; whereas task, assignments, projects, and goals represent the object of the activity.

[2]   To protect the privacy of participants, I've withheld some key information about the project. All the names of people used in the paper are pseudonyms.

Copyright © 2005, Idea Group Inc. Copying or distributing in print or electronic forms without written permission of Idea Group Inc. is prohibited.

3   I've used contradictions and breakdowns interchangeably here, although they have slightly different connotations (see Bødker, 1996).

4   It is not possible to reproduce the messages in the "Proposal Discussion" thread because of length restrictions for the chapter.

5   Michael Cole (1996) proposed the idea of using Activity Theory to design learning environments. My focus here is on studies specific to technology-supported and CSCL environments.

Copyright © 2005, Idea Group Inc. Copying or distributing in print or electronic forms without written permission of Idea Group Inc. is prohibited.

About the Authors

Tim S. Roberts (t.roberts@cqu.edu.au) is a Senior Lecturer with the Faculty of Informatics and Communication at the Bundaberg Campus of Central Queensland University. He has taught a variety of computer science subjects, including courses to over 1,000 students located throughout Australia and overseas, many of them studying entirely online. In 2001, together with Lissa McNamee and Sallyanne Williams, he developed the Online Collaborative Learning in Higher Education website at *http://clp.cqu.edu.au*. He was awarded the Bundaberg City Council's prize for Excellence in Research in 2001, and the Dean's Award for Quality Research in 2002. He has previously edited *Online Collaborative Learning: Theory and Practice* (Information Science Publishing, 2003).

* * * * * * * * * * * * * * * * * * *

Herman Buelens obtained a Ph.D. in Psychology from the Katholieke Universiteit Leuven (1995). At the Educational Support Office of the K.U., Buelens is involved in teacher training and in the implementation of both student-oriented education and ICT. He is also an Assistant Professor at the University of Antwerp, Department of Applied Economical Sciences, where he lectures on Learning Processes and Multimedia Educational Support. His research interests include group processes and collaborative e-learning.

Thanasis Daradoumis (adaradoumis@uoc.edu) has a Master's in Computer Science from the University of Illinois and a Ph.D. in Information Sciences from the Polytechnic University of Catalonia-Spain. Since 1984, he has been a professor at several universities in the USA, Greece, and Spain. Since 1998, he has been working as a Professor in the Department of Information Sciences at the Open University of Catalonia where he

Copyright © 2005, Idea Group Inc. Copying or distributing in print or electronic forms without written permission of Idea Group Inc. is prohibited.

coordinates several online courses. His research focuses on the development of methods and strategies for modeling collaborative learning in virtual environments, methods of analysis of online collaborative interactions, and assessment tools for evaluating the processes and products of collaborative learning.

Jan Elen graduated from the University of Leuven where he also received his doctoral degree. He is currently Head of the Educational Support Office of the University of Leuven. His research pertains to theoretical aspects of instructional design, instructional conceptions of students, and the relationship between research and university education. An advanced course on Instructional Design for Educational Sciences students and an Introduction to Didactics in the Teacher Training Program are his most important teaching responsibilities.

David Gibson, Ed.D is the Co-creator and Co-Principal Investigator of simSchool. Dr. Gibson oversees all phases and aspects of the project: professional development, planning and implementation, educational research and evaluation, application development and deployment, budgeting, staff and project performance monitoring, communications, dissemination and sustainability. At Vermont Institutes, Dr. Gibson is Director of Research and Development concentrating on partnership development and new programs, systems analysis, evaluation, higher education reform and statewide professional development planning. He is also Director of Research with the National Institute for Community Innovations, concentrating on vision and project development, strategic planning, professional network building, national partners, and telecommunications in learning. His research and publications include work on complex systems analysis and modeling of education, and network-based assessment and the use of technology to personalize education for the success of all students.

Bruce Havelock has fifteen years of combined experience in research and practice surrounding technology-supported learning in K-12, higher education, and business settings. He is currently a Research Associate at RMC Research Corporation in Denver, Colorado, where he directs several program evaluation studies for technology-enhanced education projects around the country. His primary research interests are collaborative technologies in teacher learning, and technology as an accelerator for educational change. He holds a Ph.D. in educational technology and cognitive studies from the University of Washington in Seattle, and a B.A. *cum laude* in Latin American history and literature from Harvard University.

Simon B. Heilesen is Senior Lecturer in Net Media at the Department of Communication, Journalism, and Computer Science, Roskilde University, Denmark. His principal research interests are World Wide Web, HCI design and communication planning, and learning and collaboration in Net environments. For more information, visit his institutional homepage at *http://www.ruc.dk/~simonhei*.

Copyright © 2005, Idea Group Inc. Copying or distributing in print or electronic forms without written permission of Idea Group Inc. is prohibited.

Sisse Siggaard Jensen is Senior Lecturer in Net Media at the Department of Communication, Journalism, and Computer Science, Roskilde University. Her principal research interests are reflective practices, social interaction, collaboration and interpersonal communication, knowledge sharing, knowledge communication, identity, avatars, and organizational games on the Net.

Aditya Johri is a doctoral student in the Learning Sciences and Technology Design program at the School of Education, Stanford University (USA). He has a master's degree in Information, Design and Technology from Georgia Institute of Technology, Atlanta. His chapter is based on research done as part of his master's thesis at Georgia Tech. Prior to joining graduate school, he worked as a software engineer in India. His research interests are in CSCL, learning in informal and work settings, and distributed work.

W.R. Klemm is a neuroscientist who has published in a wide range of topics. He has more than 400 publications to his credit, including nine books and 45 book chapters. He has also created a web site (*http://www.cvm.tamu.edu/wklemm/collab.htm*) dealing with collaborative learning and distance education. He is co-developer of a Web-like computer conferencing and document-sharing system called Forum MATRIX, as well as a computerized self-quizzing game called Get Smart! His honors include serving on the Board of Editors of seven journals: *The Technology Source; Progress in Neuropsychopharmacology & Biological Psychiatry; Archives of Clinical Neuropsychology; American Journal of Veterinary Research; Psychopharmacology; Journal of Electrophysiological Techniques; and Communications in Behavioral Biology* (now *Behavioral & Neural Biology*). He has been listed in some 17 biographies and has Distinguished Researcher awards from Texas A&M and Sigma Xi, Texas A&M. He served on the international Board of Directors of the 70,000-member Sigma Xi, The Scientific Research Society. A more complete resume is available at his website at http://www.cvm.tamu.edu/wklemm.

Antonio Santos Moreno received his Doctorate in Education from Indiana University, has been a researcher and consultant in the field of Educational Technology in several Mexican universities. He has also been a producer of instructional materials for teachers' training at different educational and industrial organizations. He has published several articles in the areas of constructivist-learning environments and the use of technology in developing countries. He has also presented papers in world conferences on themes such as contemporary pedagogies and collaborative learning. Currently, he is working as a professor at the Universidad de las Americas Puebla in Mexico, where he has developed the pedagogical model for the University's distance education program.

From 1985 to 1988, **Amiram Moshaiov** was an Assistant Professor at MIT (USA). Since 1988, he has been with the Department of Mechanics, Materials, & Systems at Tele-Aviv University (TAU) (Israel), heading the Robotics and Mechatronics Program. Following his initiative for an Israeli-Palestinian large-scale education proposal, he was a Visiting Professor at Bogazici University, Turkey, with Professor O. Kaynak, the UNESCO Chair

Copyright © 2005, Idea Group Inc. Copying or distributing in print or electronic forms without written permission of Idea Group Inc. is prohibited.

on Mechatronics; there, he initiated and coordinated a socio-technical proposal of 39 European organizations. He has served on many international program committees of engineering and of educational conferences (e.g., Special Sessions Chair of the Third International Conference on Information Technology-Based Higher Education and Training). He is also a member of the TC on Education of IASTED.

Alexandra Lilaváti Pereira Okada has a background in Computer Science Engineering, and Communications and Marketing. She teaches technology in Dante Alighieri School, A 4-17 school for 5,500 students in Sao Paulo, and in a post-graduation education course at Mackenzie University. She is also studying for a Ph.D. on Collective Building of Knowledge in Collaborative Learning Environments at Pontifical University Catholic PUC-SP. Alexandra's current research interests include networks, virtual collaborative communities, the collective building of knowledge, and methodologies for online and Web-based teaching and learning. She has been involved with numerous online learning projects at MEC – PROINFO (Ministry of Education) and OAS (Organization of American States).

Trena M. Paulus is an Assistant Professor of Educational Psychology and Counseling at the University of Tennessee (USA), working with the Collaborative Learning and Applied Educational Psychology programs. Her research explores how language, culture, collaboration, and technology impact learning. She completed her Ph.D. in Instructional Systems Technology with a minor in Computer-Mediated Communication at Indiana University and holds an M.A. in Applied Linguistics from Ohio University. She has designed, developed, taught, and researched online distance-education courses for both Indiana University and Walden University.

Celia T. Romm Livermore is a Full Professor in the Department of Information Systems and Manufacturing at Wayne State University, Detroit (USA), and an Honorary Professor at Fujian Radio and TV University, the People's Republic of China. She received her Ph.D. from the University of Toronto, Canada (1979). She has been a lecturer, consultant, and visiting scholar in Israel, Japan, Germany, Canada, the U.S., and Australia. Dr. Romm's current research interests include telework, electronic commerce, computer-mediated communication, and IT/IS education. Dr. Romm has published three books: *Electronic Commerce: A Global Perspective* (Springer, London, 1998), *Virtual Politicking* (Hampton Press, USA, 1999), and *Doing Business on the Internet - Opportunities and Pitfalls* (Springer, London, 1999; and Tianjin People's Publishing House, China, 2001). In addition, Dr. Romm has published more than 90 papers in refereed journals and chapters in collected volumes, as well as presented her work at over seventy local and international conferences.

Lorraine Sherry, Ph.D., is a Senior Research Associate with RMC Research Corporation. She has directed or served on the evaluation teams for the national PT3 evaluation; three PT3 Catalyst Grants; has two PT3 Implementation Grants and three Technology Innovation Challenge Grants, as well as the Virtual Assistive Technology University, the

Copyright © 2005, Idea Group Inc. Copying or distributing in print or electronic forms without written permission of Idea Group Inc. is prohibited.

Boulder Valley Internet Project, and the Annenberg/CPB Math and Science Project. She authored many book chapters, journal articles, and magazine articles on instructional technology. She has developed new models for technology adoption and diffusion, teacher leadership in instructional technology, and the link between technology and student performance, and has presented her research at many national conferences.

Elizabeth Stacey teaches and researches in the School of Scientific and Developmental Studies in the Faculty of Education at Deakin University, Melbourne. She is currently coordinating and teaching in the Flexible, Online and Distance Education Specialism in the Master of Professional Education and Training program. She supervises postgraduate students researching flexible learning through information and communication technologies, particularly focusing on e-learning. Her research and publications are developed from these areas. Her doctoral research explored collaborative-group learning through computer-mediated communication with adult learners. She is convener of the Flexible and Situated Learning Faculty Research Group.

Valerie Taylor has taught Professional Development and postgraduate courses in online teaching and learning, curriculum development, instructional design, and learning technology. She is a graduate of the California State University- Hayward. There, she received a Master of Science in Education with a specialty in Online Teaching and Learning, a program offered completely online, so Ms. Taylor has considerable experience as an online learner as well as educator. She teaches web design and development online and on-campus at De Anza College in the heart of Silicon Valley. Ms. Taylor works with government, corporate, and academic clients designing and developing online courses for development of an information-proficient society.

Eddy Van Avermaet received a Ph.D. in Psychology from the University of California (Santa Barbara) (1975). Since then, he has been a Professor at the Katholieke Universiteit Leuven where he lectures on Social Cognition, Social Interaction, and the Social Psychology of Communication. At K.U.Leuven, he is Director of the Laboratory for Experimental Social Psychology, Chairman of the Council of Education, Chairman of the Educational Task Force, Advisory to the rectors of the Flemish universities, and Academic Coordinator of Quality Assurance at the Flemish universities. Eddy was also chief editor of the *European Journal of Social Psychology* (1994-1997), Chairman of the Scientific Advisory Board of the Kurt Lewin Institute (2001-2005), a member of the Accreditation Committee for Dutch Research Schools in the Social Sciences (2000-2004), and a member of the Quality Assurance Committee for Psychology Curricula in The Netherlands (2000-2001). His research interests include individual versus intergroup relations, social dilemmas, implicit processes in social perception, and minority and majority influence.

Jan Van den Bulck obtained a Ph.D. in Social Sciences (K.U. Leuven, 1996), an M.A. in Politics (Hons, University of Hull, 1989), and an M.A. in Communication Science (KULeuven, 1987). He is an Associate Professor for the Department of Communication,

Copyright © 2005, Idea Group Inc. Copying or distributing in print or electronic forms without written permission of Idea Group Inc. is prohibited.

KULeuven, and is Director of the Center for Audience Research. He specializes in long-term effects of the entertainment media on the social and cognitive construction of reality. He is responsible for the research seminar described in the chapter of which he is a co-author.

Jan Van Mierlo obtained an M.A. in Communication Science from the Katholieke Universiteit Leuven (2001). As a member of the Center for Audience Research of the Department of Communication Science (K.U.Leuven), he studies group processes and their impact on both group functioning and individual results. He is preparing a Ph.D. thesis on the cultivation approach to media effects.

Fatos Xhafa (*fatos@lsi.upc.es*) has a Ph.D. in Computer Sciences from the Polytechnic University of Catalonia, Barcelona-Spain. Since 1996, he has been working as a lecturer in the Department of Languages and Informatics Systems at the Polytechnic University of Catalonia and, since 1999, as a consulting-professor at the Open University of Catalonia. Dr. Xhafa has participated in several research projects funded by the European Community.

Copyright © 2005, Idea Group Inc. Copying or distributing in print or electronic forms without written permission of Idea Group Inc. is prohibited.

Index

Copyright © 2005, Idea Group Inc. Copying or distributing in print or electronic forms without written
permission of Idea Group Inc. is prohibited.

Copyright © 2005, Idea Group Inc. Copying or distributing in print or electronic forms without written permission of Idea Group Inc. is prohibited.

J

Jigsaw 187
journals 2

K

kairic rhythmicity 60
knowledge 259
knowledge construction 212

L

leader 266
learners' perception 128
learning communities 20, 209
learning objects 26
learning online 264
learning state 266
learning styles 23
learning task 260
Lévy, Pierre 71

M

master of computer-mediated
    communication 53
matrix of time and place 55
Maturana, Humberto 71
Mead, George Herbert 52
memory matrix 187
mentoring 202
meta-abilities 260
meta-analysis 263
meta-change 266
meta-cognition 260
meta-communicate 260
meta-communication abilities 260
meta-learning 260
metacognitive abilities 266
metamorphosis 78
metaphorical space 65
methodological framework 219
models of task analysis 135
motivation 162
multiplicity 78

N

negative effects 127
negotiated time 56
negotiation 101, 270
networked learning environment 218
networked mediated collaboration 259
non-verbal communication 267

O

online collaborative learning 141, 282
online communication 259
online dialogue 203
online learning environment 274
oral communication 274
organizing strategies 67

P

pair note-taking 185
passivity 272
pauses 59
pedagogical mediation 92
pedagogical strategies 260, 265
personal and professional learning
    portfolio 201
personal punctuation 272
place 54
positive interdependence 262
problem-based learning 189
problem-oriented project work 53
problems 1
project development 189
proximity 78

Q

questionnaire 130

R

radical model 162
Rämö, Hans 52
reflection 206
Reid collaborative learning model 25
relational cognition 259
relational process of cognition 263
relations 259

Copyright © 2005, Idea Group Inc. Copying or distributing in print or electronic forms without written permission of Idea Group Inc. is prohibited.

Copyright © 2005, Idea Group Inc. Copying or distributing in print or electronic forms without written permission of Idea Group Inc. is prohibited.

Instant access to the latest offerings of Idea Group, Inc. in the fields of
INFORMATION SCIENCE, TECHNOLOGY AND MANAGEMENT!

InfoSci-Online Database

- BOOK CHAPTERS
- JOURNAL ARTICLES
- CONFERENCE PROCEEDINGS
- CASE STUDIES

The Bottom Line: With easy to use access to solid, current and in-demand information, InfoSci-Online, reasonably priced, is recommended for academic libraries. 〞
- Excerpted with permission from Library Journal, July 2003 Issue, Page 140

The InfoSci-Online database is the most comprehensive collection of full-text literature published by Idea Group, Inc. in:

- Distance Learning
- Knowledge Management
- Global Information Technology
- Data Mining & Warehousing
- E-Commerce & E-Government
- IT Engineering & Modeling
- Human Side of IT
- Multimedia Networking
- IT Virtual Organizations

BENEFITS
- Instant Access
- Full-Text
- Affordable
- Continuously Updated
- Advanced Searching Capabilities

Start exploring at www.infosci-online.com

Recommend to your Library Today!

Complimentary 30-Day Trial Access Available!

A product of:
Information Science Publishing*
Enhancing knowledge through information science

*A company of Idea Group, Inc.
www.idea-group.com

BROADEN YOUR IT COLLECTION
WITH IGP JOURNALS

Idea Group Publishing

is an innovative international publishing company, founded in 1987, specializing in information science, technology and management books, journals and teaching cases. As a leading academic/scholarly publisher, IGP is pleased to announce the introduction of 14 new technology-based research journals, in addition to its existing 11 journals published since 1987, which began with its renowned Information Resources Management Journal.

Free Sample Journal Copy

Should you be interested in receiving a **free sample copy** of any of IGP's existing or upcoming journals please mark the list below and provide your mailing information in the space provided, attach a business card, or email IGP at journals@idea-group.com.

Upcoming IGP Journals

January 2005

Int. Journal of Data Warehousing & Mining	Int. Journal of Enterprise Information Systems
Int. Journal of Business Data Comm. & Networking	Int. Journal of Intelligent Information Technologies
International Journal of Cases on E-Commerce	Int. Journal of Knowledge Management
International Journal of E-Business Research	Int. Journal of Mobile Computing & Commerce
International Journal of E-Collaboration	Int. Journal of Technology & Human Interaction
Int. Journal of Electronic Government Research	Int. J. of Web-Based Learning & Teaching Tech.'s
Int. Journal of Info. & Comm. Technology Education	

Established IGP Journals

Annals of Cases on Information Technology	International Journal of Web Services Research
Information Management	Journal of Database Management
Information Resources Management Journal	Journal of Electronic Commerce in Organizations
Information Technology Newsletter	Journal of Global Information Management
Int. Journal of Distance Education Technologies	Journal of Organizational and End User Computing
Int. Journal of IT Standards and Standardization Research	

Name:_____ Affiliation: _____

Address: _____

_____

E-mail:_____ Fax: _____

**Visit the IGI website for more information on
these journals at www.idea-group.com/journals/**

IDEA GROUP PUBLISHING
A company of Idea Group Inc.
701 East Chocolate Avenue, Hershey, PA 17033-1240, USA
Tel: 717-533-8845; 866-342-6657 • 717-533-8661 (fax)

Journals@idea-group.com   www.idea-group.com

NEW RELEASE

Online Collaborative Learning:
Theory and Practice

Tim S. Roberts, PhD, Central Queensland University, Australia

Online Collaborative Learning: Theory and Practice
provides a resource for researchers and practitioners in the
area of online collaborative learning (also known as CSCL,
computer-supported collaborative learning), particularly those
working within a tertiary education environment. It includes
articles of relevance to those interested in both theory and
practice in this area. It attempts to answer such important
current questions as: how can groups with shared goals
work collaboratively using the new technologies? What
problems can be expected, and what are the benefits? In
what ways does online group work differ from face-to-face
group work? And what implications are there for both
educators and students seeking to work in this area?

ISBN 1-59140-174-7 (h/c) • US$74.95 • ISBN 1-59140-227-1(s/c) • US$59.95
• 320 pages • Copyright © 2004

*"The new environment is one in which students are more likely to come from a
diverse range of backgrounds, have differing levels of technical and language
abilities, and desire to study at times and in places of their own choosing. "*

- Tim S. Roberts
Central Queensland University, Australia

**It's Easy to Order! Order online at www.idea-group.com or
call 717/533-8845 x10**
Mon-Fri 8:30 am-5:00 pm (est) or fax 24 hours a day 717/533-8661

INFOSCI Information Science Publishing
Hershey • London • Melbourne • Singapore

An excellent addition to your library

Usability Evaluation of Online Learning Programs

Claude Ghaoui,
Liverpool John Moores University, UK

Successful use of information and communication technologies depends on usable designs that do not require expensive training, accommodate the needs of diverse users and are low cost. There is a growing demand and increasing pressure for adopting innovative approaches to the design and delivery of education, hence, the use of online learning (also called E-learning) as a mode of study. This is partly due to the increasing number of learners and the limited resources available to meet a wide range of various needs, backgrounds, expectations, skills, levels, ages, abilities and disabilities. By focusing on the issues that have impact on the usability of online learning programs and their implementation, *Usability Evaluation of Online Learning Programs* specifically fills in a gap in this area, which is particularly invaluable to practitioners.

ISBN 1-59140-105-4(h/c); eISBN 1-59140-113-5• US$74.95 • 350 pages • © 2003

"There is a growing demand and increasing pressure for adopting innovative approaches to the design and delivery of education, hence, the use of online learning (also called e-learning) as a mode of study."

–Claude Ghaoui, Liverpool John Moores University, UK

**It's Easy to Order! Order online at www.idea-group.com
or call 717/533-8845 x10!**
Mon-Fri 8:30 am-5:00 pm (est) or fax 24 hours a day 717/533-8661

Information Science Publishing

Hershey • London • Melbourne • Singapore • Beijing

An excellent addition to your library